Enzymes Involved in Glycolysis, Fatty Acid and Amino Acid Biosynthesis:
Active Site Mechanism and Inhibition

Authored by

Marco Brito-Arias

Department of Basic Sciences,
National Polytechnic Institute,
Interdisciplinary Professional Unit of Biotechnology, Mexico City,
Mexico

Enzymes Involved in Glycolysis, Fatty Acid and Amino Acid Biosynthesis: Active Site Mechanism and Inhibition

Author: Marco Brito-Arias

ISBN (Online): 978-981-14-6090-6

ISBN (Print): 978-981-14-6088-3

ISBN (Paperback): 978-981-14-6089-0

need for a court order if at any point you breach any terms of this License Agreement. In no event will any delay or failure by Bentham Science Publishers in enforcing your compliance with this License Agreement constitute a waiver of any of its rights.

3. You acknowledge that you have read this License Agreement, and agree to be bound by its terms and conditions. To the extent that any other terms and conditions presented on any website of Bentham Science Publishers conflict with, or are inconsistent with, the terms and conditions set out in this License Agreement, you acknowledge that the terms and conditions set out in this License Agreement shall prevail.

Bentham Science Publishers Pte. Ltd.
80 Robinson Road #02-00
Singapore 068898
Singapore
Email: subscriptions@benthamscience.net

BENTHAM
SCIENCE

CONTENTS

PREFACE

The present book entitled *"Enzymes Involved in Glycolysis, Fatty Acid and Amino Acid Biosynthesis: Active Site Mechanism and Inhibition"* comprehends a thorough revision about the known reaction mechanism occurring between the enzymes related to the mentioned biosynthetic pathways with their substrates, cofactors and residues. Different disciplines provide a wealth of knowledge including crystallographic studies, kinetic studies, docking, genetic mutagenesis and biochemistry. The knowledge about the reaction mechanism is primordial to understand in a better way the normal functioning of the cell process, which is used as a starting point for preventing or to correct pathologies. Currently, the drug design relies strongly on the understanding of the interaction between the substrates or ligands with the amino acid residues and derived from these studies, a wealth of potent inhibitors have emerged for the treatment of several diseases such as cancer, tuberculosis, anti-parasitic, and also importantly metabolic syndrome alterations such as diabetes and obesity.

The enzymatic reaction mechanism includes aldolase, isomerase, kinase, mutase, synthase, dehydrogenase, reductase, transferase, hydrolase, lyase *etc.*, all of them widespread in all biochemical transformations.

This book pretends to serve as a tool for professionals involved in pharmaceutical, health, food and other related disciplines, providing well known, and key insights of the reaction mechanism occurring at the molecular level between the biological catalyst and the chemical ligands and how the transformation occurs within the cell.

ACKNOWLEDGEMENTS:

The author is grateful to the following for financial support:

- Estímulo al Desempeño Docente (EDD) IPN 20774

- Comisión de Operación y Fomento de Actividades Académicas del Instituto Politécnico Nacional (COFAA-IPN) 9500068

- CONACyT-SNI 14636

CONFLICT OF INTEREST:

The authors confirm that this chapter contents have no conflict of interest.

Marco Brito-Arias
Department of Basic Sciences,
National Polytechnic Institute,
Interdisciplinary Professional Unit of Biotechnology, Mexico City,
Mexico

CHAPTER 1

Glycolysis

Abstract: The biochemical process known as glycolysis is a fundamental pathway, which allow the glucose to be transformed into energy (ATP) and pyruvate. During this process the glucose is first phosphorylated at the 6th position, then converted to fructose by the phosphoglucose isomerase and phosphorylated to fructose 1,6-biphosphate. The next steps involve the bond cleavage by the enzyme fructose biphosphate aldolase to dihydroxyacetone phosphate and glyceraldehyde 3-phosphate, which can be converted from the keto to the aldehyde by the enzyme triose phosphate isomerase. A second phosphorylation of glyceraldehyde 3-phosphate takes place by the enzyme glyceraldehyde-3-phosphate dehydrogenase in the presence of NAD+ and Pi providing 1,3-bisphosphoglycerate. The next step was mediated by the enzyme phosphoglycerate kinase to produce ATP and 3-phosphoglycerate which undergoes phosphate migration producing 2-phosphoglycerate. Further dehydration mediated by the enzyme enolase produce phosphoenolpyruvate which is finally converted by the enzyme pyruvate kinase to pyruvate and the release of a second ATP molecule.

Keywords: ATP molecules, Biphosphate, Dehydration.

The biochemical process known as glycolysis is a fundamental pathway, which allows the glucose to be transformed into energy (ATP) and pyruvate. The glycolytic pathway occurs in either prokaryotic and eukaryotic cells, however in bacteria, the common pathway is known as the Entner-Doudoroff pathway (ED) while in eukaryotes, the glycolysis follows the Embden-Meyerhoff-Parnas (EMP) pathway (Fig. **1**) [1].

In cancer cells, the glycolytic pathway is affected significantly since tumor cells consume much higher amounts of glucose than normal cells, which is known as the Warburg effect [2]. The Embden-Meyerhof-Parnas pathway is the glycolytic pathway used by mammals, consisting of the initial phosphorylation of glucose at the 6th position, consuming the first ATP molecule to produce glucose-6-phosphate (G6P), being catalysed by the enzyme hexokinase (HK). Next, an isomerization process takes place under the catalysis of phosphoglucose isomerase (PGI) converting the pyranose into a furanose ring, resulting in the formation of fructose 6-phosphate (F6P). Then, a second phosphorylation reaction at position 1 occurs, requiring an ATP molecule, and catalysed by the enzyme phosphofructokinase (PFK), to provide fructose 1,6-diphosphate (F1,6DP) which is cleaved into dihydroxyacetone (DHAP) and glyceraldehyde 3-phospate (G3P) by the catalysis of fructose biphosphate aldolase (FBP).

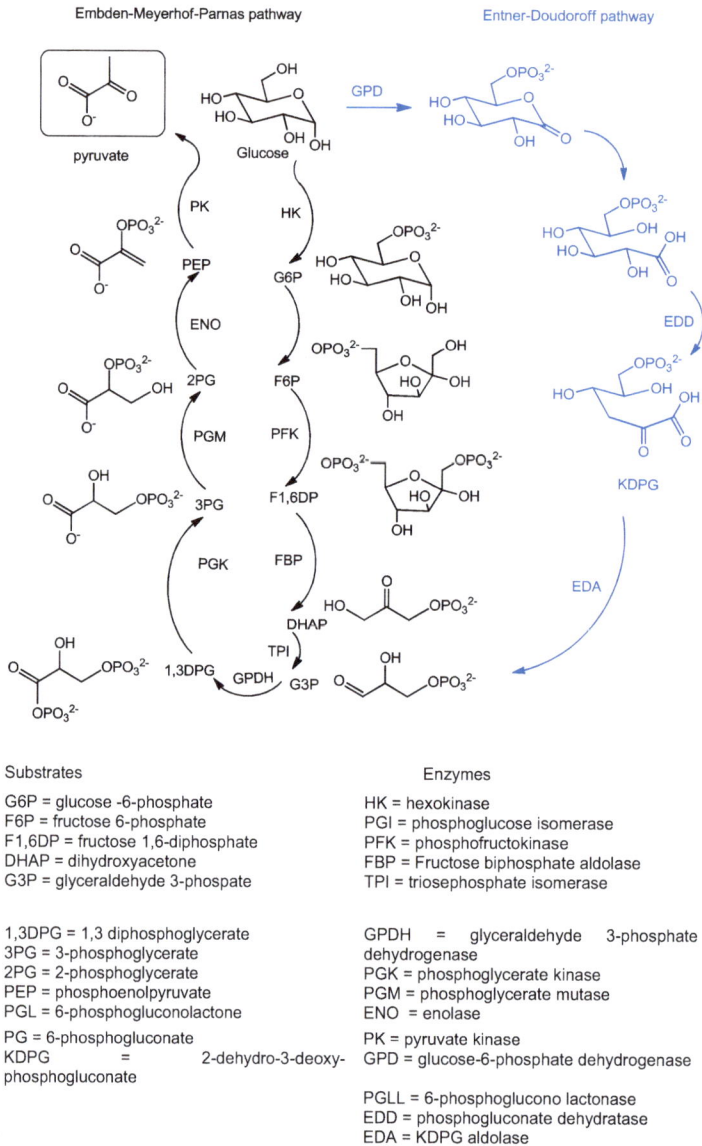

Substrates

G6P = glucose -6-phosphate
F6P = fructose 6-phosphate
F1,6DP = fructose 1,6-diphosphate
DHAP = dihydroxyacetone
G3P = glyceraldehyde 3-phospate

1,3DPG = 1,3 diphosphoglycerate
3PG = 3-phosphoglycerate
2PG = 2-phosphoglycerate
PEP = phosphoenolpyruvate
PGL = 6-phosphogluconolactone

PG = 6-phosphogluconate
KDPG = 2-dehydro-3-deoxy-phosphogluconate

Enzymes

HK = hexokinase
PGI = phosphoglucose isomerase
PFK = phosphofructokinase
FBP = Fructose biphosphate aldolase
TPI = triosephosphate isomerase

GPDH = glyceraldehyde 3-phosphate dehydrogenase
PGK = phosphoglycerate kinase
PGM = phosphoglycerate mutase
ENO = enolase

PK = pyruvate kinase
GPD = glucose-6-phosphate dehydrogenase

PGLL = 6-phosphoglucono lactonase
EDD = phosphogluconate dehydratase
EDA = KDPG aldolase

Fig. (1). The Embden-Meyerhof-Parnas and Entner-Doudoroff as the main glycolytic pathways in prokaryote and eukaryote cells.

The next steps involve inter-conversion from DHAP to the G3P by the enzyme triose phosphate isomerase (TPI). Second phosphorylation of glyceraldehyde 3-phosphate proceeds by the enzyme glyceraldehyde-3-phosphate dehydrogenase (G3P) in the presence of NAD^+ and Pi providing 1,3 diphosphoglycerate (1,3DPG). The next step is mediated by the enzyme phosphoglycerate kinase to produce the first ATP molecule and 3-phosphoglycerate which undergoes phosphate migration producing 2-phosphoglycerate. Further, dehydration mediated by the enzyme enolase produces phosphoenolpyruvate which is finally converted by the enzyme pyruvate kinase to pyruvate, leading to the generation of a second ATP molecule. It is important to notice that the overall energy yield is two ATPs, considering that one glucose consumes two ATPs to form fructose-1,6-diphosphate (F1,6DP), and then the cleavage of each F1,6DP produces two three-carbon molecules G3P and DHAP, each of them producing two ATP molecules (Fig. **2**).

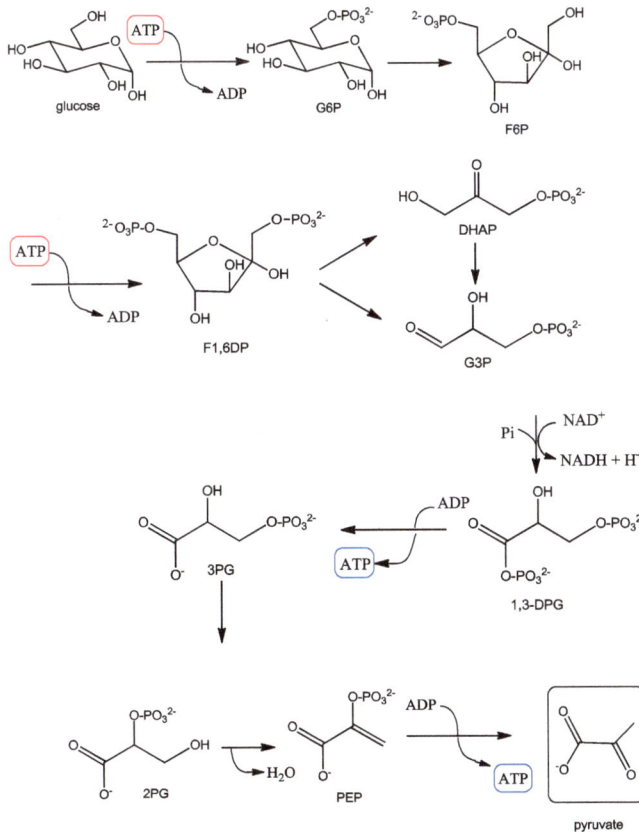

Fig. (2). The Embden-Meyerhof-Parnas pathway showing the intermediates and the steps where ATP is consumed and produced.

1. ENZYMES AND THEIR INHIBITORS INVOLVED IN THE GLYCOLYSIS PATHWAY

1.1. Hexokinase (HK)

Hexokinase is a phosphorylating enzyme widespread in microorganisms, plants and vertebrates, being responsible for the first step during glycolysis and its deficiency in humans is associated with haemolytic anemia. Thus, hexokinase uses glucose as a substrate which is irreversibly phosphorylated at the 6th position to produce glucose 6-phosphate (GLc-6-P) and ADP. The phosphorylated glucose becomes an HK natural inhibitor, and if Glc-6-P is not rapidly consumed, the enzyme decreases its activity. Besides the need for ATP, the reaction requires magnesium ion (Fig. **3**).

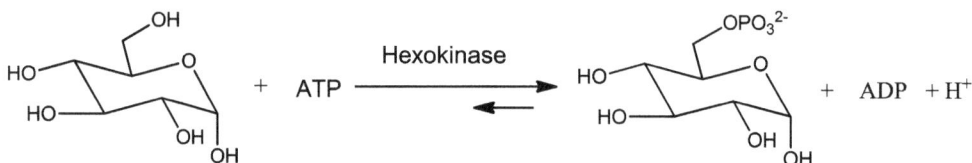

Fig. (3). The hexokinase reaction catalysing the conversion of glucose and ATP into glucose-6-phosphate and ADP.

A general mechanism for hexokinase phosphorylation reaction requires Mg^{2+}, and proposes an initial capture of hydrogen at position 6th by aspartate residue generating an alcoxide which attacks the ATP terminal phosphate group producing an O-P cleavage bond, resulting in the formation of glucose--phosphate and ADP (Fig. **4**) [3].

Different hexokinases have been analysed by crystallographic studies, and their structure determined, such as human hexokinase (PDB: 1HKB), *Bacillus subtilis* (PDB: 1XC3), and *Escherichia coli* (PDB: 1RKD), and *Sulfolobus tokodaii* (PDB: 2E2Q) in a complex with xylose and ADP (Fig. **5**). The ribbon structure shows a large α/β and a small α/β domain with the active site in a deep cleft at the central part shown by the spheres. At the catalytic site, it is possible to see the glucose as a stick model in chair conformation, aspartic acid D95 residue, the ATP molecule in orange and the Mg^{2+} ion as a green sphere [3, 4].

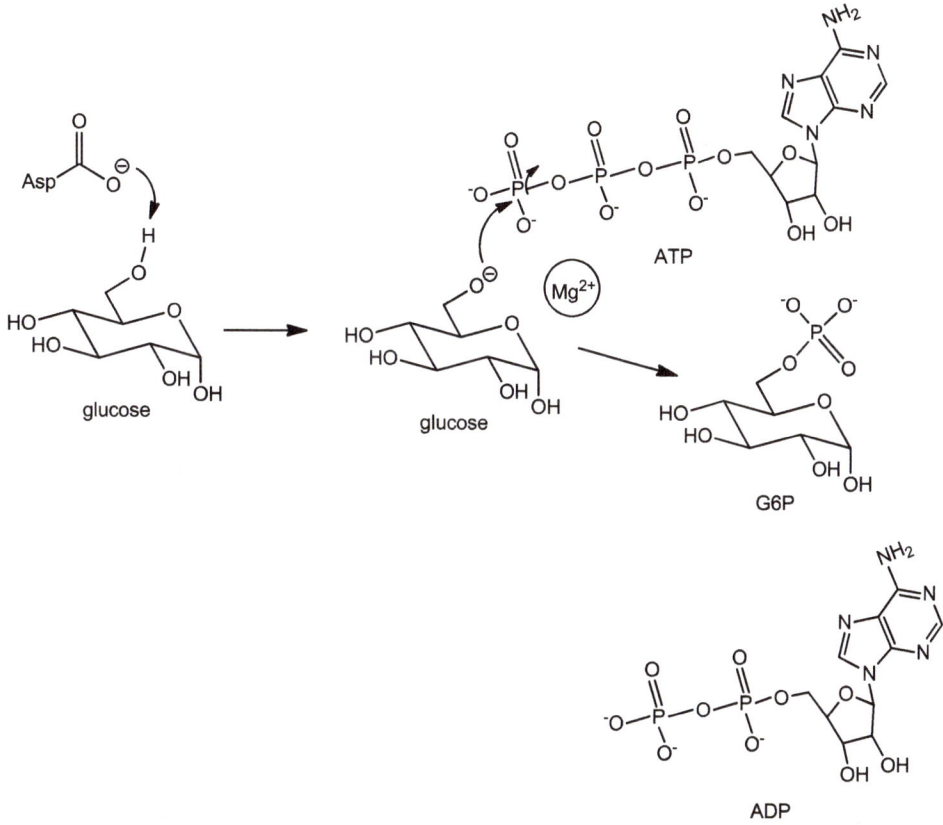

Fig. (4). General mechanism of carbohydrate kinases involving aspartate residue, glucose and ATP in the presence of Mg^{2+} to produce glucose-6-phosphate and ADP.

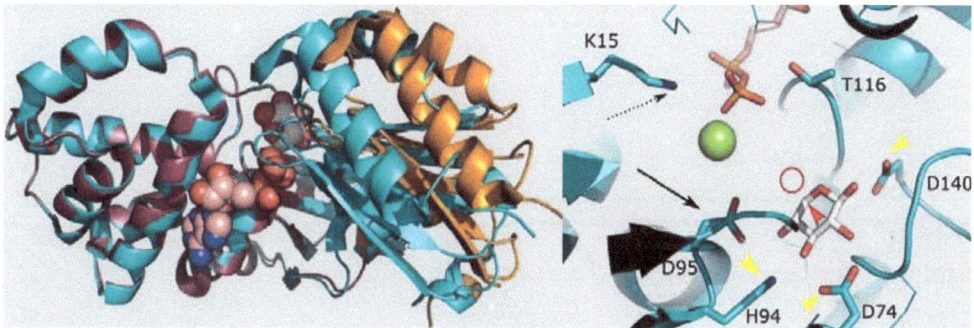

Fig. (5). Crystal structure of apo hexokinase from *Sulfolobus tokodaii* (PDB: 2E2Q), showing the large and small domains with the active site at the interphase and active site approach with the residues interacting with xylose, and ATP in the presence of Mg^{2+} ion.

1.2. Hexokinase Inhibitors

Lonidamine is an indazole-3-carboxilic acid (Fig. **6**) and it works as a mitochondrial hexokinase inhibitor effective as anti-tumour drug affecting the bioenergetics of the cells by inhibiting glycolysis, reducing ATP levels and mitochondrial respiration [5].

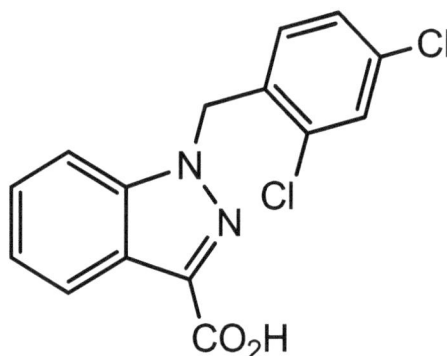

Fig. (6). Chemical structure of Lonidamine.

2-Bromo pyruvate (Fig. **7**) is an alkylating agent with an anti-glycolytic effect, due to its ability to inhibit mitochondrial hexokinase, inducing cell death in cultured tumour cells [6].

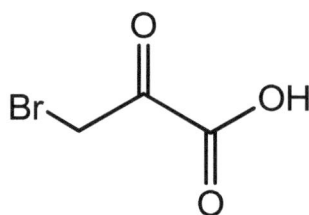

Fig. (7). Chemical structure of 2-bromo pyruvate.

2-Deoxyglucose (Fig. **8**) is a deoxyglucose derivative which can be phosphorylated by hexokinase, inducing the accumulation of 2-deoxyglucos--phosphate within the cell, leading to inhibition of the enzyme. Studies on cancer cells have shown that 2-DG significantly suppresses proliferation, causing apoptosis and reducing migration of murine endothelial cells [7].

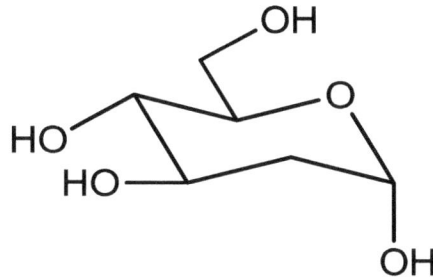

Fig. (8). Chemical structure of 2-deoxyglucose.

1.3. Phosphoglucose Isomerase (PGI)

Phosphoglucose isomerase (Fig. **9**) is a cytosolic enzyme that catalyses the second step of glycolysis, consisting of the reversible conversion of D-glucose-6-phosphate (G6P) into D-fructose-6-phosphate (F6P). Its deficiency leads to the Satoyoshi disease characterised by haemolytic anemia and muscle pain.

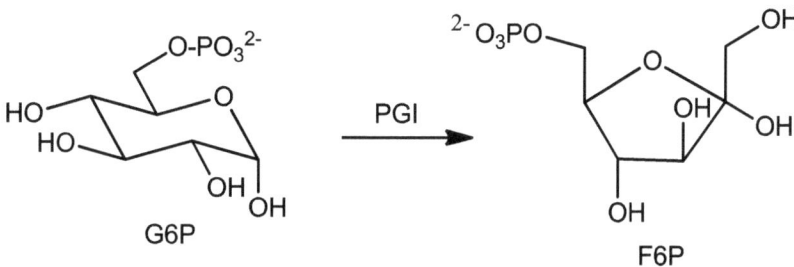

Fig. (9). The phosphoglucose isomerase reaction that catalyses the conversion of glucose-6-phosphate to fructose-6-phosphate.

The role of this enzyme is to promote the interconversion from pyranoside to furanoside rings, by series of steps involving ring-opening, isomerization and ring closure. In order to explain how the conversion takes place, a proposed mechanism considers ring opening assisted by lysine residue, followed by the formation of a cis-enediol intermediate, which isomerises to the ketone form at the C-1 position. Subsequent nucleophilic addition to the ketone at C-1 will result in the ribose ring formation according to Fig. **(10)** [8].

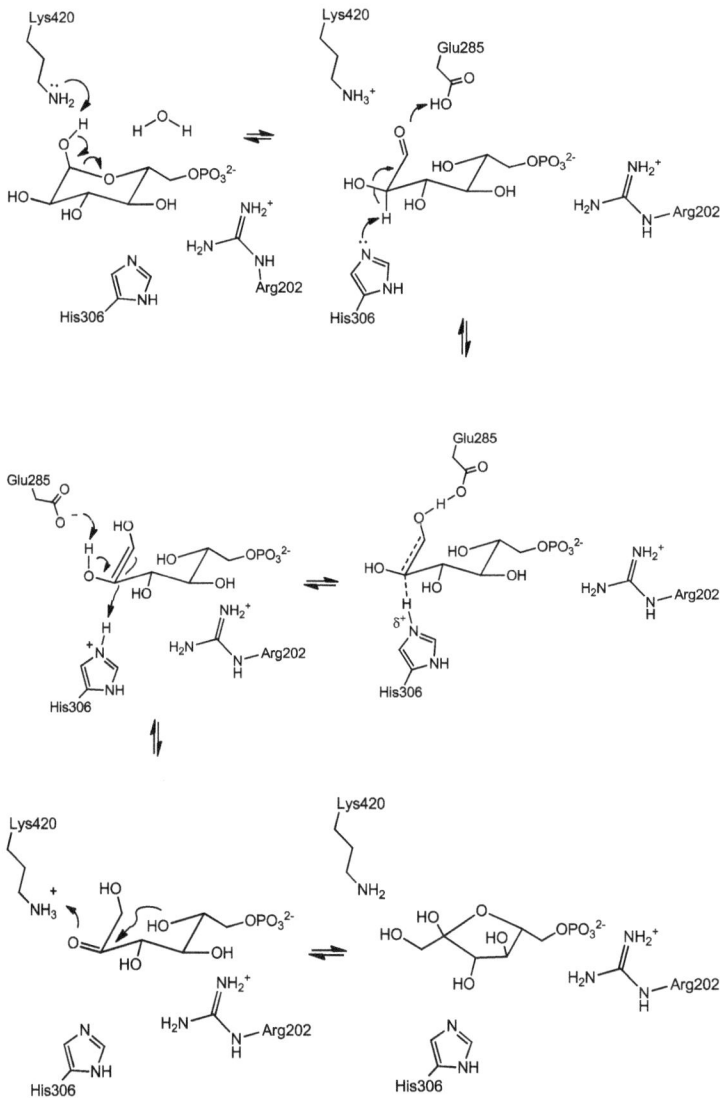

Fig. (10). Proposed phosphoglucose isomerase reaction mechanism involving amino acid residues to convert D-glucose-6-phosphate (G6P) to D-fructose-6-phosphate (F6P).

Different phosphoglucose isomerase (PGIs) structures have been reported, among them, pig (PDB: 1GZD), *Trypanosoma brucei* (PDB: 2O2C), and human (PDB: 1NUH), exhibiting high homology. The human PGI is a homodimeric structure composed by a large N-terminal domain, a second smaller domain, an extended

C-terminal arm, with the active site located between the large and the small domains (Fig. **11**). Inhibition studies comparing human (*h*PGI) and *trypanosome brucei* (*t*PGI) demonstrate that inhibitors suramin, agaricic acid and 5PAH inhibit more strongly the parasitic enzyme than human, and to clarify such inhibitory variability, a superimposed analysis on the active site was performed showing the residues and conformations with high overlapping (parasite in black and human in white) although the residues 258-264 and 299-313 in *t*PGI are in closed conformation [8 - 10].

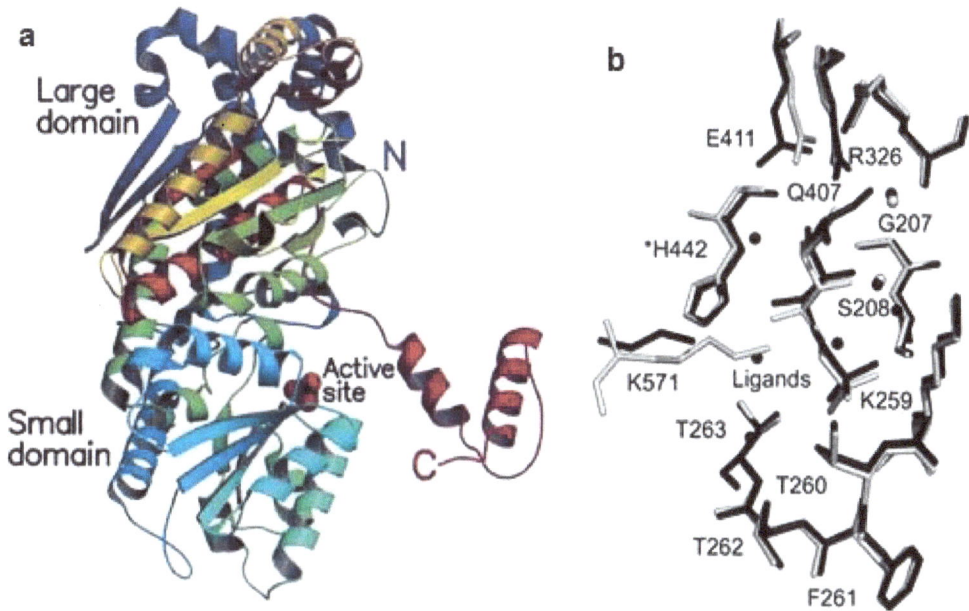

Fig. (11). a) Crystal structure of human phosphoglucose isomerase as a homodimer with a large, small side arm domains and the active site (PDB: 1NUH) b) Superimposed active sites of *Trypanosoma brucei* and human phosphoglucose isomerase.

1.4. Phosphoglucose isomerase Inhibitors

Structurally diverse compounds have shown to be potent inhibitors such as suramin K_i = 0.29 mM, agaricic acid IC_{50} value of 10 μM, 5-phosphoarabinonohydroxamic acid (5PAH) K_i = 0.05 μM, cis-enediol 6-phosphoarabinate with K_i = 0.20 μM, D-arabinonohydroxamic acid 5-phosphate with K_i = 0.50 nM, and D-arabinonate 5-phosphate with K_i = 0.23 μM (Fig. **12**). Despite their effectiveness in protozoan infections, some side effects and toxicity may appear since they can also inhibit other enzymes such as phosphoglycerate kinase, tyrosine phosphatase and glucose 6-phosphate dehydrogenase [10, 11].

suramin

agaricic acid

5PAH   cis-enediol 6-phosphoarabinate

R = NHOH D-arabinonhydroxamic acid 5-phosphate

R = O⁻ D-arabinonate 5-phosphate

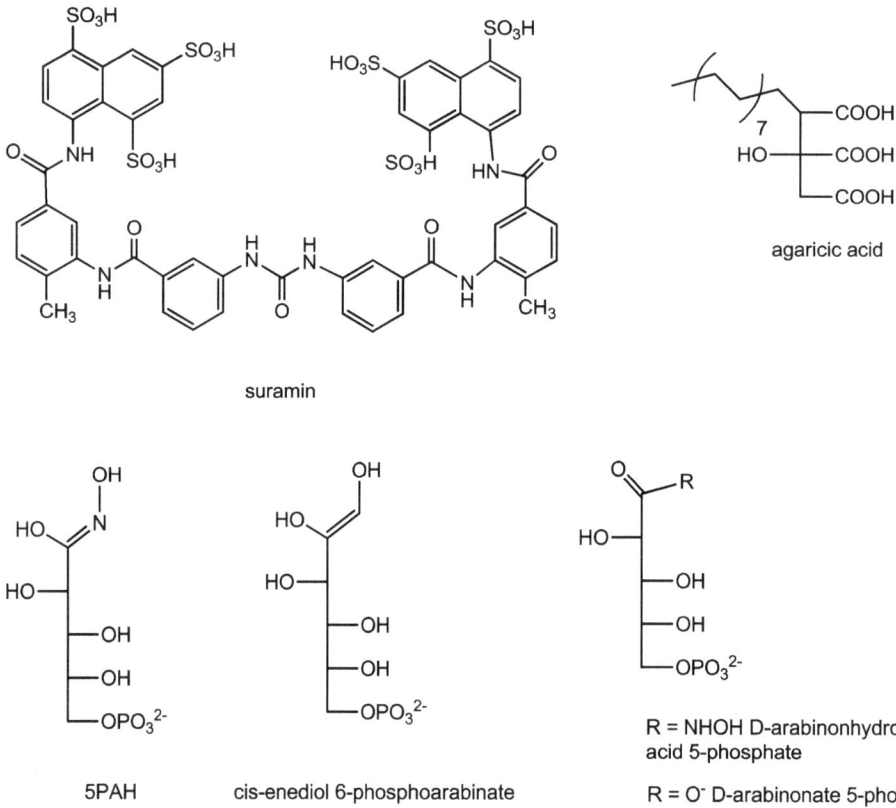

Fig. (12). Chemical structure of phosphoglucose isomerase inhibitors suramin, agaricic acid and phosphoarabinate analogues.

1.5. Phosphofructokinase (PFK)

Phosphofructokinase is an allosteric enzyme considered the main regulator of glycolysis and catalyses the formation of fructose 1,6-bisphosphate from fructose 6-phosphate (Fig. **13**). It is controlled by ATP and ADP or AMP in a way that high ATP concentration reduces the enzyme activity, and low ADP or AMP increases. Its deficiency is associated with muscle disorders such as the Tarui disease characterised by fatigue, exercise intolerance and muscle cramps.

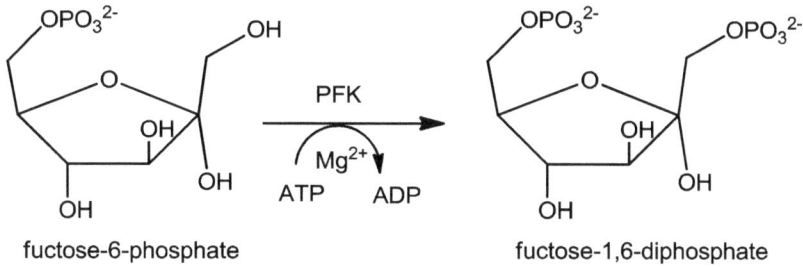

Fig. (13). The phosphofructokinase reaction.

A basic mechanism considers a proton transfer from F-6P to aspartate residue (Asp256) and the alkoxide formed at position 1 attacking the ATPs terminal phosphate group stabilized by Mg^{2+}, to yield fructose 1,6 diphosphate and ADP (Fig. **14**) [12].

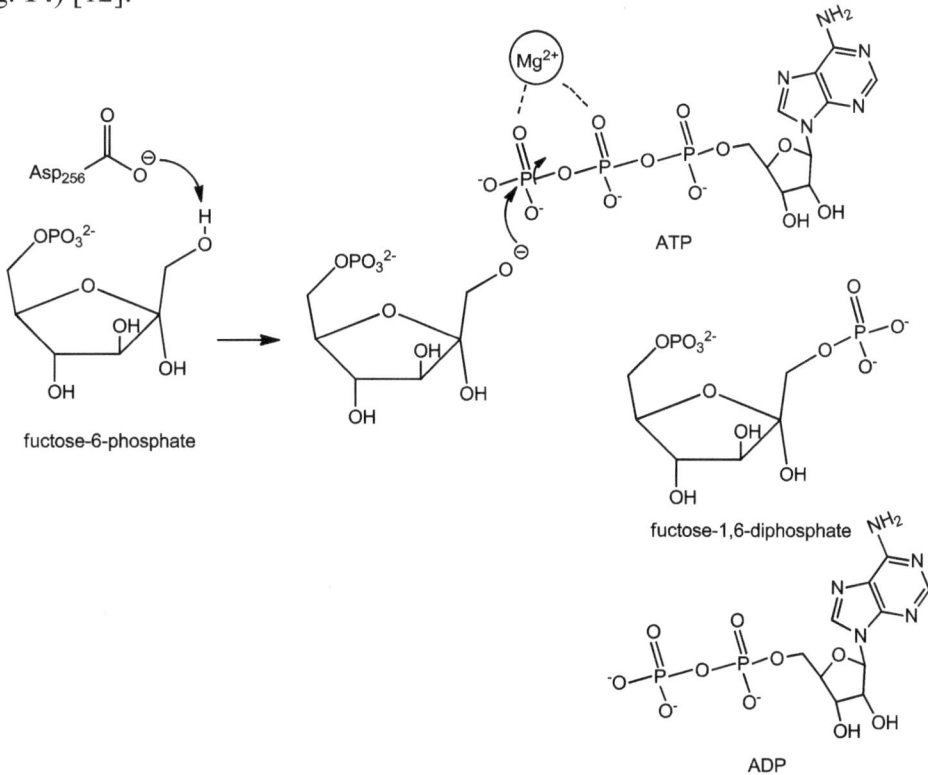

Fig. (14). Proposed reaction mechanism for the conversion of fructose-6-phosphate to fructose-1,6-diphosphate.

The preliminary crystallographic analysis of human muscle phosphofructokinase has been described showing a tetrameric enzyme dissociated into a dimer [13]. The packing and quaternary structure of one monomer were compared with rabbit muscle PFK (PDB 3O8N) obtaining more than 90% sequence identity. Superimposed active site regions of *Saccharomyces cereviciae* (*Sc*PFK) and *Escherichia coli* (*Ec*PFK) showing the key residues participating in the binding with Fru6-P and the cofactor (Fig. **15**) [14].

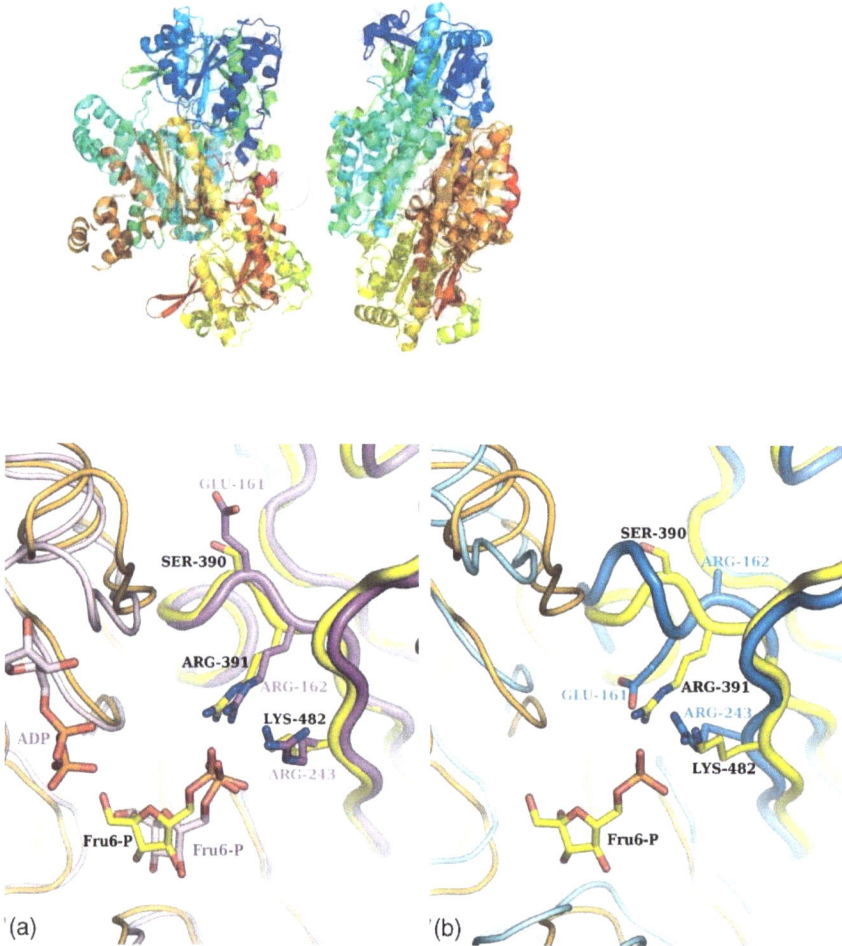

Fig. (15). Crystal structure of human muscle phosphofructokinase (PDB: 4OMT) and superimposed active site regions of *Saccharomyces cereviciae* and *Escherichia coli* showing the residues interacting with fructose 6-phosphate.

1.6. Phosphofructokinase (PFK) Inhibitors

A sulfonamide isoxazole and thiadiazole analogues (Fig. **16**) were described as potent inhibitors, displaying IC_{50} values of 370 nM, although some toxicity against cultured *T. brucei* ED_{50} = 16.3 μM and K_i of 52 nM were encountered [15].

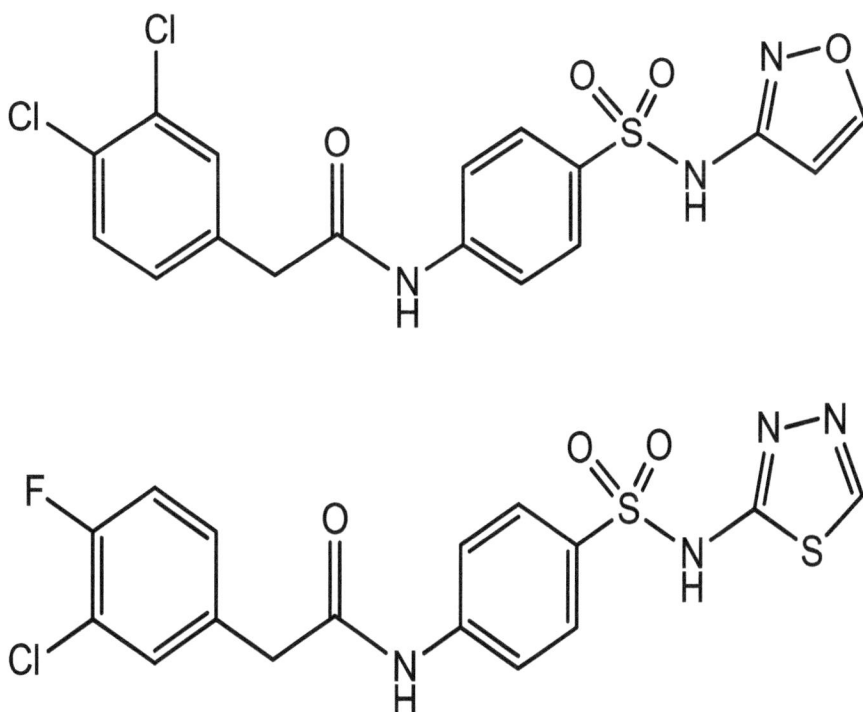

Fig. (16). Chemical structure of sulfonamide isoxazole and thiadiazole analogues as phosphofructokinase inhibitors.

1.7. Fructose Biphosphate Aldolase (FBP Aldolase)

Human aldolase FBP exist in three isoforms aldolase A, B and C present in muscle and blood cells, liver and brain respectively. It is responsible for the cleavage reaction of fructose 1,6-biphosphate to yield D-glyceraldehyde 3-phosphate, and dihydroxyacetone phosphate (Fig. **17**). Its deficiency has been correlated to hereditary fructose intolerance which can trigger haemolytic anemia and myopathy.

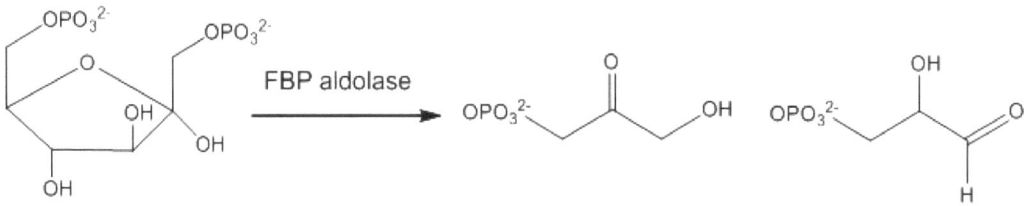

Fig. (17). The fructose biphosphate aldolase reaction leading to the formation of dihydroxyacetone phosphate and D-glyceraldehyde 3-phosphate.

The proposed mechanism (Fig. **18**) involves the furanose ring opening at the anomeric position to furnish a ketone which suffers a nucleophilic attack from a lysine residue producing and amino acid imine. Next a retro Claisen type reaction lead to the diol cleavage giving D-glyceraldehyde 3-phosphate and the enamine which tautomerize to the iminiun salt providing after hydrolysis dihydroxyacetone phosphate [16].

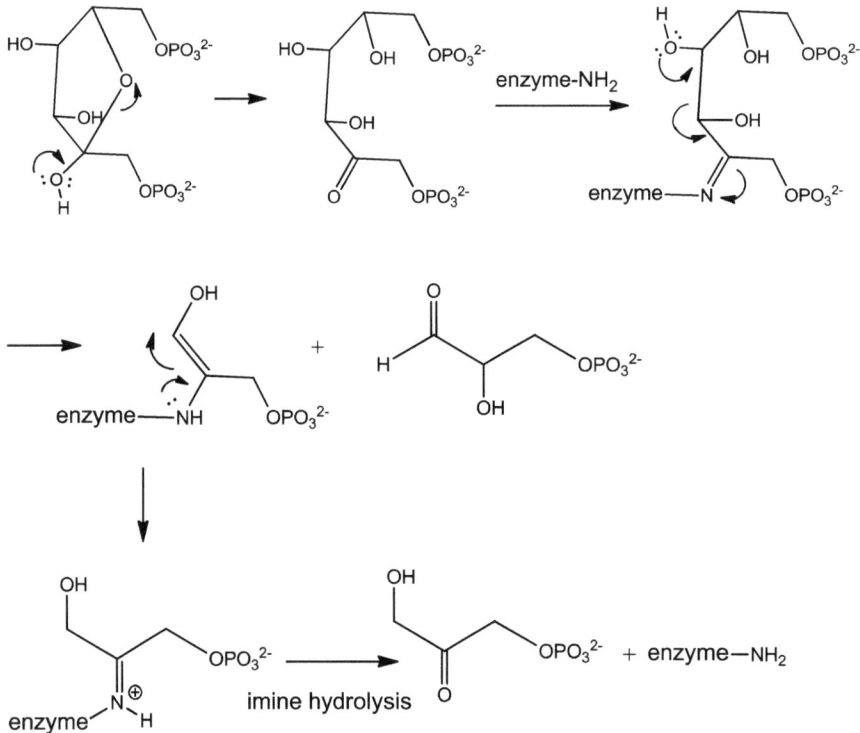

Fig. (18). Proposed reaction mechanism catalysed by fructose biphosphate aldolase leading to ring opening, imine formation and retro Claise aldol reaction to yield dihydroxyacetone phosphate and D-glyceraldehyde 3-phosphate.

Mammalian FBP aldolase from muscle tissue has higher substrate specificity for fructose 1,6-bisphosphate over fructose 1-phosphate. Human muscle aldolase bisphosphate complexed with fructose 1,6-bisphosphate was determined by crystallographic studies, and the residues localized at the catalytic site assigned. The residues involved in the interactions with the substrate are Glu187 in the proximity with C-3, Lys146 establishing contacts with C-3 and C-4, Glu189, Lys146 and Arg148 close to the cleavage point of the substrate and Lys229 was determined to be responsible for the Schiff′s formation with the ketone group (Fig. **19**) [17].

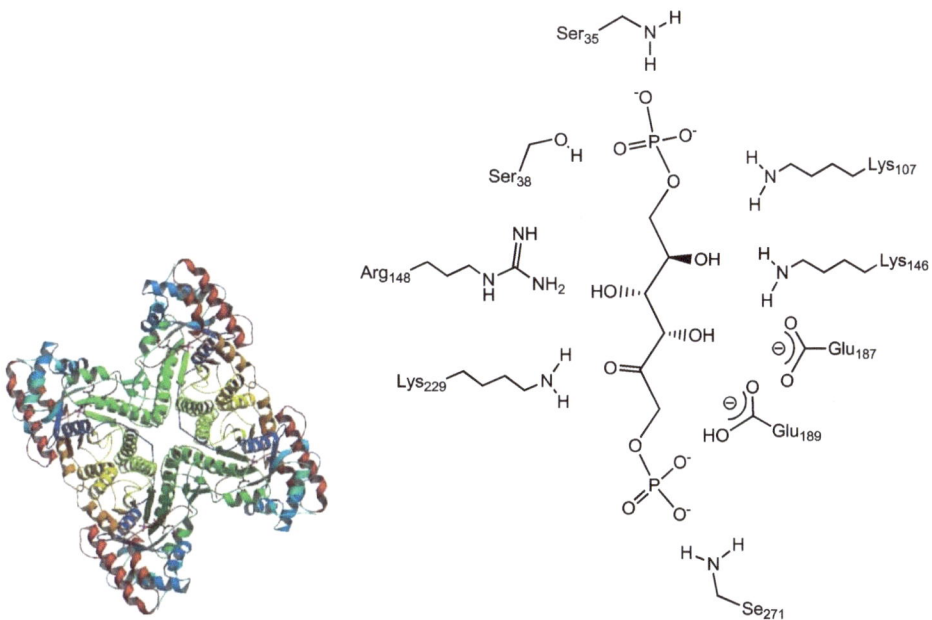

Fig. (19). Crystal structure of mammalian FBP aldolase from muscle tissue (PDB: 4ALD) and residue interacting with fructose 1,6-bisphosphate.

1.8. Fructose Biphosphate Aldolase Inhibitors

Hydroxynaphthaldehyde phosphate (Fig. **20**) has been evaluated as selective inhibitor against the glycolytic fructose 1,6-bisphosphate aldolase from *Trypanosoma brucei*, observing K_i values of 23.03 ± 2.31 µM [18].

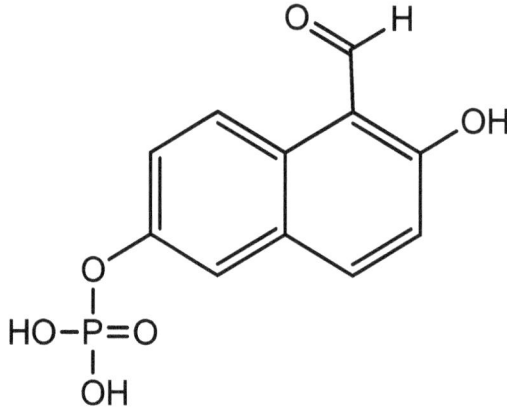

Fig. (20). Chemical structure of hydroxynaphthaldehyde phosphate.

A potent FBP aldolase class II inhibitor is the DHAP analogue phosphoglycolic hydroxamic acid (Fig. **21**) with Ki of 10 nM in yeast and much less active in FBP aldolase class I (mammalian aldolase) with K_i of 1 μM [19].

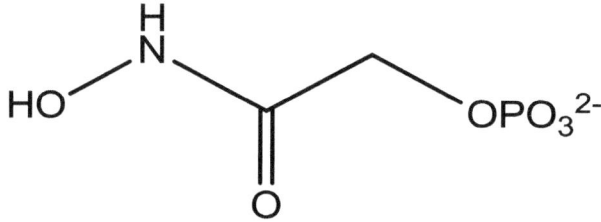

Fig. (21). Chemical structure of phosphoglycolic hydroxamic acid.

1.9. Triosephosphate Isomerase (TIM)

This enzyme is considered the most active in glycolysis and catalyses an isomerization process, converting dihydroxyacetone phosphate into glyceraldehyde 3-phosphate (Fig. **22**). Its malfunction produce accumulation of dihydroxyacetone phosphate mainly in red blood cells resulting in serious damage including haemolytic anemia and neurologic dysfunction.

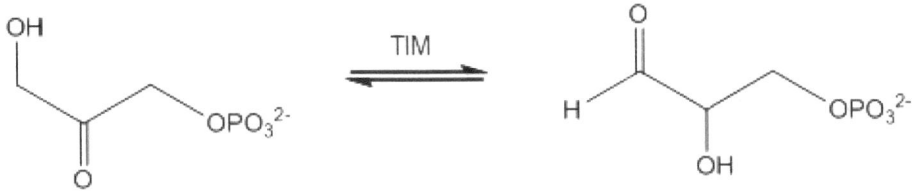

Fig. (22). The triosephosphate isomerase reaction converting dihydroxyacetone phosphate into glyceraldehyde 3-phsphate.

The reaction mechanism involves an initial attack of glutamate residue to the α-carbonyl position to produce an enediol intermediate, which further interacts with a histidine residue following a hydrogen transfer step to produce glyceraldehyde 3-phosphate (Fig. **23**).

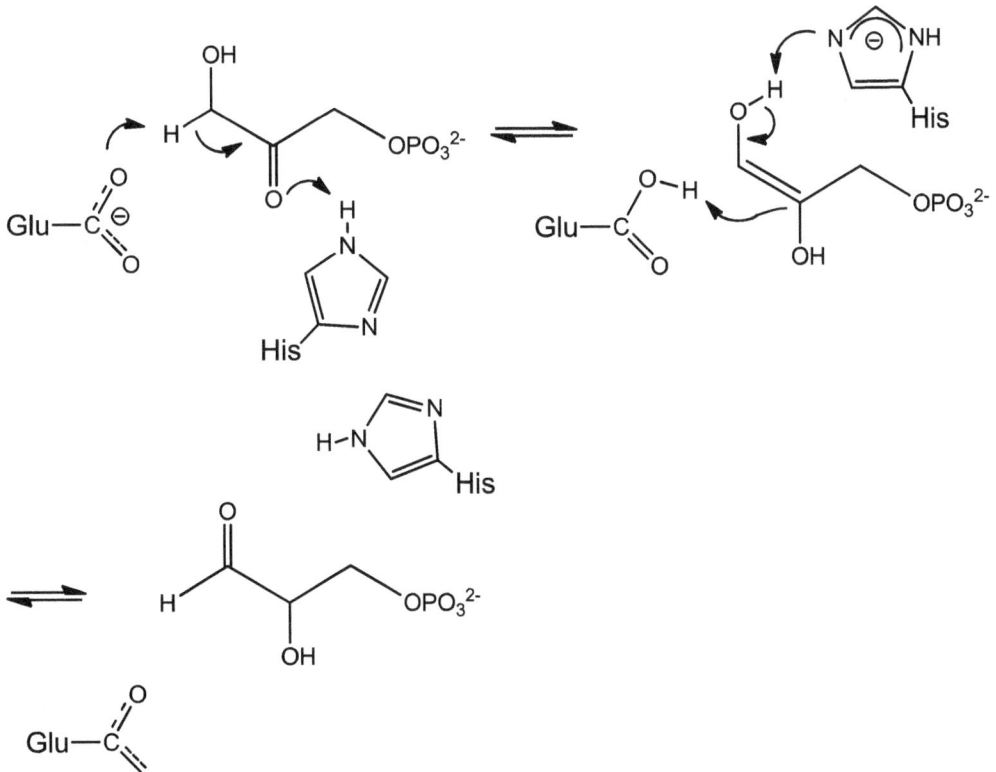

Fig. (23). Proposed reaction mechanism for the conversion of dihydroxyacetone phosphate into glyceraldehyde-3-phosphate.

Crystallographic studies of triosephosphate isomerase from different species have been described, among them *Giardia lamblia* in an effort to find novel therapeutic agents against giardiasis. The ribbon structure shows a homodimer with each monomer formed by eight β-strands and eight α-helices. Directed mutagenesis on cysteine222 residue has been directed to understand the stability and structure of TIM enzyme, leading to the conclusion that Cys222 residue is required for activity. On the other hand the active site region has been identified showing the residues in the proximity of 2-phosphoglycolate as ligand are Glu98, His96, Glu174, Ile175, Ser170, Gly214, Val236, Leu235 and Lys13 (Fig. **24**) [20].

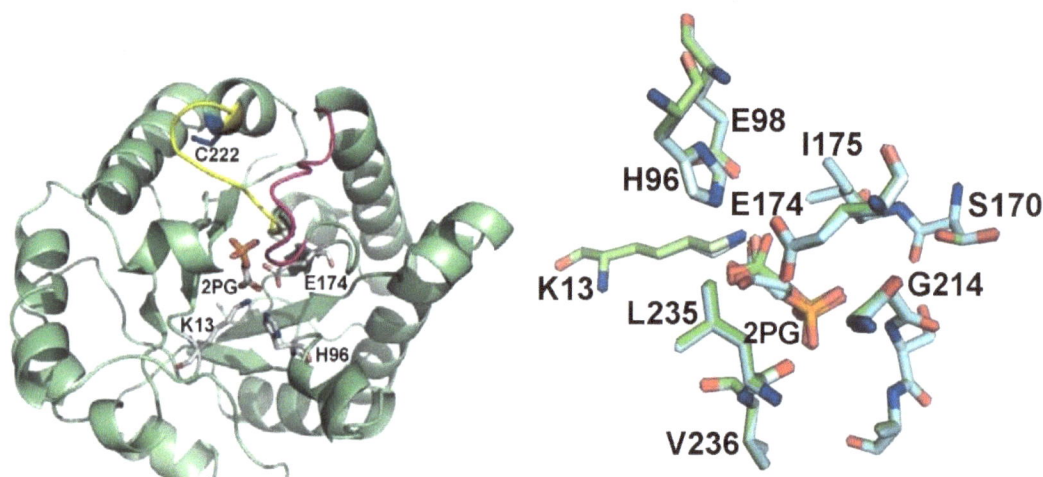

Fig. (24). Crystal structure of triosephosphate isomerase from *Giardia lamblia* and active site region of interacting residues with the ligand (PBD: 4B15).

1.10. Triosephosphate Isomerase (TPI) Inhibitors

A recent study on glycolysis describes the natural pyruvate kinase (PK) substrate phosphoenolpyruvate (Fig. **25**) as an important TPI inhibitor, which can alter oxidative stress in cancer cells and actively respiring cells. Crystal studies reveals that PEP competitively inhibits the interconversion of glyceraldehyde-3-phosphate and dihydroxyacetone phosphate [21].

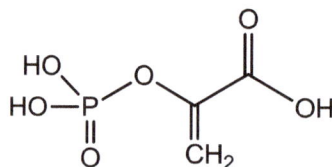

Fig. (25). Chemical structure of phosphoenolpyruvate.

1.11. Glyceraldehyde Phosphate Dehydrogenase (GAPDH)

This enzyme is involved in the 6th step of glycolysis consisting in the oxidative phosphorylation of D-glyceraldehyde 3-phosphate (D-G3P) to give 1,3-diphosphoglycerate (1,3-DPG), requiring NAD(P)$^+$ or NAD$^+$ as cofactor and phosphate (Fig. **26**).

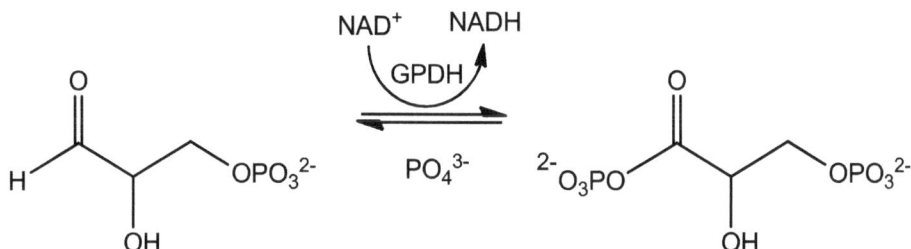

Fig. (26). The glyceraldehyde phosphate dehydrogenase reaction converting D-glyceraldehyde 3-phosphate to 1,3-diphosphoglycerate.

The mechanism involves a two-step sequence, the first implying the binding of D-G3P with the binary complex GAPDH·NAD, through the thiol group to form a thiohemiacetal and after hydride transfer assisted by NAD$^+$ to give a thioester. The second step consisting in the nucleophilic attack of the phosphate group to the thioacylenzyme intermediate to yield after enzyme displacement the substrate 1,3-dPG (Fig. **27**) [22].

Chloroplast GAPDH has been solved as hetero tetramer A$_2$B$_2$, with B subunits having a small C-terminal extension, and 80% identical in sequence to A subunits. The active site amplification allows identifying in detail the residues located around NADP in green, sulphate ions as gold spheres, and a disulfide bond between Cys349 and Cys358 in yellow. A schematic illustration shows the interactions between the residues of GAPDH with NADP+ and sulphate ion (Fig. **28**) [23].

1.12. Glyceraldehyde Phosphate Dehydrogenase (GAPDH) Inhibitors

The preparation of the ψGAPDH peptide (Fig. **29**) with sequence Leu-Gly-Glu Val-Ile-Gly that selectively inhibits the phosphorylation of GAPDH has been described, observing after inhibition assays the reduction of GAPDH activity *in vitro* [24].

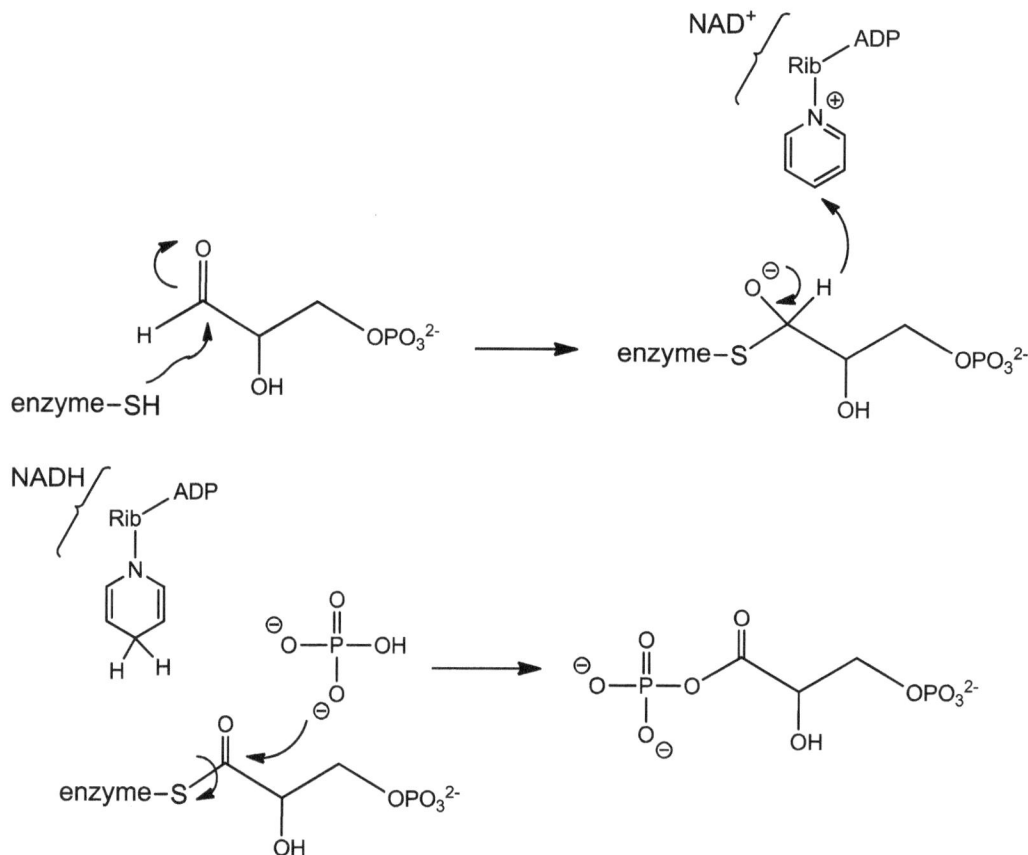

Fig. (27). Proposed reaction mechanism of D-glyceraldehyde 3-phosphate (D-G3P) phosphorylation.

Pentalenolactone and koningic acid (Fig. **30**) are known antibiotics having potent inhibition of the glycolytic enzyme glyceraldehyde-3-phosphate dehydrogenase, and therefore were used as synthons for preparing glyceraldehyde-3-phosphate analogues mimicking some of the functionalities present in the natural products [25].

2-Phenoxy naphthoquinone (Fig. **31**) has been described as a potent GAPDH inhibitor and cytotoxic compound, displaying its inhibitory activity against *Trypanosoma brucei* with EC_{50} values of 80nM [26].

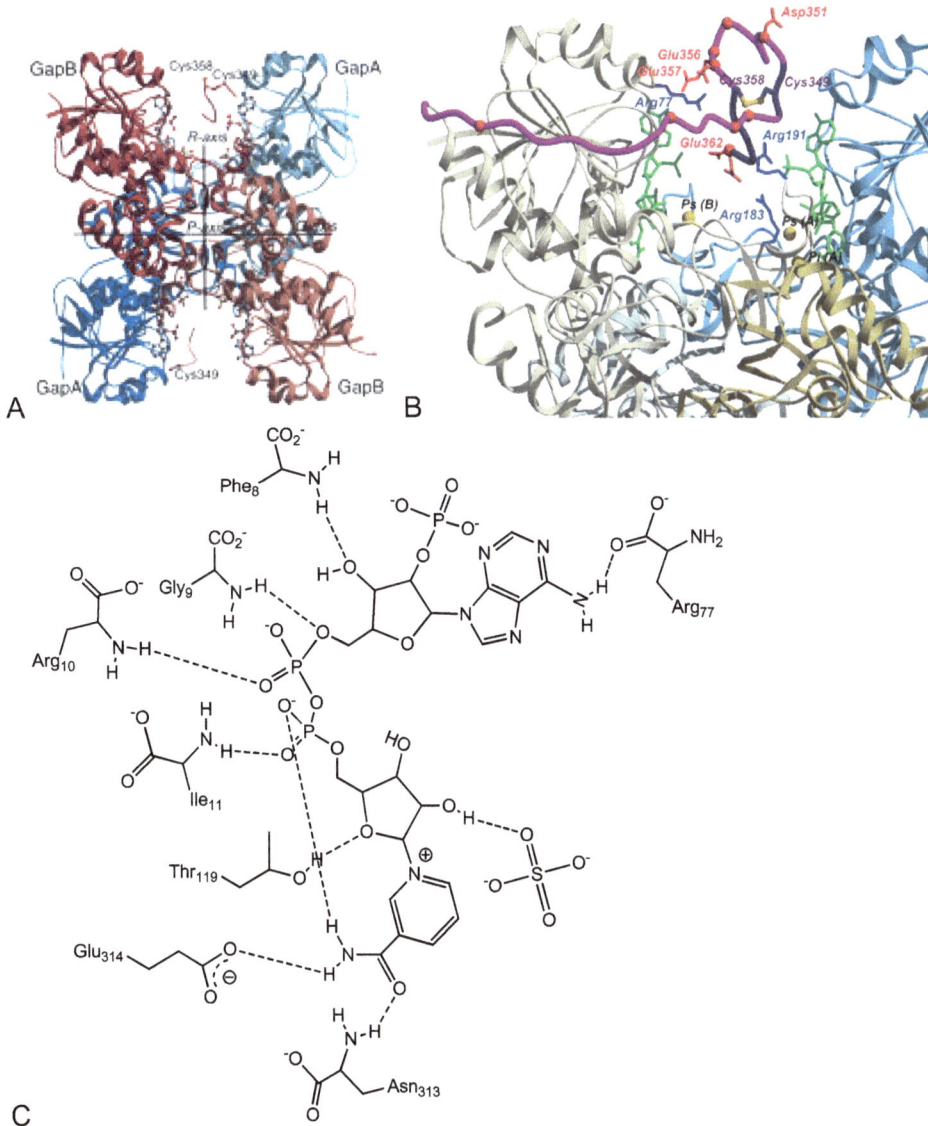

Fig. (28). A) Crystal structure of chloroplast glyceraldehyde phosphate dehydrogenase as hetero tetramer A_2B_2 (PDB: 2PKR). B) Model diagram of active site region with cofactor NADP. C) Interaction diagram between the ligands and cofactor NADP.

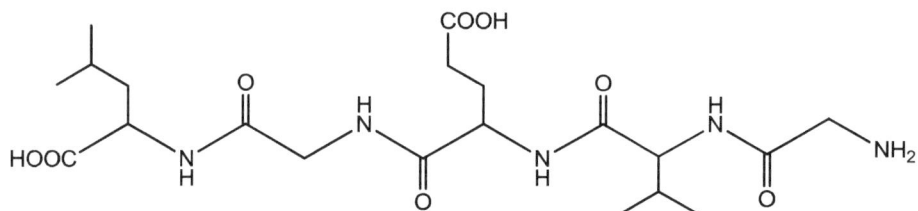

Fig. (29). Chemical structure of the ψGAPDH peptide.

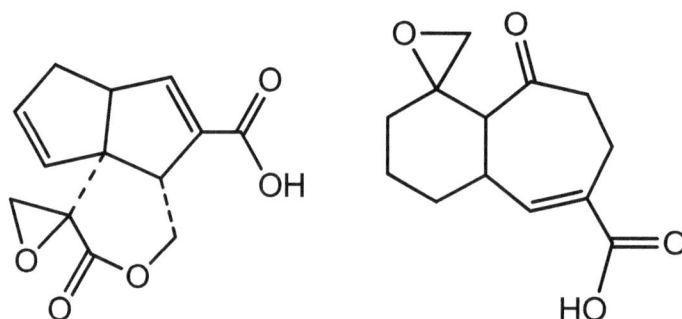

Fig. (30). Chemical structure of entalenolactone and koningic acid.

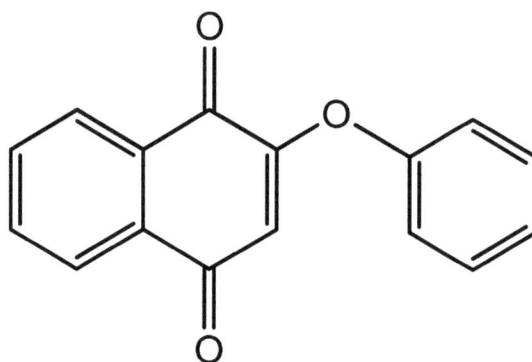

Fig. (31). Chemical structure of 2-phenoxy naphthoquinone.

A combination of antimycin A and sodium iodoacetate (Fig. **32**) has been used to suppress ATP synthesis through irreversible inhibition of glyceraldehyde phosphate dehydrogenase [27].

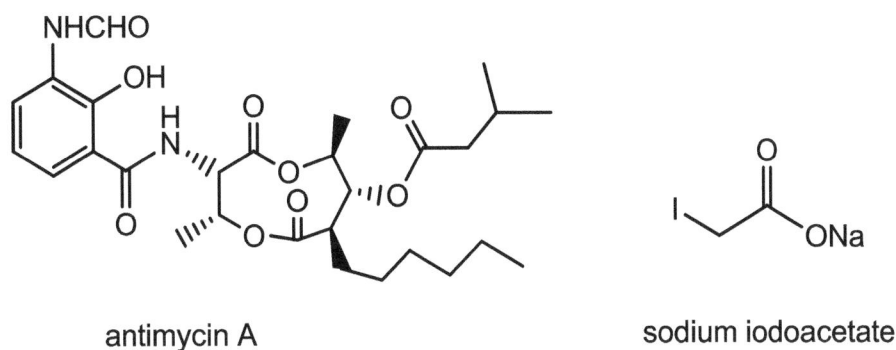

Fig. (32). Chemical structure of antimycin A and sodium iodoacetate.

1.13. Phosphoglycerate Kinase (PGK)

The PGK enzyme is a transferase responsible for the conversion of 1,3-diphosphoglycerate to 3-phosphoglycerate generating the first ATP molecule produced during the glycolytic pathway (Fig. **33**).

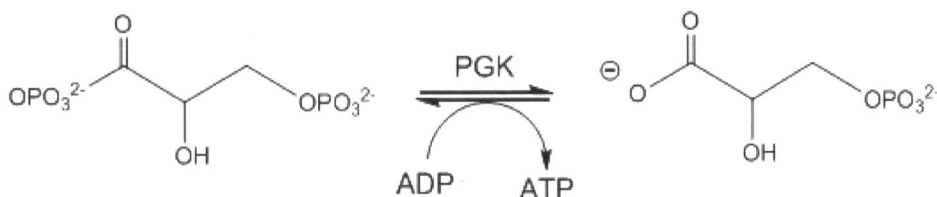

Fig. (33). The phosphoglycerate kinase reaction that converts 1,3-diphosphoglycerate to 3-phosphoglycerate.

This enzyme exists in two isoforms PGK1 and PGK2 (Fig. **34**), which are identical in about 97% in sequence and the active sites are essentially identical [28]. Human PGK is also responsible for the final phosphorylation of the non-natural nucleosides to the triphosphate homologues and the sequence identified under this study are Leu256, Lys215, Asp218, Lys219, Ala214, Gly238, Phe291, Leu313, Gly312, Val341, Glu343, Pro338 [29].

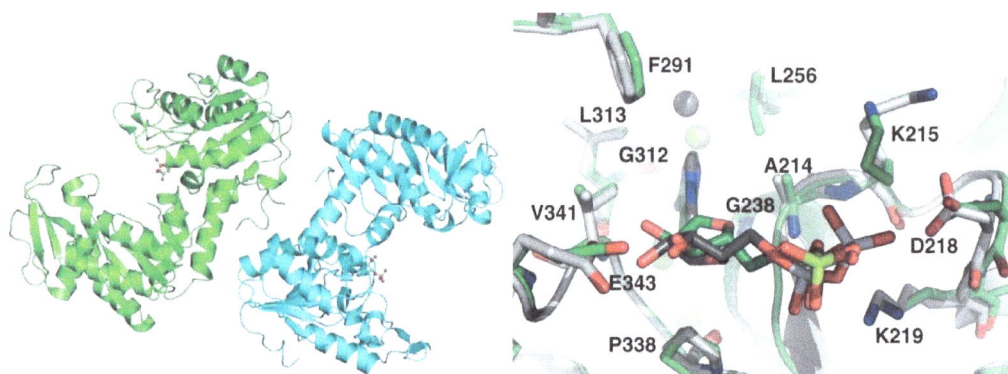

Fig. (34). Crystal structure of phosphoglycerate kinase (PDB: 3C39).

1.14. Phosphoglycerate Kinase Inhibitors

Moderate inhibition of this enzyme can cause large changes in the glycolytic supply in parasites such as *Trypanosoma brucei*, *Trypanosoma cruzi*, and *Leishmania spp* without affecting host red blood cells.

Aromatic and aliphatic fluorobisphosphonates (Fig. **35**) were subjected to structure-activity relationship studies including comparative molecular field analysis (CoMFA) and docking analysis presenting as result different degrees of human PGK enzyme inhibition [30].

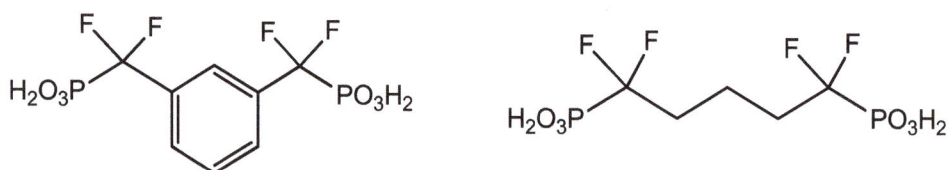

Fig. (35). Chemical structure of aromatic and aliphatic fluorobisphosphonates.

1.15. Bisphosphoglycerate Mutase (BPGM)

This is an enzyme present in erythrocytes and placental cells and catalyses the transformation of 1,3-bisphosphoglycerate to 2,3-bisphosphoglycerate (Fig. **36**), which acts as allosteric effector of haemoglobin shifting the equilibrium between the oxy and deoxy conformations of haemoglobins favouring the unligated form [31].

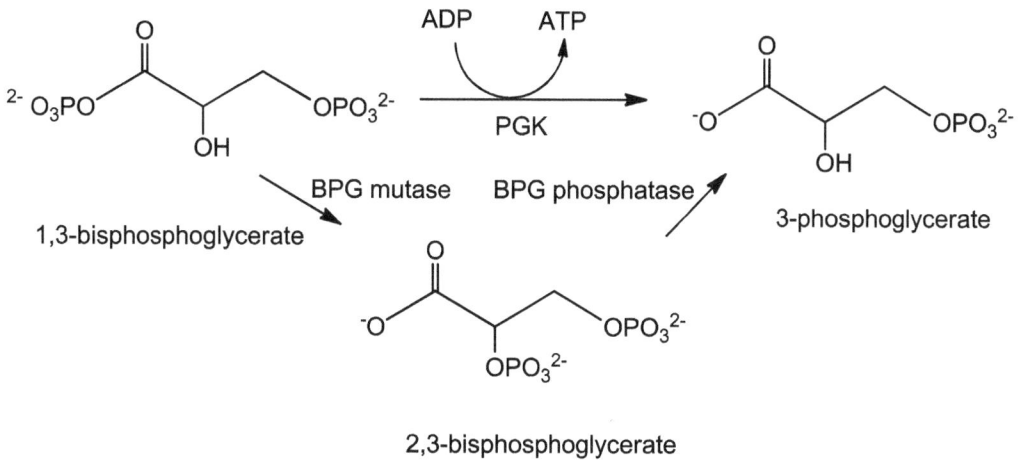

Fig. (36). The bisphosphoglycerate mutase reaction.

1.16. Phosphoglycerate Mutase (PGM)

These enzyme catalyses the internal transfer of a phosphate group from C-3 to C-2 to convert 3-phosphoglycerate to 2-phosphoglycerate (Fig. **37**). There are broadly two classes of PGMs: cofactor dependent (dPGM) and cofactor independent (iPGM), and there have been reports addressing that in tumor cells the activity of dPGM is higher, suggesting tumor progression. Also its deficiency reduce muscle activity and causes premature fatigue and contractures, and is consider a metabolic myopathie aside of myophosphorylase, muscle phosphofructokinase and lactate dehydrogenase deficiencies [32].

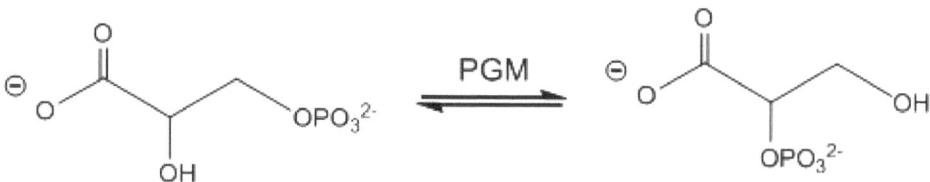

Fig. (37). The phosphoglycerate mutase reaction.

In human B type phosphoglycerate mutase (Fig. **38**) the residues found at the active site are Arg116, Arg117, Arg110, Cys23, His188, Asn17, Gly189, Glu89, Asn190, and Tyr92, interacting with 2,3-bisphosphoglycerate [33].

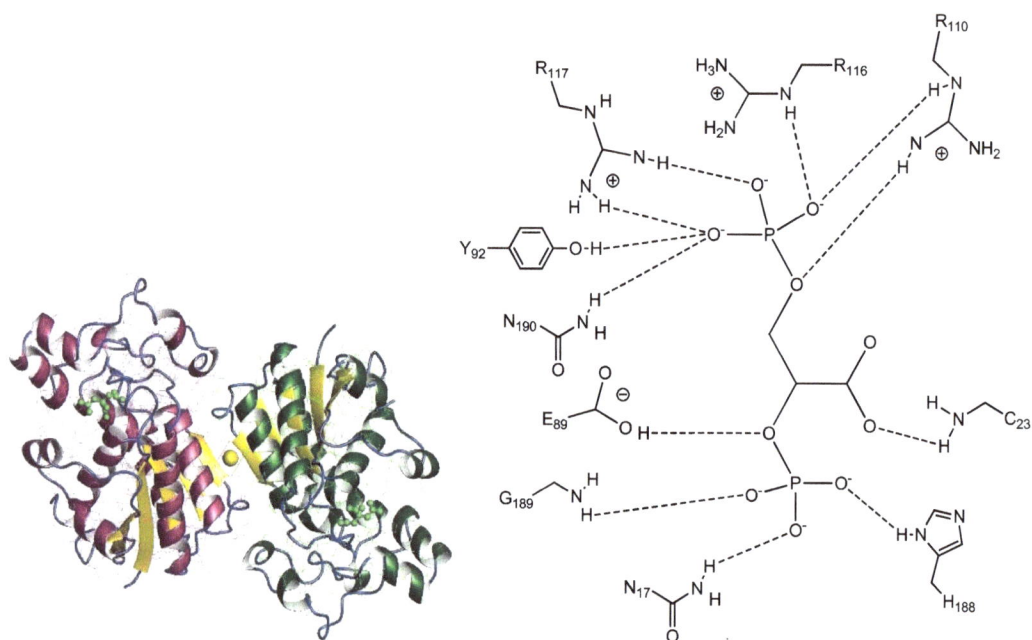

Fig. (38). Crystal structure of human B type phosphoglycerate mutase (PDB: 1YFK) and interacting diagram with 2,3-bisphosphoglycerate.

1.17. Phosphoglycerate Mutase Inhibitors

Phosphoglycerate mutase has been identified as a potential anticancer and anti-parasitic target. Some candidates resulting as effective inhibitors are represented in Fig. (**39**), displaying the following IC_{50} values: MJE3 (IC_{50} = 33 μM), PGMI-004A (IC_{50} =13.1 μM), and EGCG (IC_{50} = 0.49 μM) [34]. This enzyme appear to be a good drug target, because the parasite cofactor-independent enzyme (iPGAM) and the human cofactor-dependent enzyme are not homologous and thus share no common features [35]. (-)-Epigallocatechin-3-gallate (EGCG), the major natural catechins of green tea extract, has been identified as potent PGAM1 inhibitor [36].

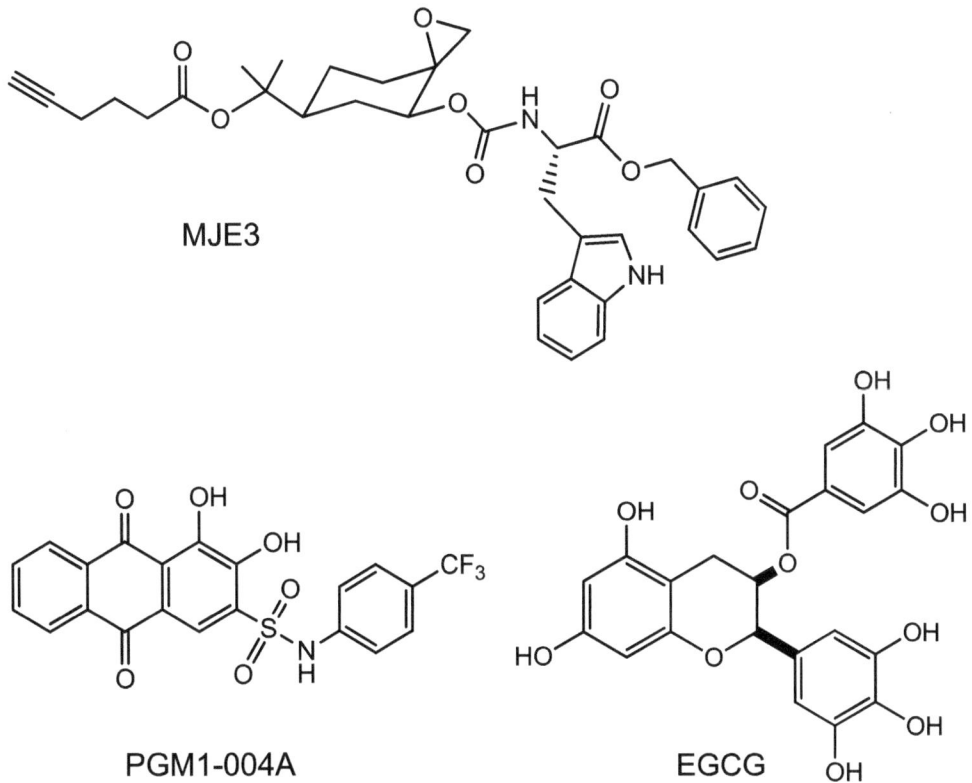

Fig. (39). Chemical structure of phosphoglycerate mutase inhibitors.

1.18. Enolase (ENO)

This metalloenzyme converts 2-phosphoglycerate to phosphoenolpyruvate (Fig. **40**), and the single water molecule produced during the glycolysis. The Mg^{2+} and Zn^{2+} are ions that strongly bind to the active site, and activate the enzyme. There are three enolase isotypes α-enolase found in a variety of tissues, β-enolase in muscle y γ-enolase present in neurons. Enolase has been implicated in numerous diseases such as cancer, autoimmune disorders and bacterial infections [37].

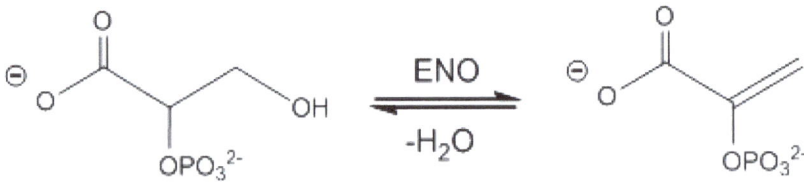

Fig. (40). The enolase reaction.

The proposed mechanism involves the participation of lysine residue, which serves as a base to withdraw the C-2 hydrogen, forming an enol structure in resonance with the carboxylate form. During the last step the enol will promote the exit of the hydroxyl group to generate finally phosphoenolpyruvate (Fig. **41**).

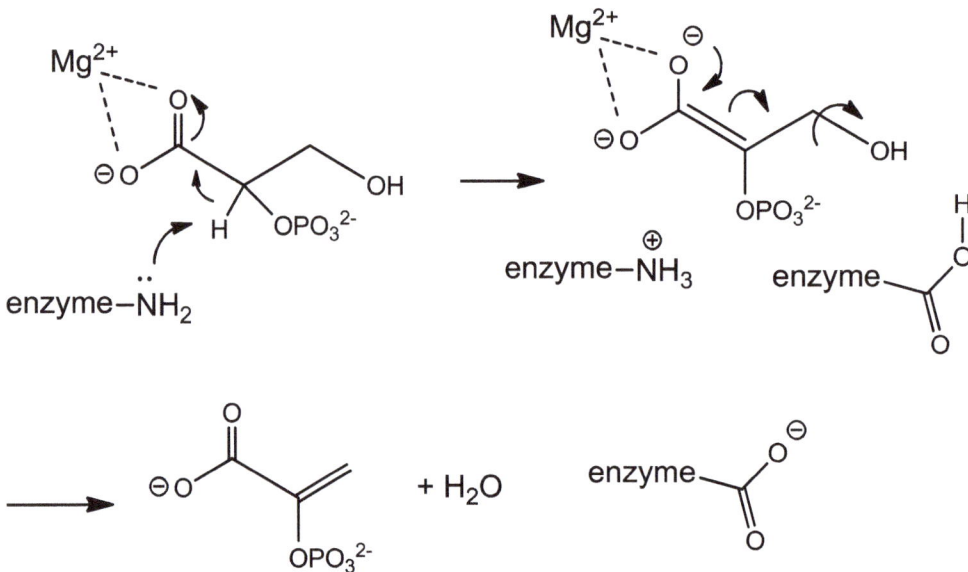

Fig. (41). Proposed mechanism for the phosphoenolpyruvate formation.

The crystal structure of human α-enolase was solved as a dimer having a N-terminal containing three stranded β-sheet and three α-helices, while the C-terminal domain an eightfold β-α barrel. The residues are localised in regions L1 to L3 loops with Mg^{2+} ion bound to Glu292, Asp317, Asp244, Ser39 and water molecules (Fig. **42**) [37].

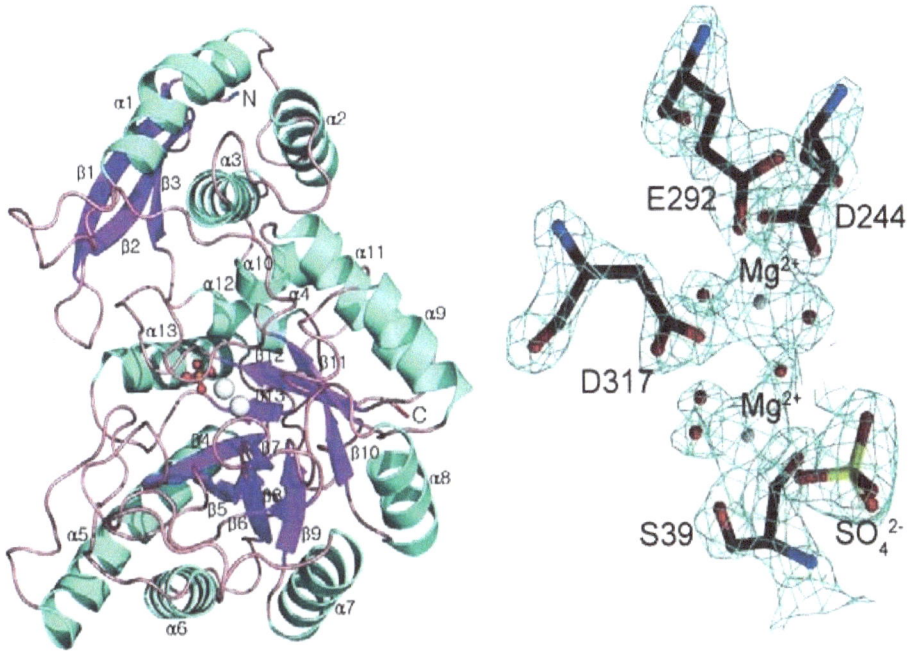

Fig. (42). Crystal structure of human α-enolase (PDB: 3B97) and active site electron density map with the residues around Mg^{2+} ions.

1.19. Enolase Inhibitors

Fluorine ion (F^{1-}) has been reported to produce weak inhibition on enolase enzyme although in the presence of Pi the inhibition increase [38]. On the other hand, phosphonoacetohydroxamate and the antibiotic SF2312 (Fig. **43**) produced by actinomycete *Micromonospora*, were evaluated as enolase ENO2 resulting in potent inhibitors at low nanomolar concentration [39].

Methylglyoxal (Fig. **44**) was evaluated as inhibitor and as effective glycating factor of human muscle-specific enolase, observing maximum inhibition at 4.34 µM with 82% of enzymatic decrease [40].

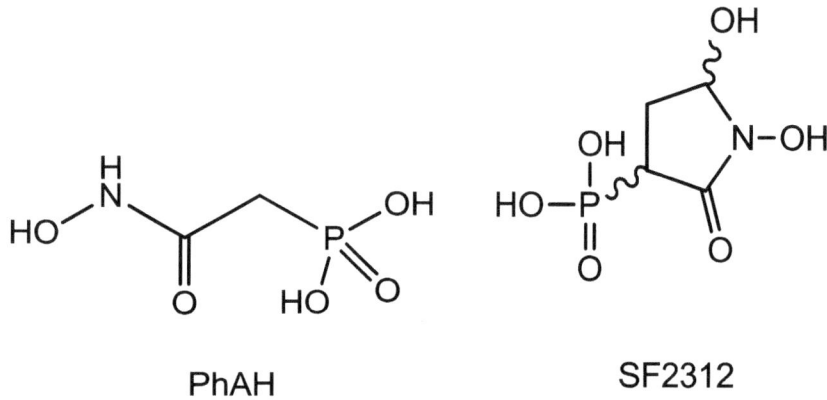

Fig. (43). Chemical structure of phosphonoacetohydroxamate and the antibiotic SF2312.

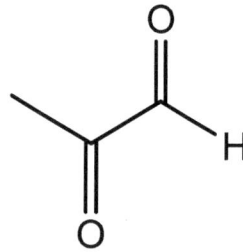

Fig. (44). Chemical structure of methylglyoxal.

1.20. Pyruvate Kinase (PK)

The last step in the glycolytic pathway catalysed by pyruvate kinase consisted in the pyruvate formation from phosphoenolpyruvate (Fig. **45**). The process requires K^+, Mg^{2+} (or Mn^{2+}) for activity, and implies the transfer of the phosphate linked to phosphoenolpyruvate to ADP, producing the second ATP during the glycolysis.

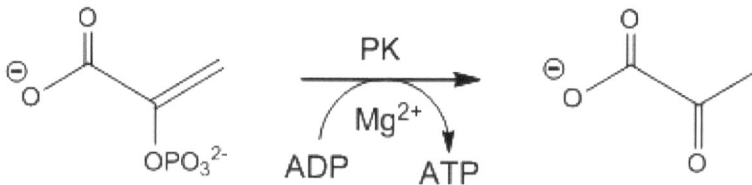

Fig. (45). The pyruvate kinase reaction.

The reaction mechanism occurs in two steps involving the formation of an enolate intermediate by attachment of MgADP with the phosphate group, followed by an assisted protonation of the enolate (Fig. **46**).

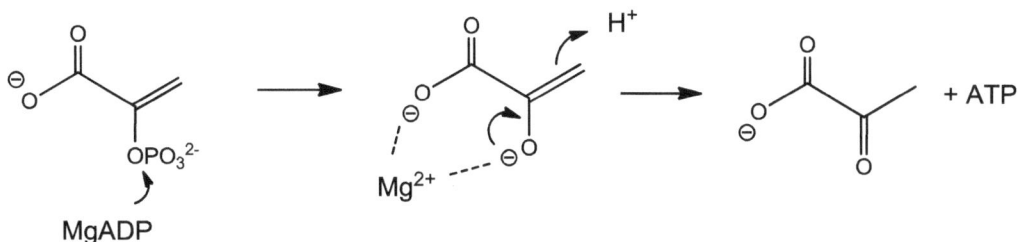

Fig. (46). Simplified reaction mechanism for the pyruvate formation.

Pyruvate kinase is important during exercising since the reduction in its activity affects the production of energy leading to premature fatigue. The deficiency of this enzyme also produces the blood disorder haemolytic anemia [41].

Human pyruvate kinase is present in 4 isozymes (M1 in muscle, M2 fetal, R and L), and M2 isozyme (hPKM2), which are also expressed in early foetal tissues and in most cancer cells are activated to promote cell proliferation due the energy required to assure cancer develop and progression [42] The allosteric and catalytic site of human erythrocyte pyruvate kinase (Fig. **47**) has been also determine, attached to phosphoglycolate in the presence of Mn^{2+} and K^+ interacting with residues Gly(G)338, Asp(D)339, Glu(E)315, Thr(T)371, and Arg(R)116 [41].

1.21. Piruvate Kinase Inhibitors

Cancer cells systematically express the M2 isoform of the glycolytic enzyme pyruvate kinase (PKM2) whose expression is needed for aerobic glycolysis and cell proliferation *in vivo*. Therefore PK enzyme became a good target for controlling cancer cell proliferation, and in an effort to find good candidates a number of small molecule have been synthesized and evaluated. Among them the pyrrole salicylate and rhodanine derivative (Fig. **48**) expressing high potency for PKM2 with IC_{50} values of 10 μM and 20 μM, respectively [43].

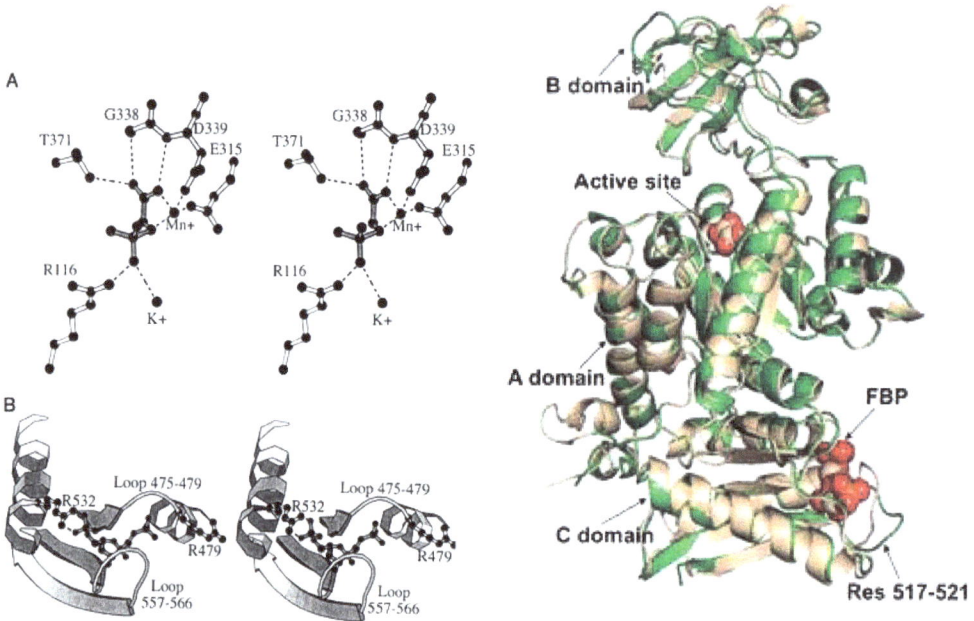

Fig. (47). The allosteric and catalytic site of human erythrocyte pyruvate kinase (PDB: 2VGB) and crystal structure of chain B of PKM2 variant showing the fructose 1,6-bisphosphate (FBP) binding site (PDB: 6B6U).

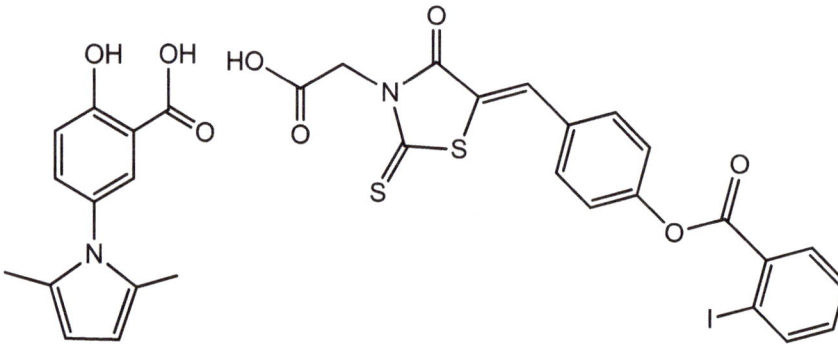

Fig. (48). Chemical structure of pyrrole salicylate and rhodanine derivative.

On the other hand, amino acid tryptophan (Fig. **49**) administrated at higher concentration (about 10 times) on the cerebral cortex in rats induces pyruvate kinase inhibition by approximately 20% [44].

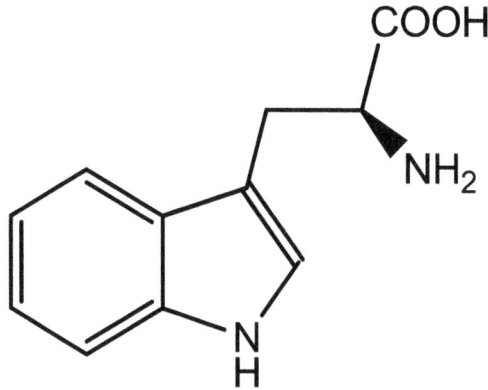

Fig. (49). Chemical structure of tryptophan.

Pyruvate kinase (PK) in many organisms, from bacteria to humans, is inhibited by reactive oxygen species (ROS) superoxide (O^{2-}), hydrogen peroxide (H_2O_2) produced by reduction of O^{2-} through dismutation. Hydroxyl radical (OH^-) arises from electron exchange between O^{2-} and H_2O_2 *via* the Harber–Weiss reaction or it is also generated by the reduction of H_2O_2 by the Fenton reaction [45].

Pyruvate Kinase M2 has been connected to different inflammatory disorders such as Crohn's disease and rheumatoid arthritis, and therefore anti-inflammatory drugs became potential PKM2 inhibitors [46].

The isomeric natural products alkannin and shikonin (Fig. **50**) are described as inhibitors of dimeric PKM2 at concentrations that resulted in over 50% inhibition without affecting PKM1 activity [47].

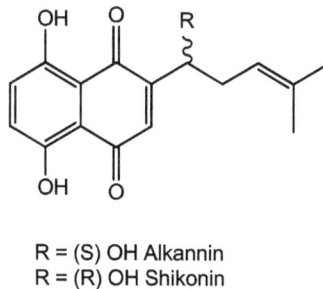

R = (S) OH Alkannin
R = (R) OH Shikonin

Fig. (50). Chemical structure natural products alkannin and shikonin.

CHAPTER 2

Citric Acid Cycle (Krebs)

This is an essential process considered a second step on the respiratory chain, consisting of a series of eight reactions that will produce 8 electrons transported by 3 NADH/H$^+$ and 1 FADH$_2$ molecules, aside from an ATP molecule.

The cycle begins with the condensation of acetyl-CoA with oxaloacetate catalysed by citrate synthase to form citrate, a six-carbon molecule and CoA. Next, a dehydration reaction by aconitase occurs to produce cis-aconitate, which by the addition of a water molecule is converted to D-isocitrate. This citrate isomer is oxidized by NAD$^+$ and transformed to α-ketoglutarate by isocitrate dehydrogenase, along with the formation of a second molecule of NADH/H$^+$ and CO$_2$. A condensing reaction between α-ketoglutarate and CoA catalysed by α-ketoglutarate dehydrogenase resulted in the formation of succinyl-CoA, NADH/H$^+$ and CO$_2$. The succinyl-CoA is hydrolysed by succinyl-CoA synthase to give succinate, CoA, and one ATP or GTP molecule. The succinate is oxidized to fumarate by succinate dehydrogenase in the presence of FAD as an electron acceptor to produce FADH$_2$. Then, fumarate is converted to malate by the enzyme fumarase, and finally, oxidation of the hydroxyl group to the ketone by malate dehydrogenase to provide oxaloacetate and the third molecule of NADH/H$^+$, completing the cycle (Fig. **51**).

1. ENZYMES INVOLVED IN THE CITRIC ACID CYCLE

1.1. Citrate Synthase (CS)

The first step of the citric acid cycle consisting of the Claisen condensation between acetyl-CoA with oxaloacetate catalysed by citrate synthase to give the intermediate citryl CoA which is hydrolysed to citrate and coenzyme A (Fig. **52**) [48].

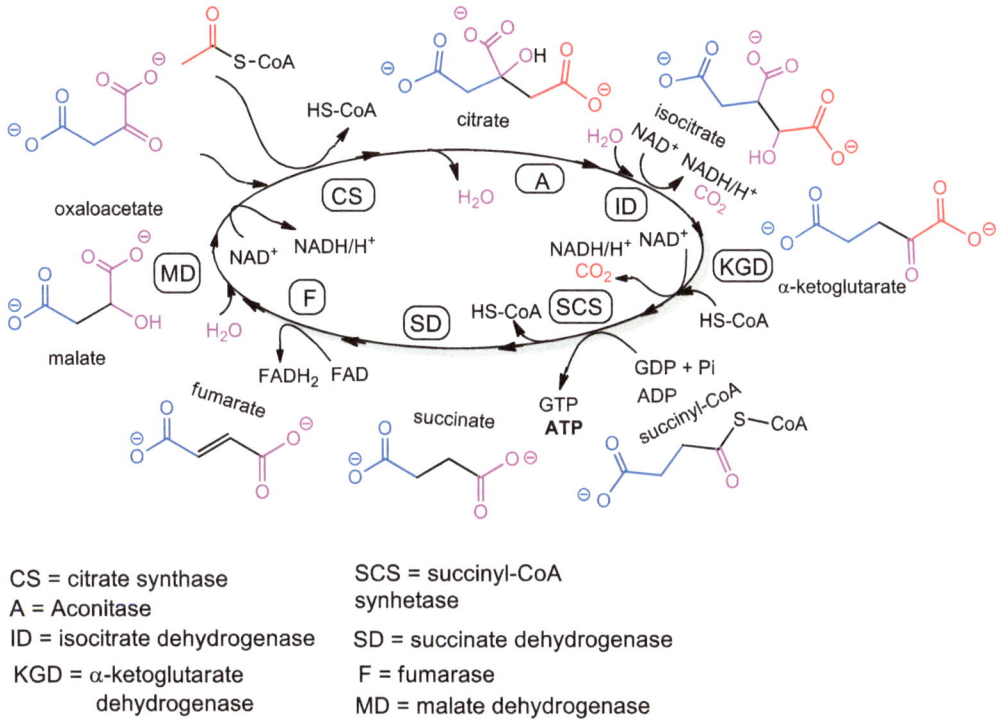

CS = citrate synthase
A = Aconitase
ID = isocitrate dehydrogenase
KGD = α-ketoglutarate
          dehydrogenase

SCS = succinyl-CoA
synhetase
SD = succinate dehydrogenase
F = fumarase
MD = malate dehydrogenase

Fig. (51). The citric acid cycle.

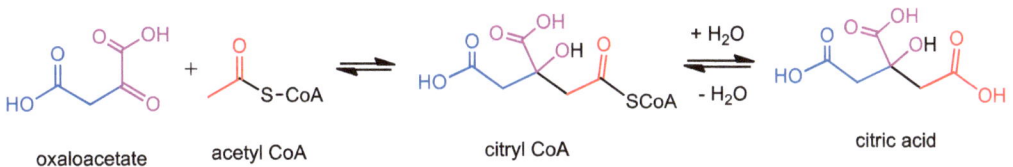

Fig. (52). The citrate synthase reaction.

The crystal structure of CS from different species such as pig, chicken, eukaryotes, bacteria gram-positive, and archaea have been solved as dimeric structures, except from gram-negative bacteria characterized as hexameric structure. The citrate synthase from *Antarctic bacterium* strain DS2-3R (*Ds*CS) was compared with Hyperthermophilic archaeon *Pyrococcus furiosus* (*Pf*CS) showing 40% identity. At the binding site, *Pf*CS citrate ligand is bound to His223, His262, and Arg271, Arg337, Arg356), while the binding of CoA to Lys254, Lys256, Lys305, Ile257, Ala260, Gly259, Asn310, Arg263, Arg353 [49]. The surface representation of the active site for *Ds*CS with citrate and CoA in the active site are represented in Fig. (**53**) [50].

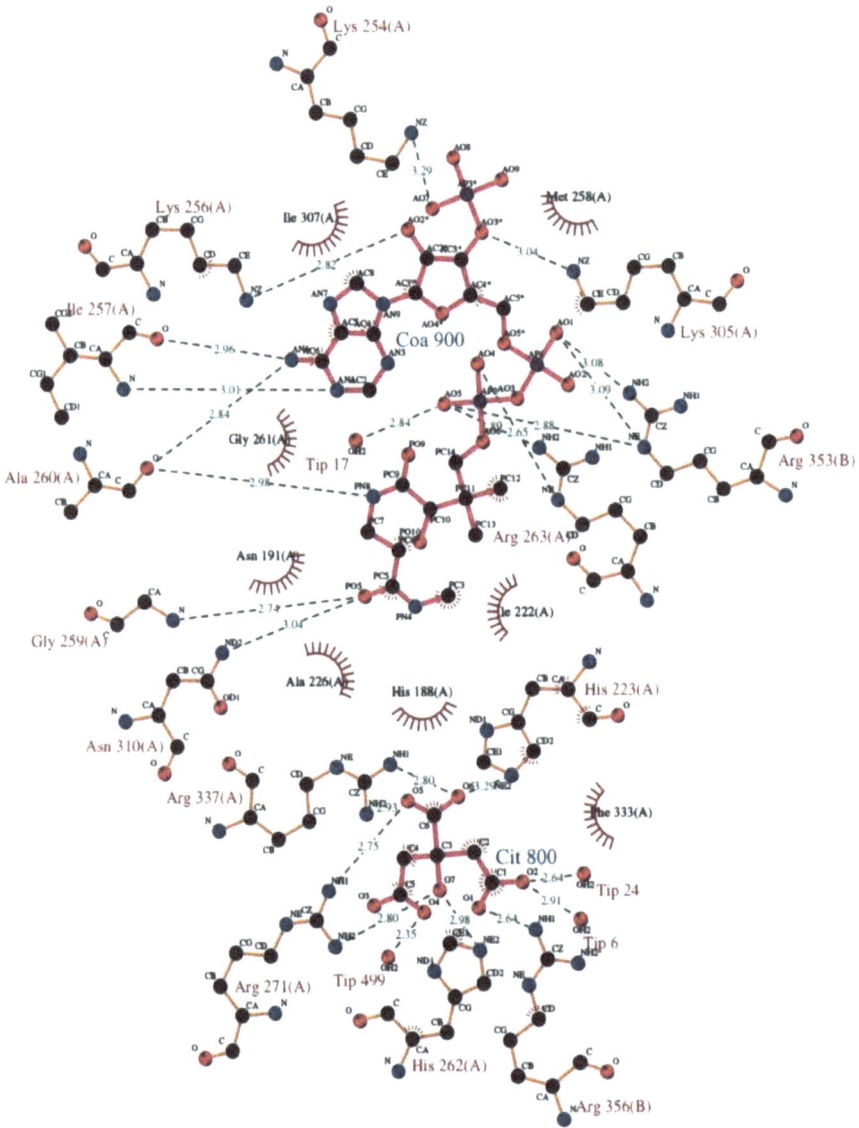

Key

- Ligand bond
- Non-ligand bond
- Hydrogen bond and its length
- Non-ligand residues involved in hydrophobic contact(s)
- Corresponding atoms involved in hydrophobic contact(s)

Fig. 53 cont.....

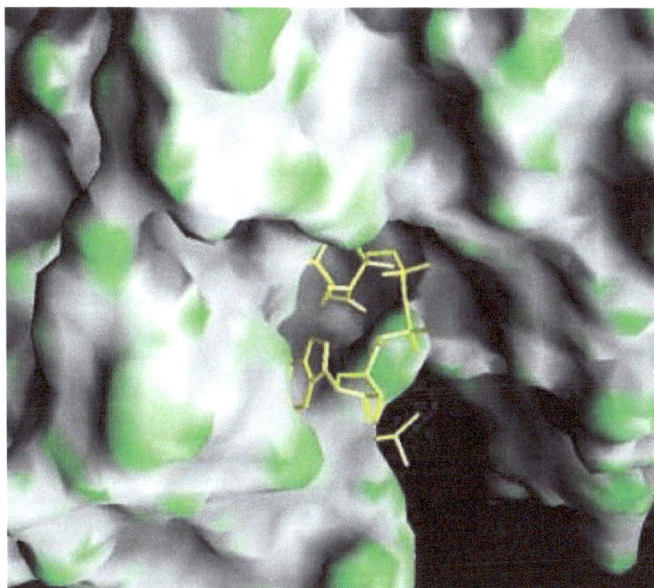

Fig. (53). Diagram of residues interacting with citrate and CoA of *Pyrococcus furiosus* and surface representation of citrate synthase from *Antarctic bacterium* (PDB: 1A59).

1.2. Citrate Synthase Inhibitors

Carboxymethyl-CoA (Fig. **54**) was identified as a potent inhibitor using citrate synthase from pig liver as the target enzyme, displaying $K_i = 0.08$ µM and affinity higher than acetyl-CoA by 100-fold. It was also determined that the affinity toward the enzyme alone was low ($K_s = 230$ µM), although in the presence of oxaloacetate increase substantially (Ks = 0.07 µM) [51].

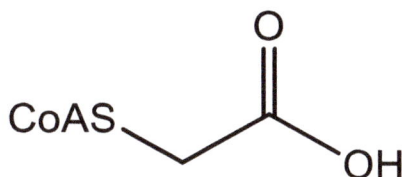

Fig. (54). Chemical structure of Carboxymethyl-CoA.

1.3. Aconitase (Citrate-isocitrate Hydroxylase)

This enzyme catalysed the isomerization conversion from citrate to isocitrate following a two steps sequence: dehydration of citrate to form cis-aconitate and rehydration to produce isocitrate (Fig. **55**). The overall process requires a [4Fe-

4S] cluster involving the formation and release of cis-aconitate and rebind to form (2R,3S) isocitrate [52].

Fig. (55). The isomerization conversion of citrate to isocitrate

Crystal structure of human cytosolic aconitase was determined showing α-helices, a long central helix as cylinder, β-strands as arrows and at the centre the [4Fe-4S] box showing its residue environment (Fig. **56**) [53].

Fig. (56). Crystal structure of human cytosolic aconitase showing α-helices and β-strands, and [4Fe-4S] cluster environment (PDB: 2B3X).

1.4. Aconitase Inhibitors

Nitric oxide free radical (NO·) and other reactive oxygen species (ROS) affect aconitase activity under low oxygen concentrations, giving as result the accumulation of citrate [54].

Fluoroacetate and fluorocitrate (Fig. **57**) when metabolized are highly toxic due to the inactivation of aconitase. The mechanism for inhibition implies the formation of cis-aconitate and then after the flip of the molecule, the addition of the hydroxyl group promotes the exit of fluorine and as a result 4-hydroxy-tras-s-aconitate tightly bound to the enzyme [55].

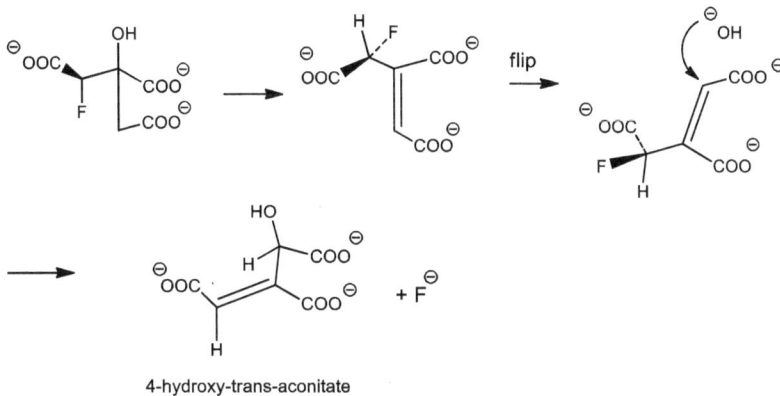

4-hydroxy-trans-aconitate

Fig. (57). Chemical structure of fluorocitrate and proposed inactivation mechanism.

Alloxan (Fig. **58**) is a diabetogenic agent affecting mitochondria and endocrine pancreas and its pathology includes the induced formation of reactive oxygen species promoting selective necrosis of beta cells [56]. The alloxan concentration needed to induce 50% aconitase inhibition is 6.9 x 10-6 M using citrate as a substrate [57].

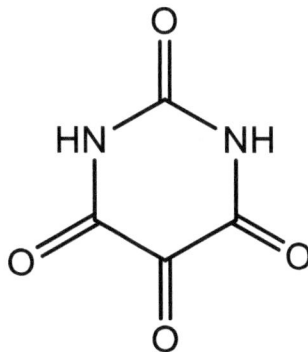

Fig. (58). Chemical structure of alloxan.

1.5. Isocitrate Dehydrogenase (IDH)

This enzyme is involved in the irreversible oxidative decarboxylation of isocitrate to give α-keto glutarate (AKG) and NADH/H$^+$ (Fig. **59**).

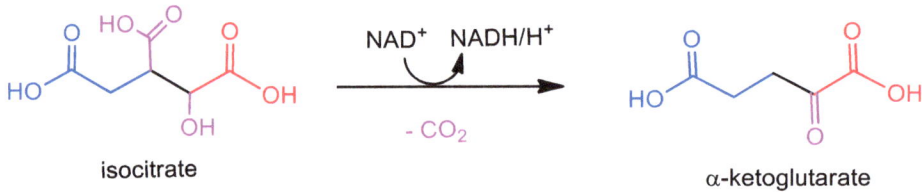

Fig. (59). The isocitrate dehydrogenase reaction.

This reaction gained special attention due to its importance in the normal functioning of the cells, but also because in tumor cells the citrate cycle can be reached from glutamine coming from the glutaminase pathway providing an extra supply of αKG (Fig. **60**) [58, 59].

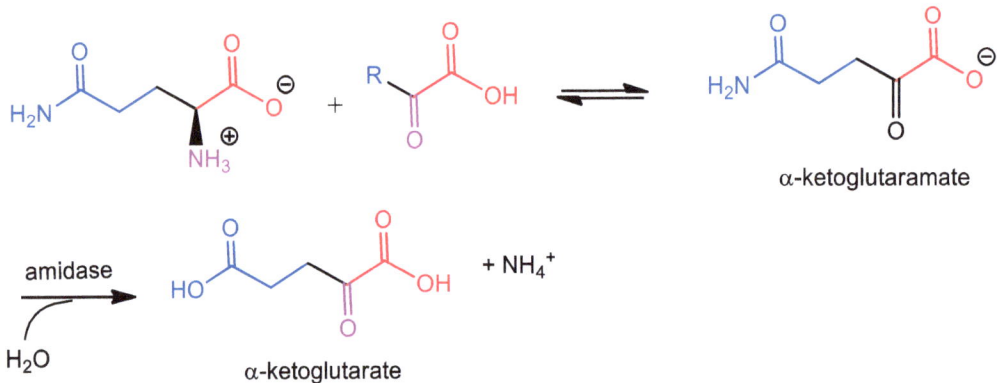

Fig. (60). α-Ketoglutarate formation from the glutamine pathway.

Isocitrate dehydrogenase from yeast (Fig. **61**) is a hetero-octamer composed of IDH1 and four IDH2 containing the catalytic binding site for isocitrate/Mg^{2+}, and NAD$^+$. The regulatory site is formed by the small domains IDH1 and IDH2 forming a four-helix bundle at the heterodimer interface, which binds to citrate at residues 78-92, and to AMP at residues 275-283, producing a conformational change, from unwind to extended loop [60].

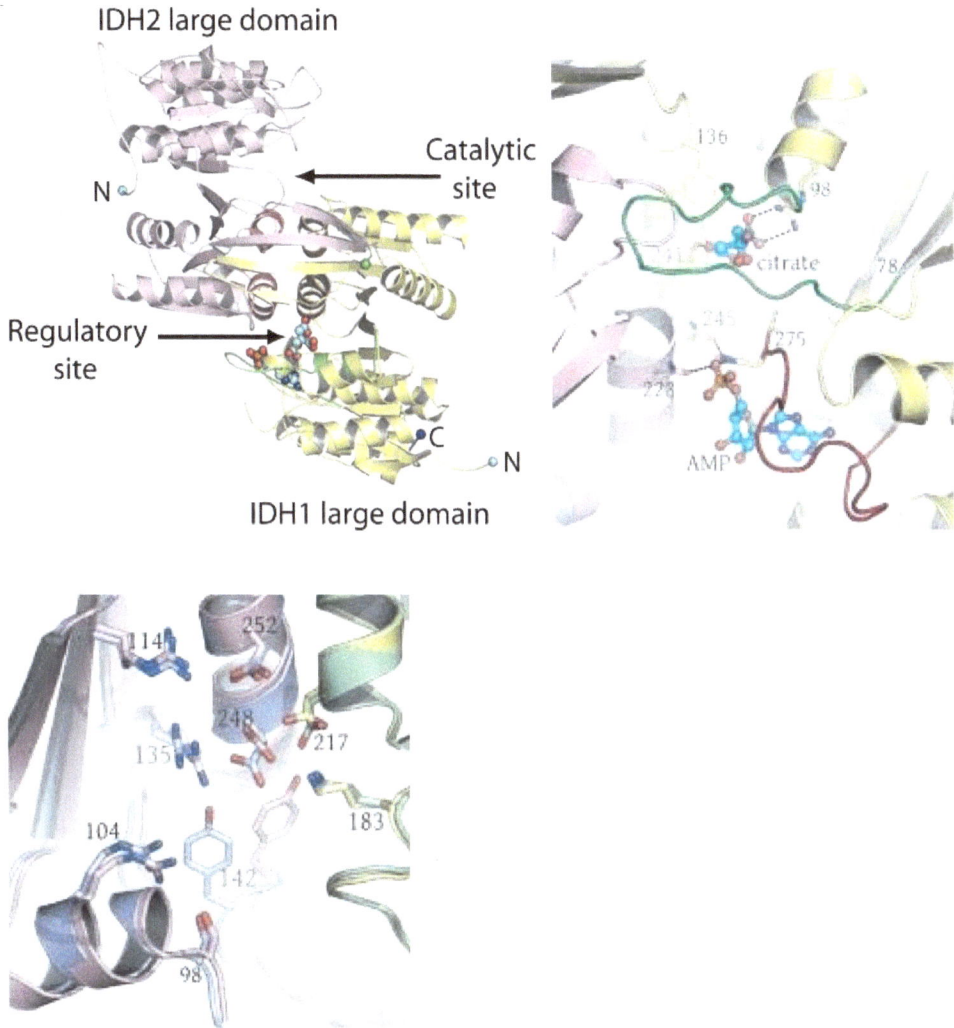

Fig. (61). Isocitrate dehydrogenase IDH1 and IDH2 subunits from yeast, and the regulatory and catalytic sites (PDB: 3BLV).

The catalytic site of *Thermus thermophiles* IDH shows the residues interacting with isocitrate ligand including nicotinamide cofactor (Fig. **62**) [61].

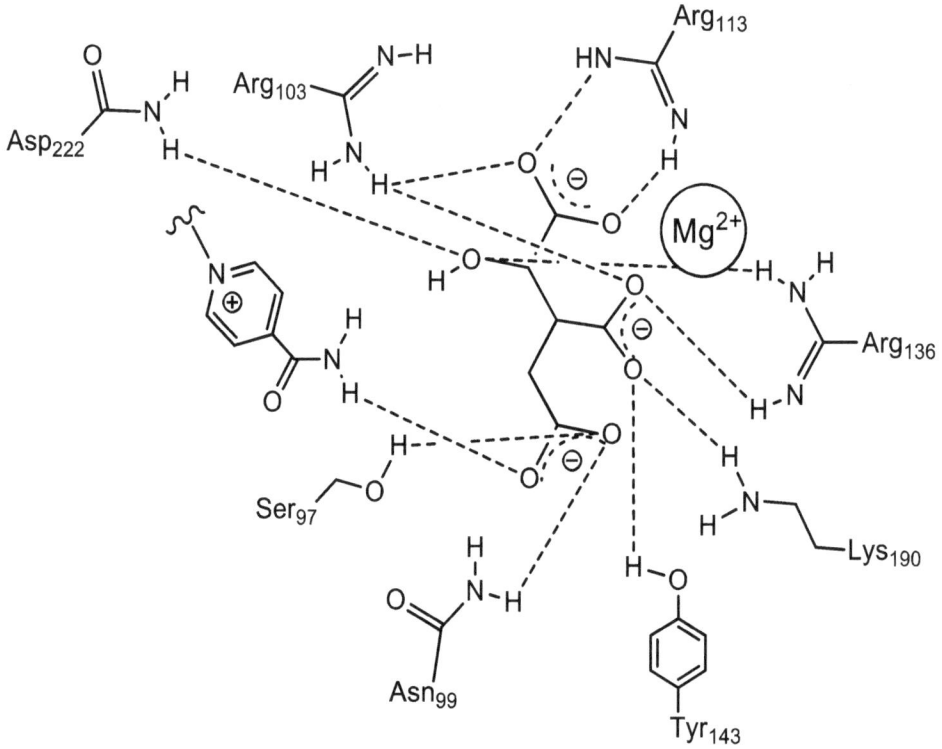

Fig. (62). Interaction diagram of isocitrate with residues from *Thermus thermophiles*.

1.6. Isocitrate Dehydrogenase Inhibitors

The inhibition of this enzyme has been a subject of intensive studies due to its important role in the control of several cancer types and therefore a number of inhibitors of IDH1 have been described (Fig. **63**) [62].

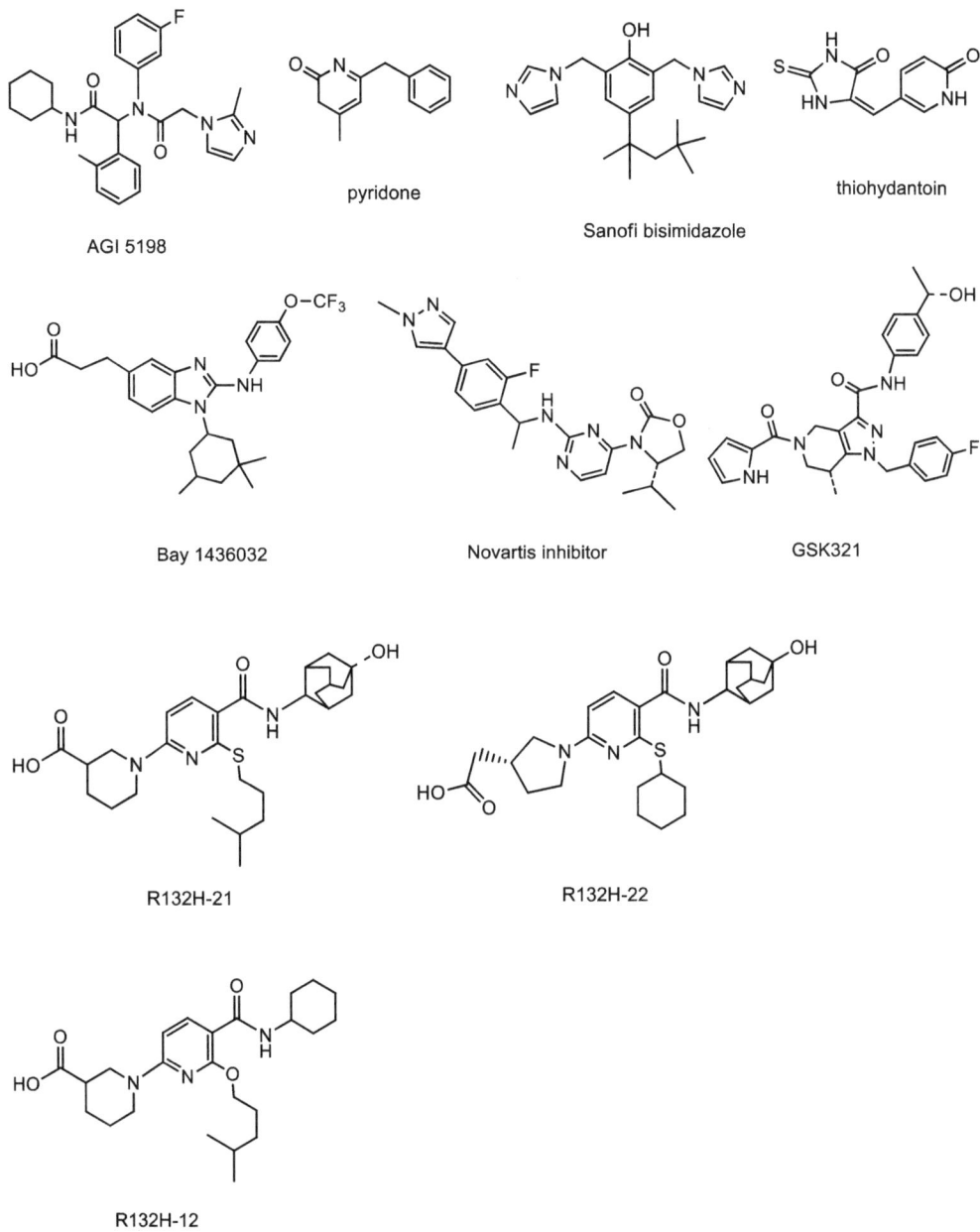

Fig. (63). Chemical structure of isocitrate dehydrogenase inhibitors.

1.7. α-Keto Glutarate Dehydrogenase (KGD)

α-Keto glutarate dehydrogenase also known as oxoglutarate dehydrogenase is a highly regulated enzyme and is a primary site of control in the metabolic flux through the Krebs cycle [63] It catalyses an oxidative decarboxylation (similar to pyruvate dehydrogenase) to convert α-keto glutarate to succinyl-CoA producing also NADH/H$^+$ and CO_2, and requires thiamine pyrophosphate (PLP) as a cofactor (Fig. **64**).

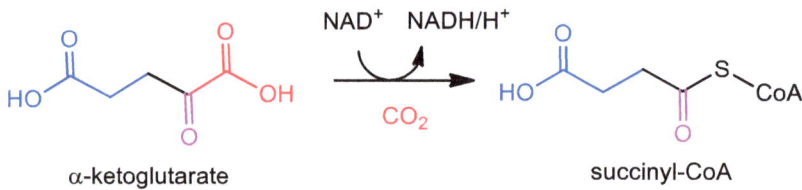

Fig. (64). The α-Keto glutarate dehydrogenase reaction.

α-keto glutarate (AKG) can be alternatively be converted in L-glutamine which is an amino acid with important implications in the cell growth, by the catalysis of glutamine dehydrogenase (GDH) and glutamine synthetase (GS). Conversely L-glutamine after crossing the plasma membrane can produce AKG through the glutaminase pathway I and II.

The catalytic mechanism combines a multi-subunit system composed of E1 (α-ketoacid decarboxylase), E2 (dihydrolipoyl transacetylase), and E3 (dihydrolipoamide dehydrogenase). The transformation sequences involve thiamine diphosphate, CoASH, lipoic acid, FAD$^+$, and NAD$^+$ being attached to the three sub-units (Fig. **65**) [64, 65]

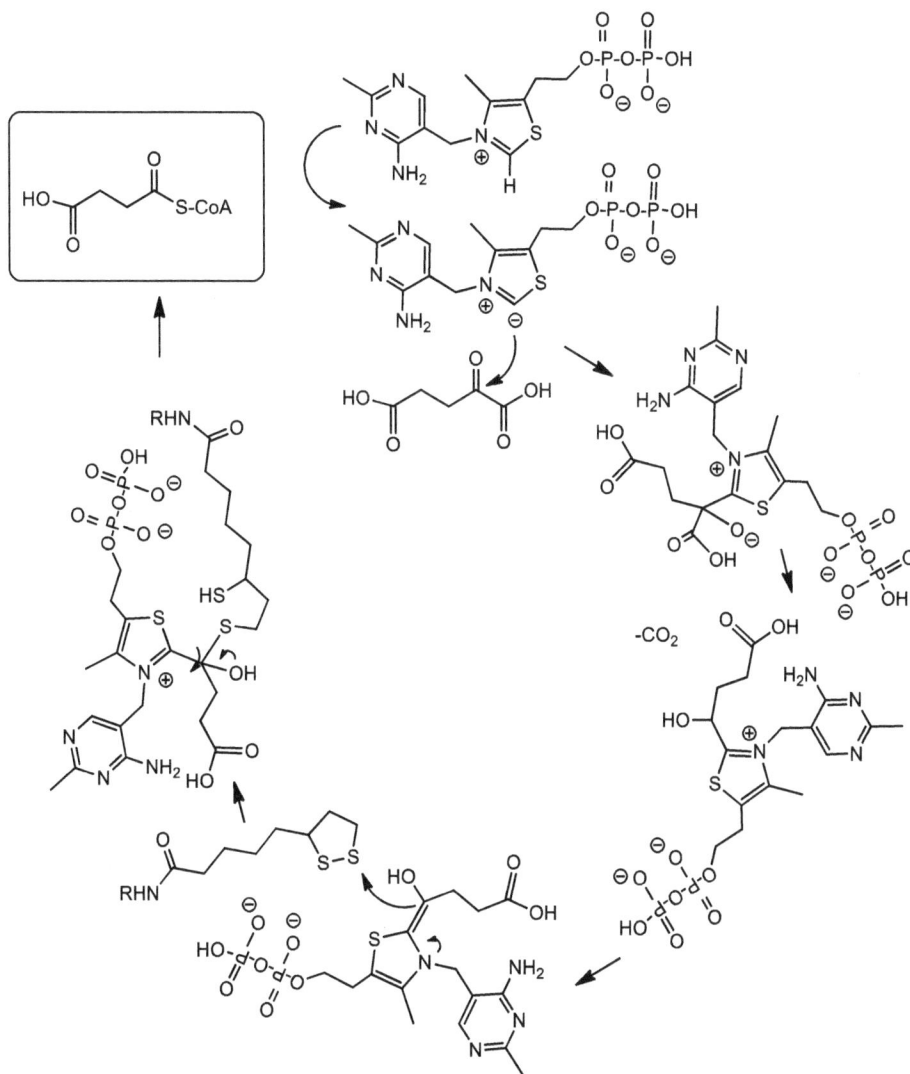

Fig. (65). The proposed mechanism for the succinyl-CoA formation.

The thiamine dependent α-ketoacid decarboxylase E1o component from *Escherichia coli* was determined by X-ray diffraction, showing at the active site some of the ligands involved in combination with thiamine diphosphate, Mg^{+2} and oxaloacetate as natural inhibitor located at positions where the substrate is expected to bind (Fig. **66**) [66].

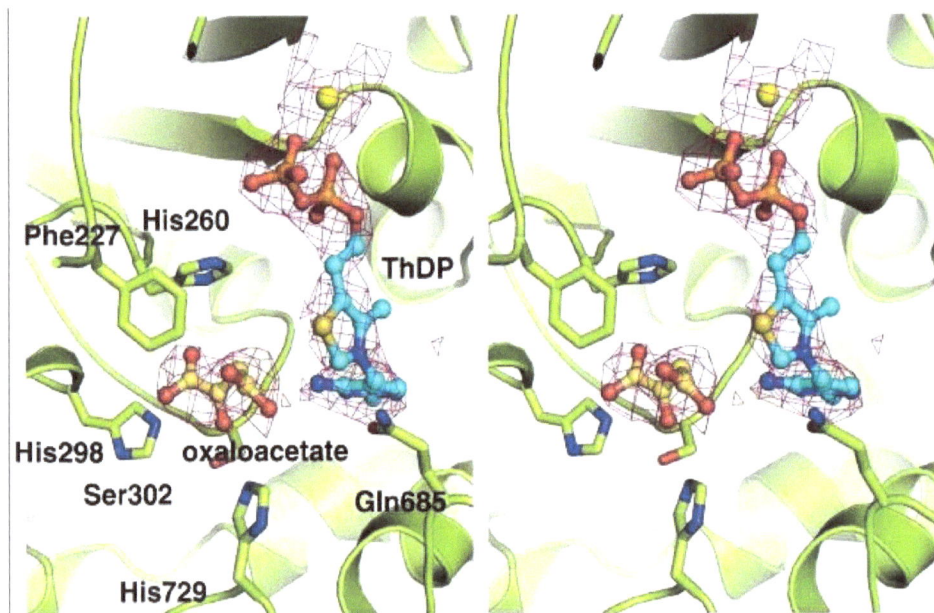

Fig. (66). The crystal structure of 2-oxoglutarate dehydrogenase from *E. coli* (PDB: 2JGD).

1.8. α-Keto Glutarate Dehydrogenase Inhibition

α-Keto glutarate dehydrogenase is highly sensitive to reactive oxygen species such as superoxide anion $O_2^{\cdot-}$, hydrogen peroxide H_2O_2, hydroxyl radical, and peroxynitrate anion $ONOO^-$ [67].

A lipoate analog CPI-613 (Fig. **67**) was designed as a strategy for controlling tumor cell growth through the redox modification of KGD leading to inactivation. For instance, it was observed that CPI-613 increases the production of reactive oxygen species in H460 human lung carcinoma [68].

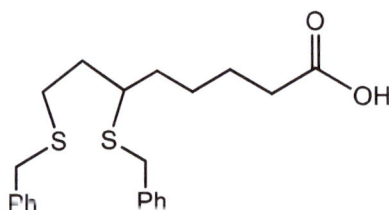

Fig. (67). Chemical structure of lipoate analog CPI-613.

The α-ketoglutarate dehydrogenase inhibitor is (S)-2-[(2,6-dichlorobenzoyl) amino] succinic acid (AA6) as depicted in Fig. (**68**), has been evaluated against breast cancer and additional correlation studies were carried out to correlate enzyme inhibition with metastasis progression *in vivo* using 4T1 orthotopic mouse model. As a result, the report suggests that AA6 does not directly interfere with metastasis although it could induce a delay in the process [69].

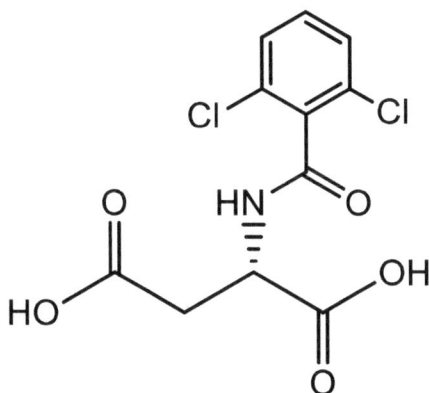

Fig. (68). Chemical structure of succinic acid derivative AA6.

1.9. Succinyl-CoA Synthetase or Succinyl-CoA Ligase

This enzyme is responsible for the conversion of succinyl-CoA to succinate, giving additionally free CoA and ATP or GTP depending on the subunit (Fig. **69**). The reaction occurs following two events well characterized, the first being the formation of succinyl-phosphate intermediate, with the release of CoA, and the second the succinate formation and ATP or GTP [70].

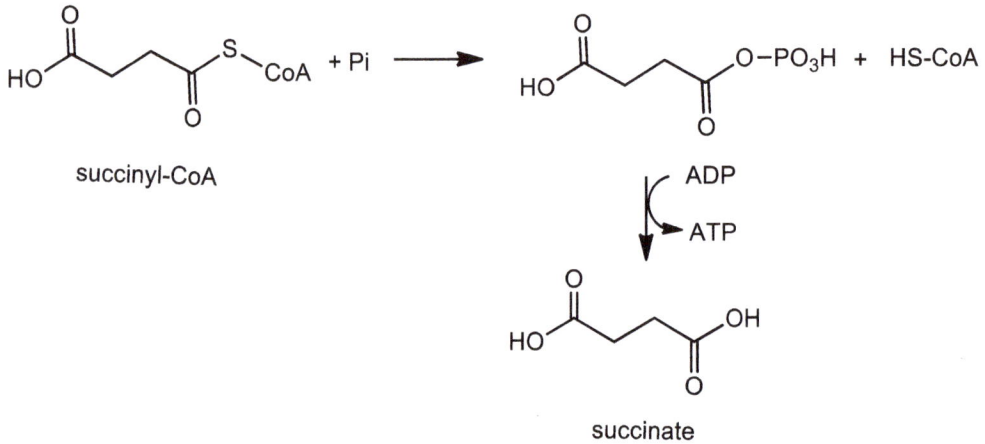

Fig. (69). The succinyl-CoA synthetase reaction.

Succinyl-CoA synthetase has been characterized from different species, including *E.coli* and mammals. The structure of pig GTP-specific succinyl-CoA synthetase isoform was described as αβ-heterodimer, with α- and β- subunits in yellow and green, respectively. The active site residues of GTP-specific succinyl-CoA (GTPSCS) were identified and the superposed conformation determined with and without binding to CoA aside from water molecules present at the interacting region (Fig. **70**) [71].

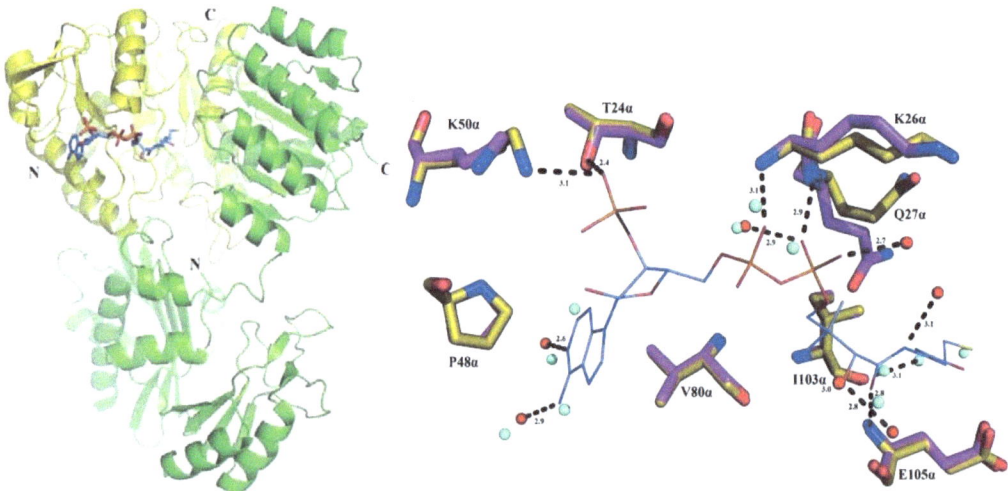

Fig. (70). The crystal structure of pig GTP-specific succinyl-CoA synthetase isoform and active site residues showing superposed conformation when bound and in absence of CoA (PDB: 4XX0).

1.10. Succinyl-CoA Synthetase Inhibitors

Isothiazolone derivative LY266500 (Fig. **71**) has been described as a potent inhibitor of histidine phosphorylation of mitochondrial succinyl CoA synthetase in the unicellular eukaryotic parasite *Trypanosoma brucei*. Thus, the proliferation of procyclic *T. brucei* depends on mitochondrial activity inhibited with LY266500, displaying an IC_{50} value of 0.6 μM, which made the compound a promising candidate for lead development [72].

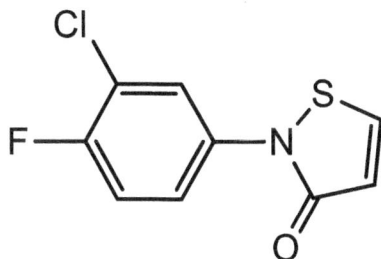

Fig. (71). Chemical structure of isothiazolone derivative LY266500.

Streptozotocin (Fig. **72**) with a full name 2-deoxy-2-(3-methyl-3-nitrosoueido)-D-glucopyranose is a known diabetogenic inducer with proved succinyl-CoA synthetase inhibition capacity, producing 50% enzyme inhibition at 10^{-8} M from mouse liver and kidney [73].

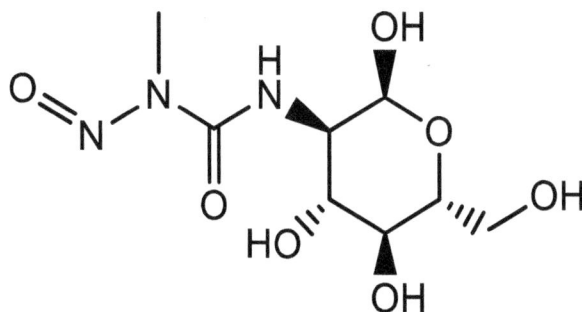

Fig. (72). Chemical structure of streptozotocin.

1.11. Succinate Dehydrogenase (SDH) or Succinate-coenzyme Q Reductase (SQR)

Succinate dehydrogenase is a membrane-bound dehydrogenase complex attached to the respiratory chain and Krebs cycle. It contains non-heme iron, labile sulphur, and FAD cofactor [74]. It is responsible for the redox transformation of succinate to fumarate, and $FADH_2$ (Fig. **73**) which is transferred to the respiratory chain ubiquinone pool (UQ). The reduction of SDH activity can induce increased oxidative stress, leading to notorious cellular damage [75].

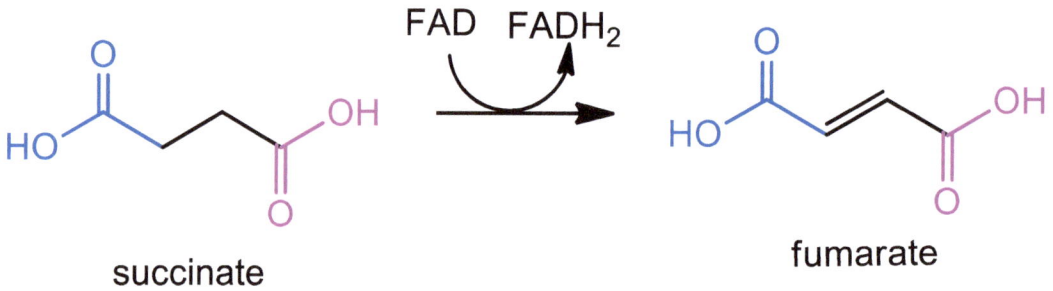

Fig. (73). The succinate dehydrogenase reaction.

SDHs mammalian mitochondrial is composed of two subunits, a flavoprotein (SdhA) and an iron-sulfur protein (SdhB) subunit, and two hydrophobic membrane anchor subunits, SdhC and SdhD. The *E. coli* monomeric structure is represented with FAD (gold), oxaloacetate (green), the heme b (magenta), and ubiquinone (yellow) as shown in Fig. (**74**) [76].

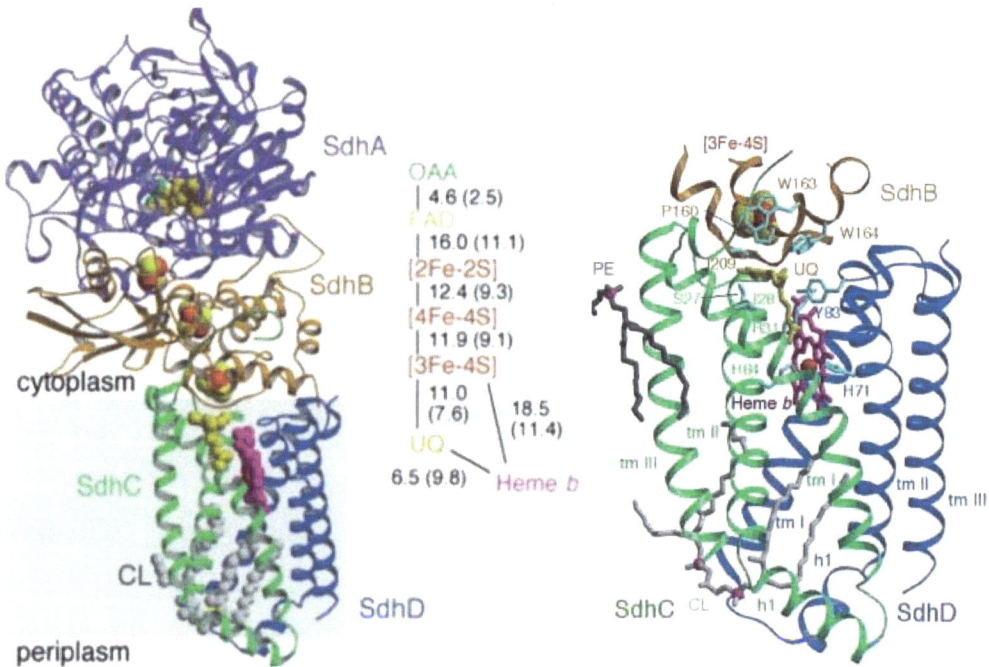

Fig. (74). The crystal structure of succinate dehydrogenase from *E. coli* (PDB: 1NEK).

1.12. Succinate Dehydrogenase Inhibitors

Succinate dehydrogenase might be inhibited by natural substrate malonate, oxaloacetate [77], and by sulfonamide derivative diazoxide (Fig. **75**) having the former, the highest inhibition of SDH (79%) and the later (47%) in wild type mice [78].

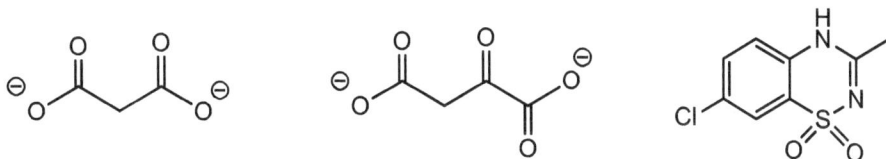

Fig. (75). Chemical structure of malonate, oxaloacetate and diazoxide.

Tumor necrosis factor receptor-associated protein 1 (TRAP1) considered a chaperone protein with important properties on mitochondrial bioenergetics in tumor cells, has been identified as succinate dehydrogenase inhibitor abolishing succinate oxidation and ROS generation [79]. Alternatively, thenoyltri-fluoroacetone (TTFA) is a thiophene derivative (Fig. **76**) identified as an SDH inhibitor with the ability for binding at the quinone binding sites B and D, producing electron arrest from succinate to coenzyme Q [80].

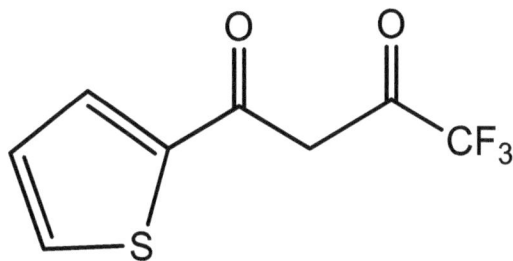

Fig. (76). Chemical structure of thenoyltrifluoroacetone.

1.13. Fumarate Hydratase or Fumarase

This enzyme performs the stereospecific reversible hydration-dehydration of fumarate to L-malate (Fig. **77**), and it is grouped in Class I for parasites with a catalytically essential [4Fe-4S] cluster and Class II for humans with Fe-independent homotetrameric enzymes [81].

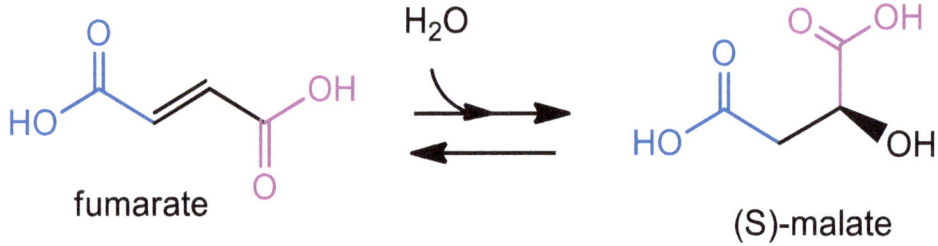

Fig. (77). The fumarate hydratase reaction.

The crystal structure of fumarate hydratase from *M. tuberculosis* showed the presence of C- and N- terminal domains as α-helix linked by the central domain, two allosteric sites, four active sites as dashed circles, and four subunits (Fig. **78**) [82].

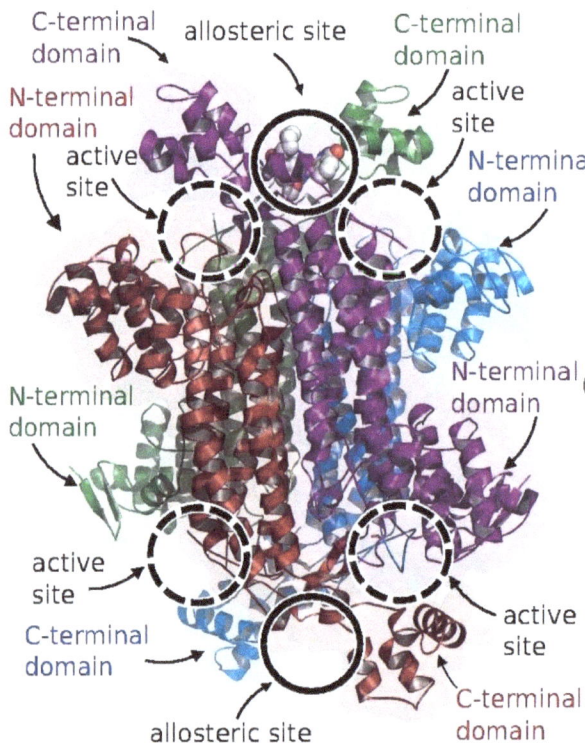

Fig. (78). The crystal structure of fumarate hydratase from *M. tuberculosis* (PDB: 5F92)

1.14. Fumarate Hydratase Inhibitors

Biphenyl pyrrolidinone derivative (Fig. **79**) was identified as competitive inhibitors of fumarate hydratase isolated from SW620 cells with $K_i = 4.5$ μM [83].

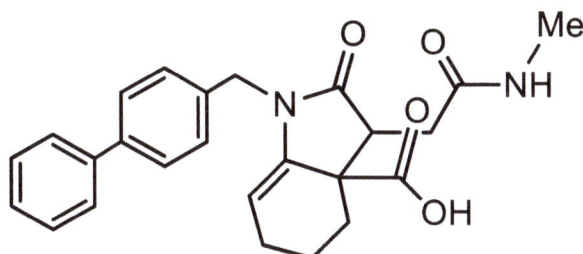

Fig. (79). Chemical structure of biphenyl pyrrolidinone derivative.

1.15. Malate Dehydrogenase (MDH)

The last enzyme involved in the citric acid cycle is MDH, having the role of performing the malate oxidation, to give oxaloacetate and reduced NADH (Fig. **80**).

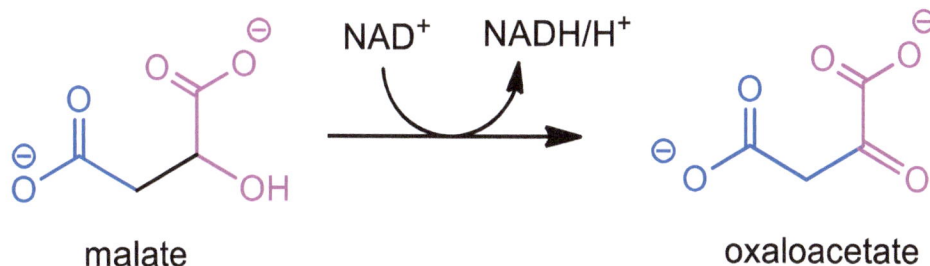

Fig. (80). The malate dehydrogenase reaction.

The reaction involves a concerted transformation involving a histidine residue serving as a base, to form an alkoxide which promotes a hydride migration to the NAD$^+$ as acceptor (Fig. **81**).

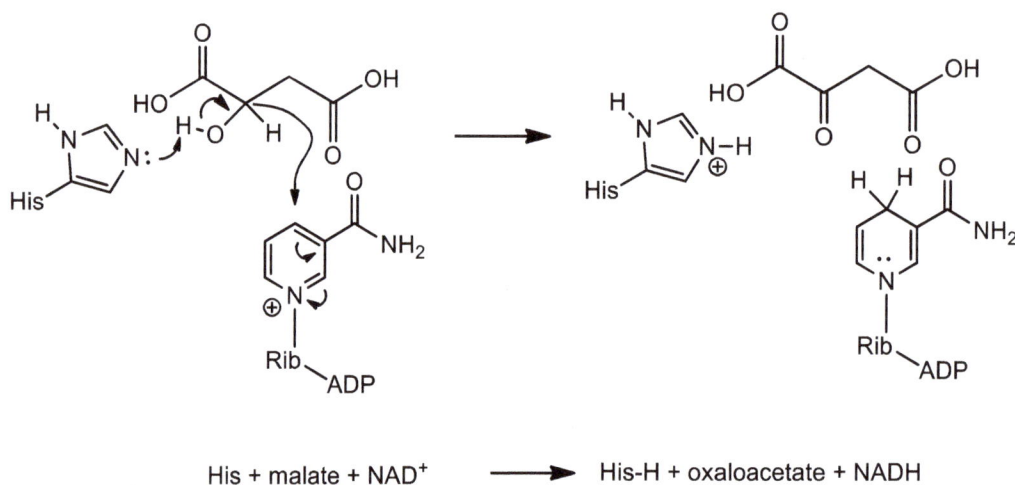

His + malate + NAD$^+$ $\longrightarrow$ His-H + oxaloacetate + NADH

Fig. (81). The proposed reaction mechanism for oxaloacetate formation.

Malate dehydrogenase is a ubiquitous enzyme and the crystal structure of different microorganism determined, among them the high resolution of *Methylobacterium extorquens* (*Mex*MDH), composed by a tetramer $(\beta\alpha\beta)_2(\alpha\beta)_2$ with an NAD$^+$ and substrate binding sites. At the active site, the residues displaying interactions with NAD+ are Hist176, Met144, Ile119, and Asp149, and with oxaloacetate Hist176, Arg89, Arg83, Arg152, and Asp149 (Fig. **82**) [84].

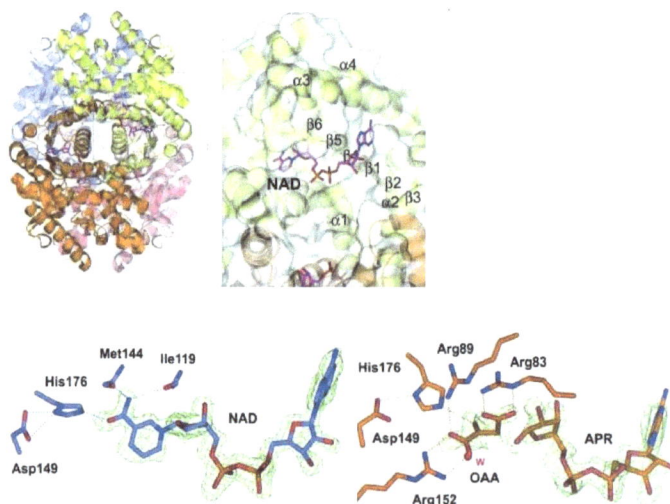

Fig. (82). Tetrameric structure of *Mex*MDH and active site interactions of residues with NAD$^+$ and oxaloacetate (PDB: 5ULV).

Additionally, the crystal structure of malate dehydrogenase from the hyperthermophilic archaeon *Archeoglobus fulgidus*, in complex with its cofactor NAD$^+$, was solved (Fig. **83**), showing the residues interacting and polar contacts establishing with the cofactor a sulphate anion [85].

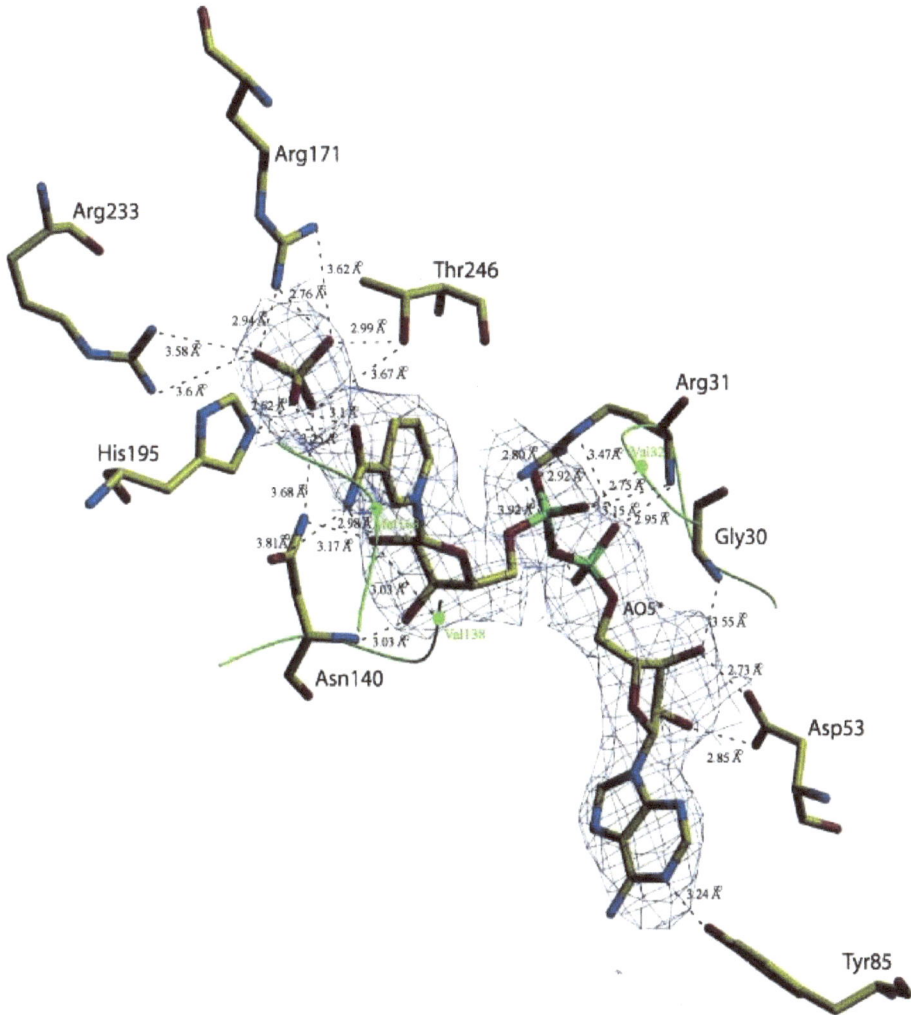

Fig. (83). The crystal structure of malate dehydrogenase from *Archeoglobus fulgidus*, in complex with its cofactor NAD$^-$ (PDB: 2X0i).

1.16. Malate Dehydrogenase Inhibitors

The aryloxyacetylamino benzoic acid derivative LW6 (Fig. **84**) has been described as malate dehydrogenase-2 inhibitor in human activated T cells and hypoxia-inducible factor (HIF-1) [86, 87]

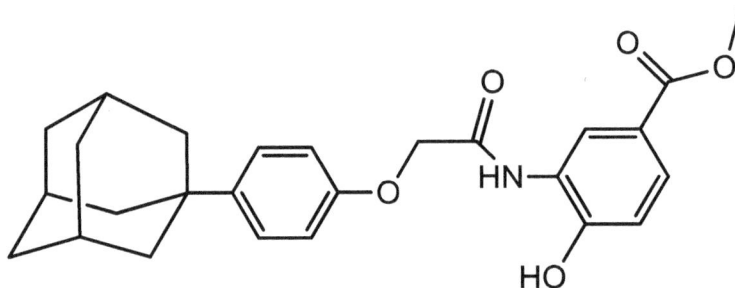

Fig. (84). The chemical structure of aryloxyacetylamino benzoic acid derivative LW6.

CHAPTER 3

Fatty Acid Biosynthesis

The fatty acid biosynthesis, also known as lipogenesis, is a process that occurs in the cytosol and its role is the production of the fatty acid palmitate from acetyl-CoA as a starting material. In humans, the process for the *de novo* biosynthesis of long-chain fatty acids is assigned to the cytosolic enzyme human fatty acid synthase (FAS) responsible for the catalysis of palmitate C16 from acetyl-coenzyme A and malonyl-coenzyme A in the presence of NADPH. During the elongation process, the chain is attached to the acyl carrier protein (ACP) and transported to the active site by enzymes known as malonyl-CoA transacylase (MAT or acetyl-CoA being used), β-ketoacyl synthase (KS), β-ketoacyl reductase (KR), dehydratase (DH) and β-enoyl reductase (ER) (Fig. **85**).

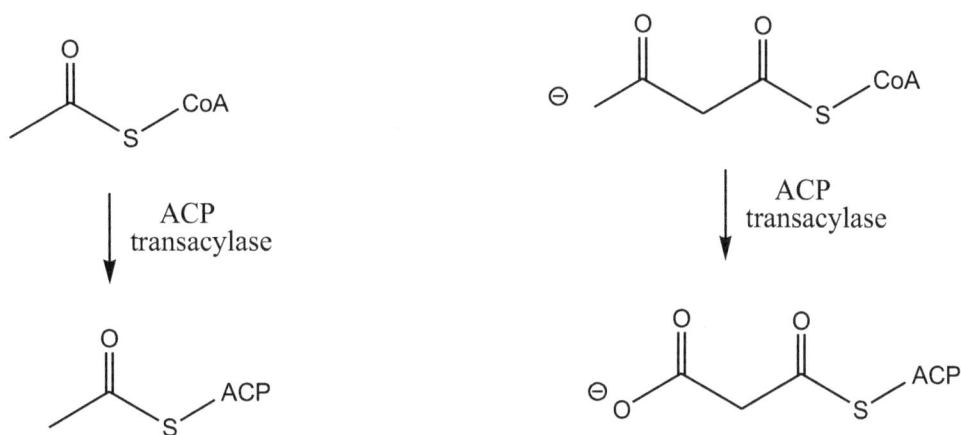

Fig. (85). Transacylation reaction by the acyl carrier protein (ACP)

The initiation process involves the parallel conversion of acetyl-CoA into acetyl-S-ACP, and malonyl-CoA to malonyl-S-ACP by the acyl carrier protein (ACP).

The phosphoryl transfer reaction to convert acetyl-CoA to acetyl-S-ACP involves the post-translational addition of the phosphopantetheine prosthetic group bearing the acyl group to the ACP through a serine residue (Fig. **86**) [88].

Acetyl-CoA

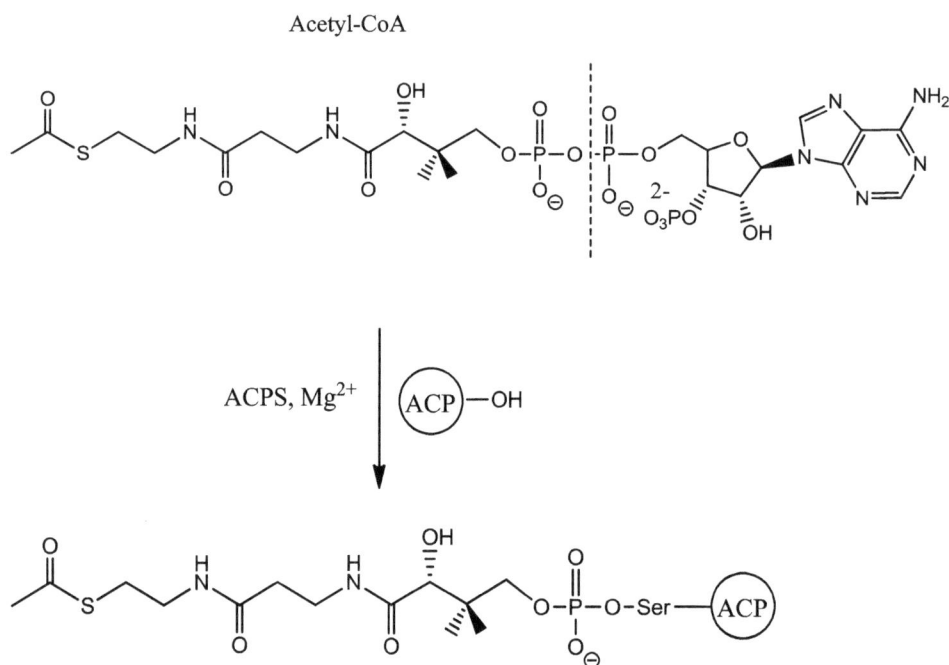

Fig. (86). The phosphoryl transfer reaction to convert acetyl-CoA to acetyl-S-ACP.

The transfer of malony-CoA to malonyl-ACP is mediated by the malonyl-CoA:ACP transacylase enzyme (MAT) according to the cycle shown in Fig. (**87**), involving histidine and serine residues, the later attacking the thioester carbonyl, forming a tetrahedral intermediate, which undergoes proton exchange followed by the attack of ACP-SH to furnish a second tetrahedral intermediate which upon releasing the serine residue provides malonyl-ACP [89].

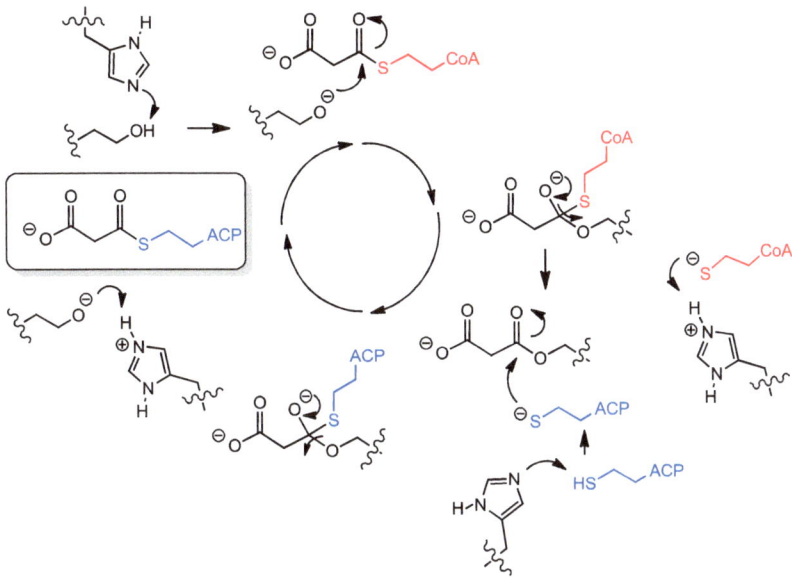

Fig. (87). Full cycle to convert malony-CoA to malonyl-ACP.

Malonyl-CoA transacylase (MAT) is a monomer composed of two subdomains, containing at the catalytic site Phe200, Thr56, Gln9, Met126, Arg122, His201, Ser97, Gln60, Val98, Ser97 residues (Fig. **88**).

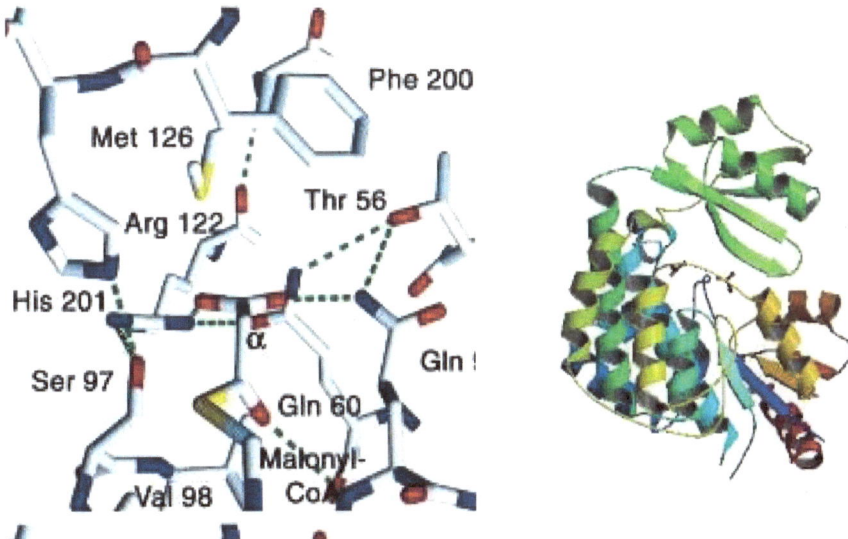

Fig. (88). Crystal structure of malonyl-CoA transacylase and the catalytic site (PDB: 1NM2).

The next step consisted in the conjugation of acetyl-S-ACP and malonyl-S-ACP mediated by the enzyme β-ketoacyl synthase to produce acetoacetyl-ACP and CO_2. The resulting ketone is reduced by the β-ketoacyl synthase in the presence of NADPH to generate 3-hydroxybutyryl ACP. Subsequent β−elimination of water by the enzyme 3-hydroxyacyldehydatase produces the unsaturated crotonyl ACP, which is finally reduced to the butyryl ACP by the enzyme enoyl reductase in the presence of NADPH (Fig. **89**).

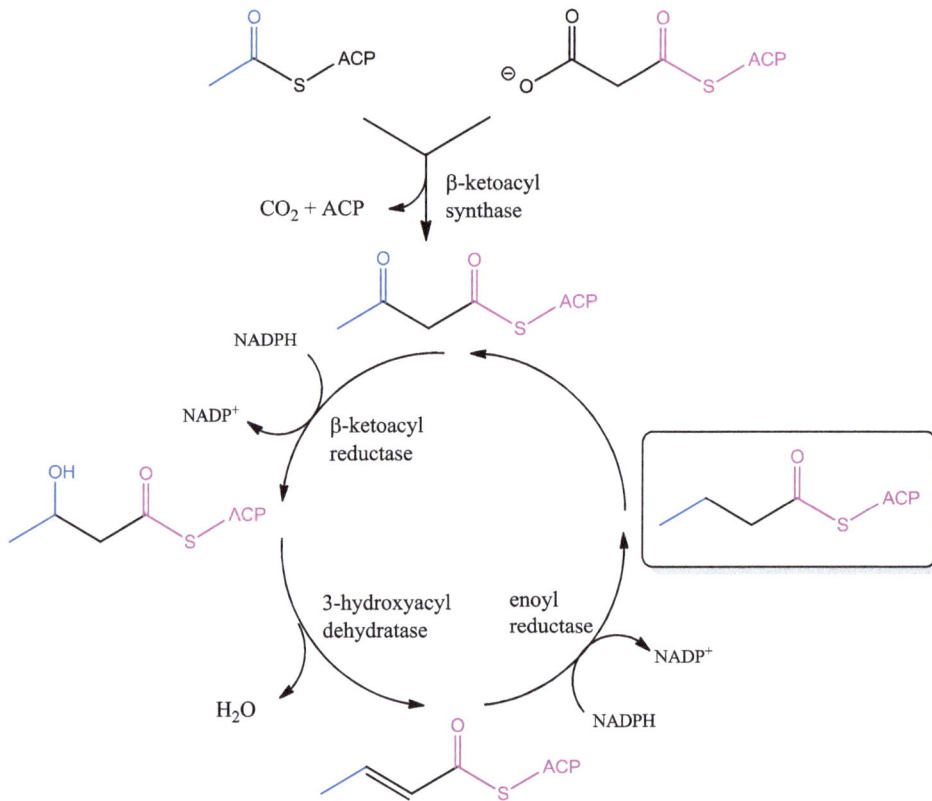

Fig. (89). Full cycle for the formation of butyryl ACP from acetyl-S-ACP and malonyl-S-ACP

The cycle is repeated for the synthesis of palmitate, consuming 7 ATP and 8 NADPH, using specific synthase for short less than C6, medium between C6 and C12 and long C16 or larger chains (Fig. **90**).

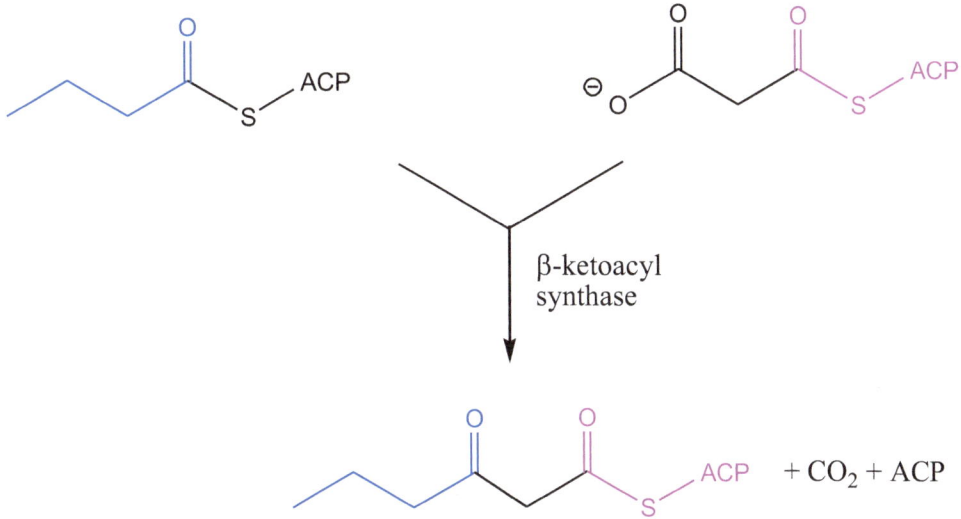

Fig. (90). Elongation process catalysed by β-ketoacyl synthase.

1. ENZYMES INVOLVED IN THE LIPOGENESIS

The pyruvate produced during glycolysis is critical since it is required for the acetyl-CoA production needed to initiate the fatty acid and cholesterol synthesis in the mitochondrion. This irreversible step is carried out by the pyruvate dehydrogenase, which catalyses the decarboxylation of pyruvate and further conjugation with coenzyme A to yield acetyl-CoA and reduced NADH (Fig. **91**).

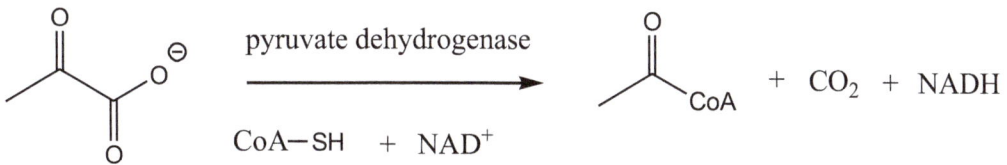

Fig. (91). The pyruvate deshydrogenase reaction.

1.1. Pyruvate Dehydrogenase Complex (PDC)

In humans PDC is composed by a complex of three enzymes: pyruvate dehydrogenase (E1), dihydrolipoyl acetyltransferase (E2) and dihydrolipoyl dehydrogenase (E3), in addition to the E3-binding protein (E3BP) and two regulatory components: pyruvate dehydrogenase kinase (PDK) and pyruvate dehydrogenase phosphatase (PDP) [90]. In humans, tissue-specific isoforms of

PDK and PDP provide valuable targets for therapeutic intervention of diabetes, heart ischemia, and cancer [91].

Fig. (**92**) represents the sequence for acetylCoA formation, starting with thiamine pyrophosphate and 2-oxoacid decarboxylase (E1) which promote the oxidative decarboxylation of pyruvate to give the E1 TPP-CHOHCH$_3$ complex. Subsequently, it is conjugated to the lipoamide attached to the dihydrolipoamide acetyltransferase (E2) through a lysine residue, affording the acylated lipoamide which will transfer the acyl group to CoA and along with reduced lipoamide which eventually is reoxidized by E3 enzyme in the presence of NAD$^+$ [92].

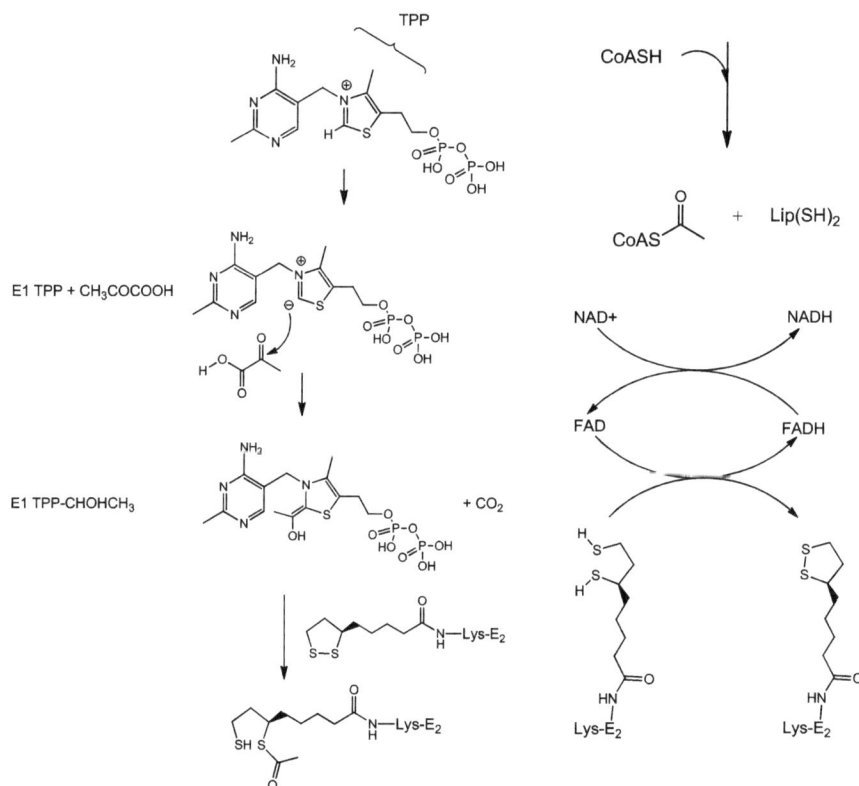

Fig. (92). Schematic representation of the pyruvate dehydrogenase complex for synthesizing acetylCoA.

The pyruvate dehydrogenase complex has been studied by a combination of crystallography, NMR spectroscopy and electron microscopy, showing a symmetrical core with a long flexible tail having several additional functional domains (Fig. **93**) [93, 94].

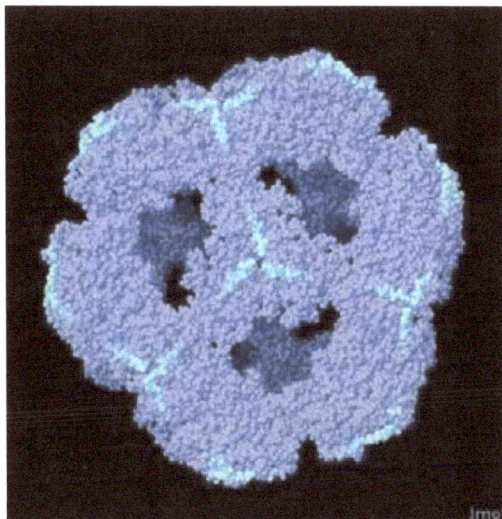

Fig. (93). Electron microscopy image of the pyruvate dehydrogenase complex.

In humans, tissue-specific isoforms of PDK (Fig. **94**) have been under study as target enzymes for diabetes, heart ischemia, and cancer [95]. In PDK1, the substrate is bound to the active site composed by Phe65, Phe62, Glu197, Leu201, Leu195, Phe202, S75, Leu79, Phe78 [96].

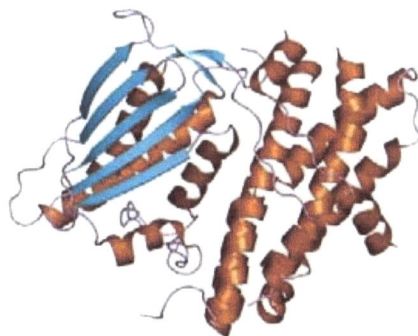

Fig. (94). Crystal structure of tissue-specific isoforms of PDK1 (PDB: 2Q8F).

1.2. Pyruvate Dehydrogenase Inhibitors

The design and synthesis of 4-aminopyrimidine derivatives as a selective *Escherichia coli*, and porcine pyruvate dehydrogenase complex E1 inhibitors were described, presenting potent inhibition for the *E. coli* E1 complex, and negligible inhibition for the porcine E1 complex (Fig. **95**) [97].

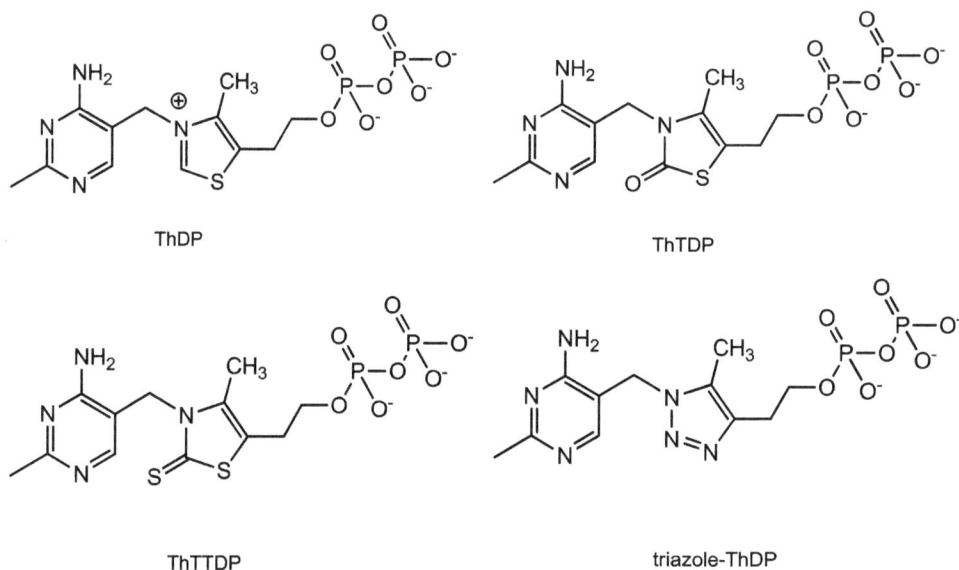

Fig. (95). Chemical structure 4-aminopyrimidine PD inhibitors.

Moreover, structurally diverse compounds either natural or synthetic have been described as PD inhibitors (Fig. **96**), such as arsenoxide (BrCH$_2$CONHPhAsO), dichloroacetate, AZD7545, natural products radicicol and moniliformin a toxin produced by *Fusarium proliferatum* and *F. subglutinans*, as well as other *Fusarium* species which contaminate food and feed having adverse effects on human and animal health [98 - 100].

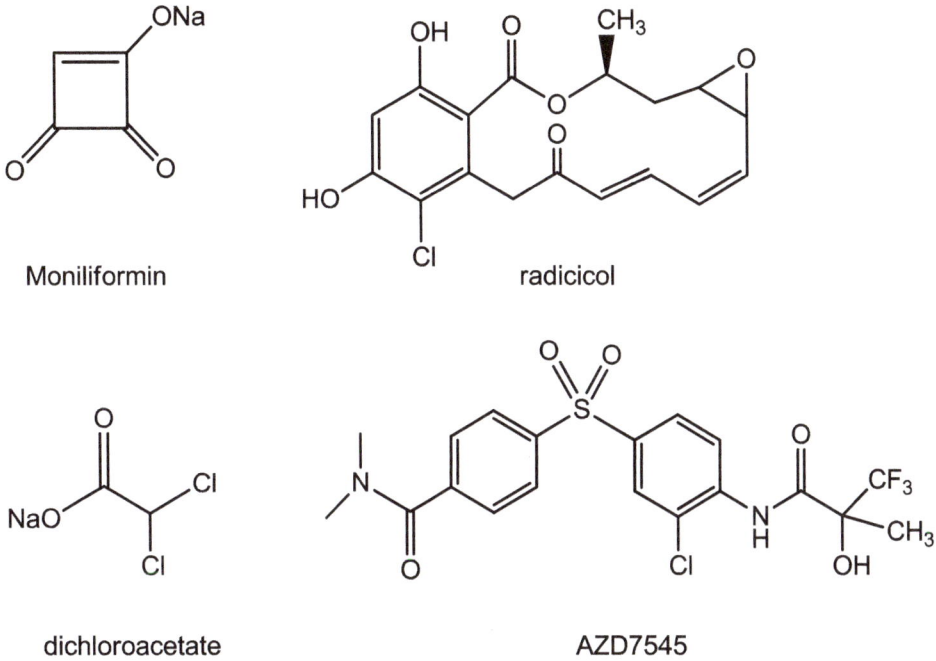

Fig. (96). Chemical structure of natural or synthetic PD inhibitors.

1.3. Citrate Synthase (CS)

Citrate synthase catalyses the aldol type reaction between acetyl-CoA coming from the glycolytic pathway and transported from the mitochondria to the cytosol, with oxaloacetate coming from the citric acid cycle to furnish citrate (Fig. **97**).

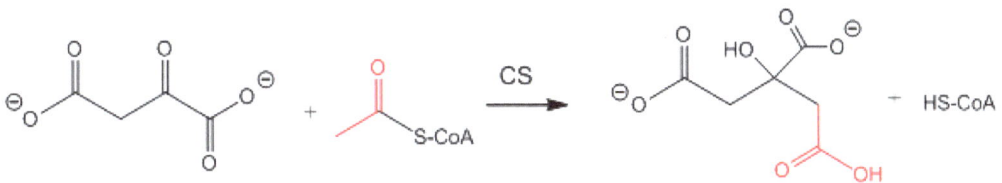

Fig. (97). The citrate synthase reaction.

The structure of pig heart citrate synthase was determined by crystallographic studies (Fig. **98**) observing a dimeric protein composed by two identical subunits. Partial resolution in the active site residues with cystamine includes Cys184, and Gly201 [101]. Also the crystal structure of citrate synthase from *Thermophilic* *archaeon* with high degree of structural homology was determined [102].

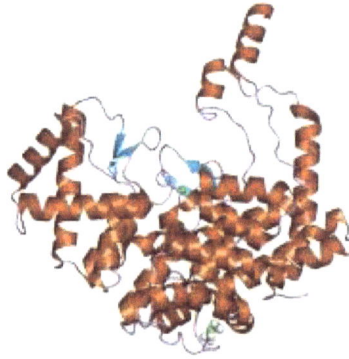

Fig. (98). The crystal structure of pig heart citrate synthase (PDB: 3ENJ).

A plausible reaction mechanism is represented in Fig. (**99**), summarizing some of the key steps which explain the condensation between acetyl-CoA and oxaloacetate, involving conserved residues His and Asp. Remarkably, the formation of acetyl-CoA enolate which directs a nucleophilic attack to the α-ketone to produce citryl-CoA, is followed by final hydrolysis furnishing citric acid [103].

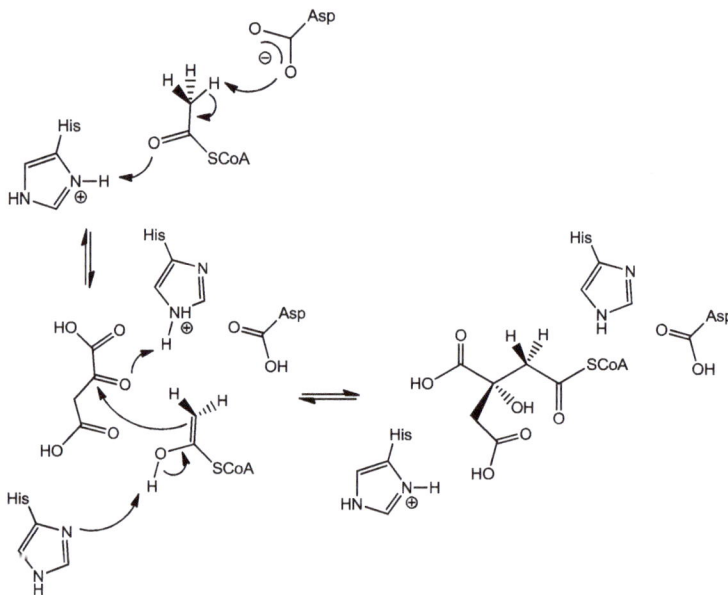

Fig. (99). Proposed reaction mechanism for the citrate formation.

Crystallographic studies at the active site of citrate synthase form *Metallosphaera sedula* MsCS were conducted, establishing as the key the residues His214, His253 and Asp307 during the catalysis (Fig. **100**) [104].

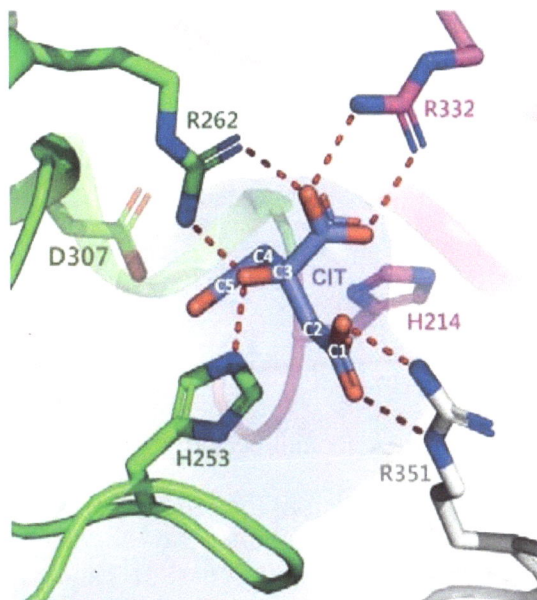

Fig. (100). The binding site residues of citrate synthase form *Metallosphaera sedula*.

1.4. Citrate Synthase Inhibitors

Naturally occurring molecules such as citrate, ATP and NADH are known to inhibit CS. Based on the inhibition kinetics of citrate synthase, the Km values increase while Vmax values keep constant what is typical behaviour for competitive inhibitors [105].

1.5. ATP-citrate Lyase (ACLY)

This enzyme catalyses the reversible transformation from citrate to acetyl-CoA and oxaloacetate, requiring ATP and water (Fig. **101**).

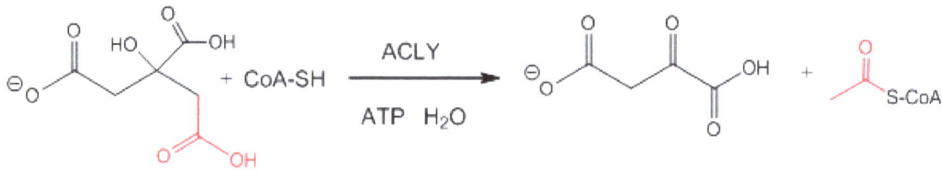

Fig. (101). The ATP-citrate lyase reaction.

A detailed study based on sequence similarity between enzymes proposed that the reaction catalysed by ACLY actually occurs in four steps [106].

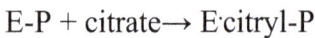

ATP + E → E-P + ADP

E-P + citrate→ E·citryl-P

E·citryl-P + CoA → E·citryl-CoA + Pi

E·citryl-CoA → E + acetyl-CoA + oxaloacetate

ATP-citrate lyase is tightly involved in the process known as *de novo* lipogenesis as the main route for the production of fatty acids and cholesterol, and it has been found to be overexpressed in several types of cancers since the fatty acid biosynthesis is enhanced substantially. Therefore ACLY inhibitors have been evaluated [107, 108].

Truncated human ATP-citrate lyase in a complex with citrate and with strong inhibitor NDI-091143 at the ACLY citrate domain allowed to observe the interactions with their respective residues (Fig. **102**) [109, 110].

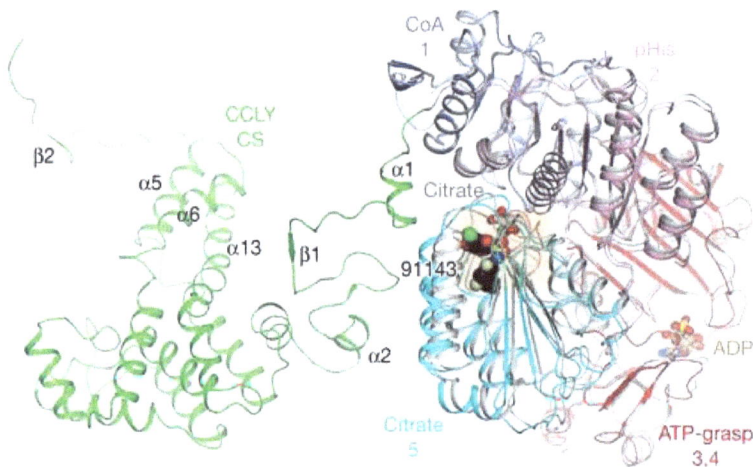

Fig. 102 cont.....

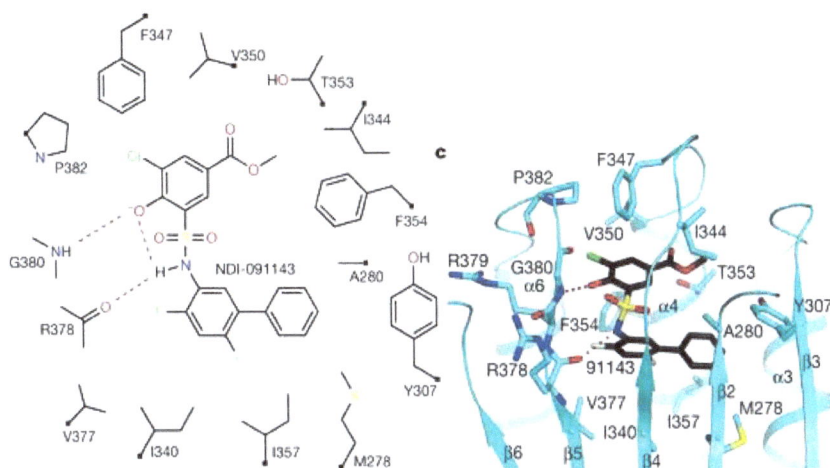

Fig. (102). Crystal structure of human ATP-citrate lyase in a complex with citrate and interactions at the binding site with strong inhibitor NDI-091143 (PDB: 6O0H).

1.6. ATP-citrate Lyase Inhibitors

As mentioned, ATP citrate lyase is an important target enzyme for inhibition because of representing a starting point in the fatty acid, cholesterol, acetylation-prenylation of proteins, and in cancer cells ACLY is overexpressed. Also in patients with high cholesterol, dyslipidaemia and hepatic steatosis ACLY reveals as key enzyme for prevention and treatment.

Some of the small molecule inhibitors reported are: (-)-hydroxycitrate 1 with K_i value of 0.15 μM, succinic acid derivative 2 with K_i = 1.0 μM, benzenesulfonamide derivatives 3 and 4 with IC_{50} of 0.13 μM and K_i value of 7.0 nM, respectively (Fig. **103**) [111, 112].

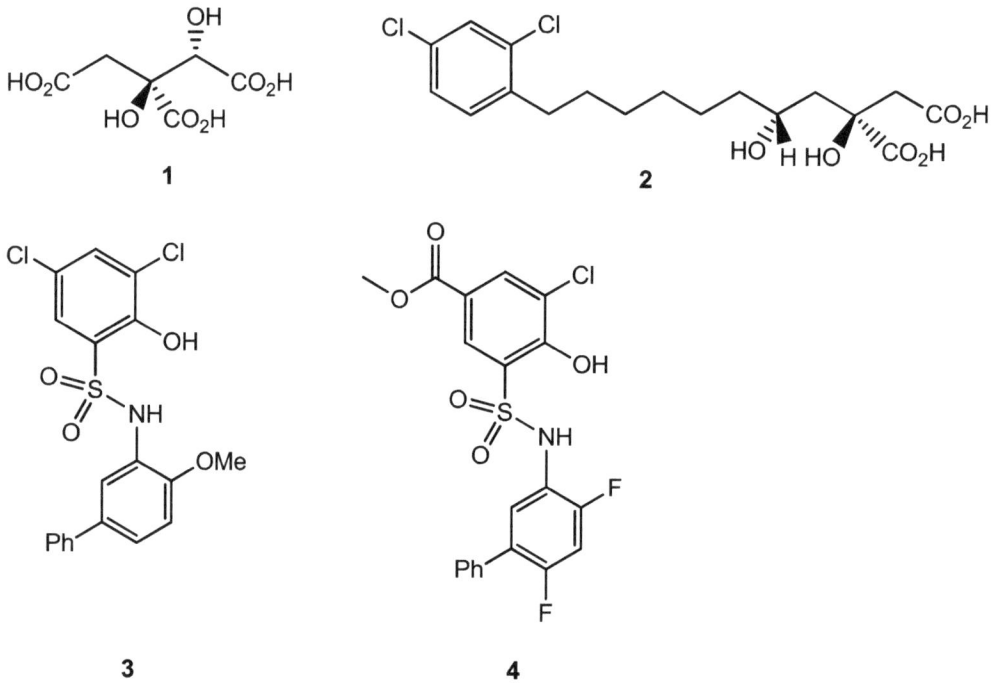

Fig. (103). Chemical structure of ATP citrate lyase inhibitors.

1.7. Ketoacyl Synthase (KAS)

This is a condensing enzyme responsible for the coupling reaction between acyl-CoA with malonyl-CoA, acyl-ACP with malonyl-ACP or combinations to produce 3-ketoacyl-CoA or 3-ketoacyl-ACP. The Claisen type condensation occurs in three step sequence summarized as: trans-thioesterification with cysteine residue, decarboxylation and nucleophilic substitution with enolate to provide β-ketoacyl-ACP (Fig. **104**).

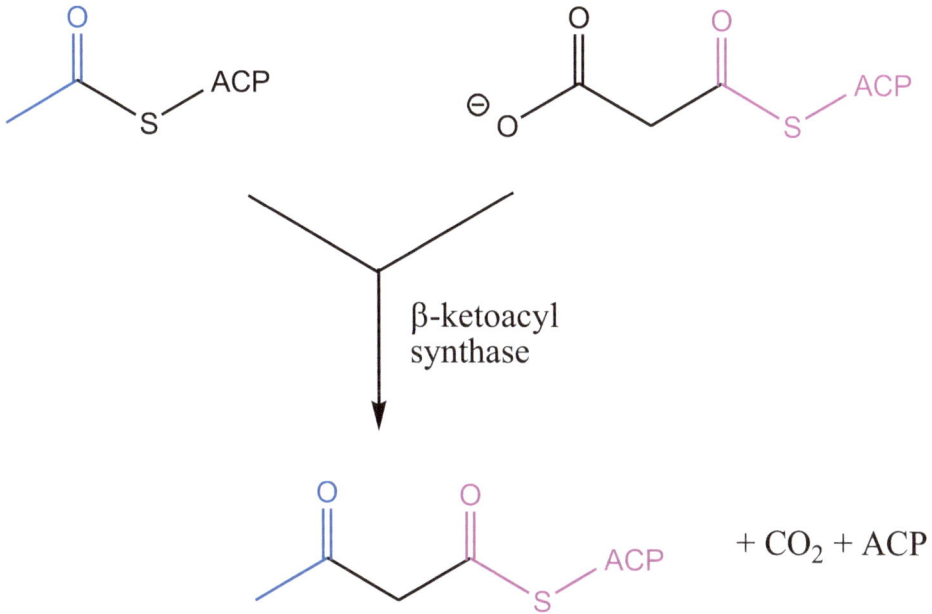

Fig. (104). The ketoacyl synthase reaction.

A mechanism proposes a nucleophilic attack by cysteine to acetyl-CoA producing a tetrahedral intermediate stabilized by Gly306, Cys112 and His244. Next the alcoxide generated promotes the exit of the CoA-S⁻ group negatively charged and final hydrogen transfer from histidine leads to the acylated product (Fig. **105**) [113].

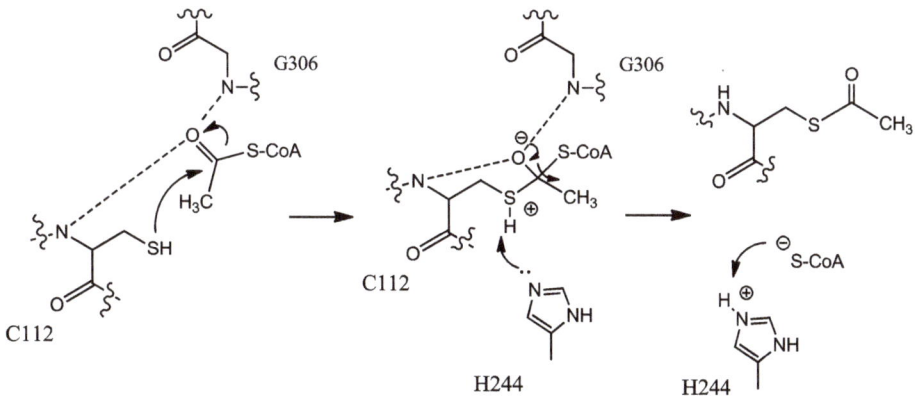

Fig. (105). The ketoacyl synthase reaction mechanism.

Mutagenesis of *E. coli* KAS I demonstrates that His298 and His333 are needed for the malonyl decarboxylation reaction, however the mechanism is not conclusive, since it has been proposed the formation of CO_2 and alternatively the bicarbonate anion formation as it is shown in Fig. (**106**) [114].

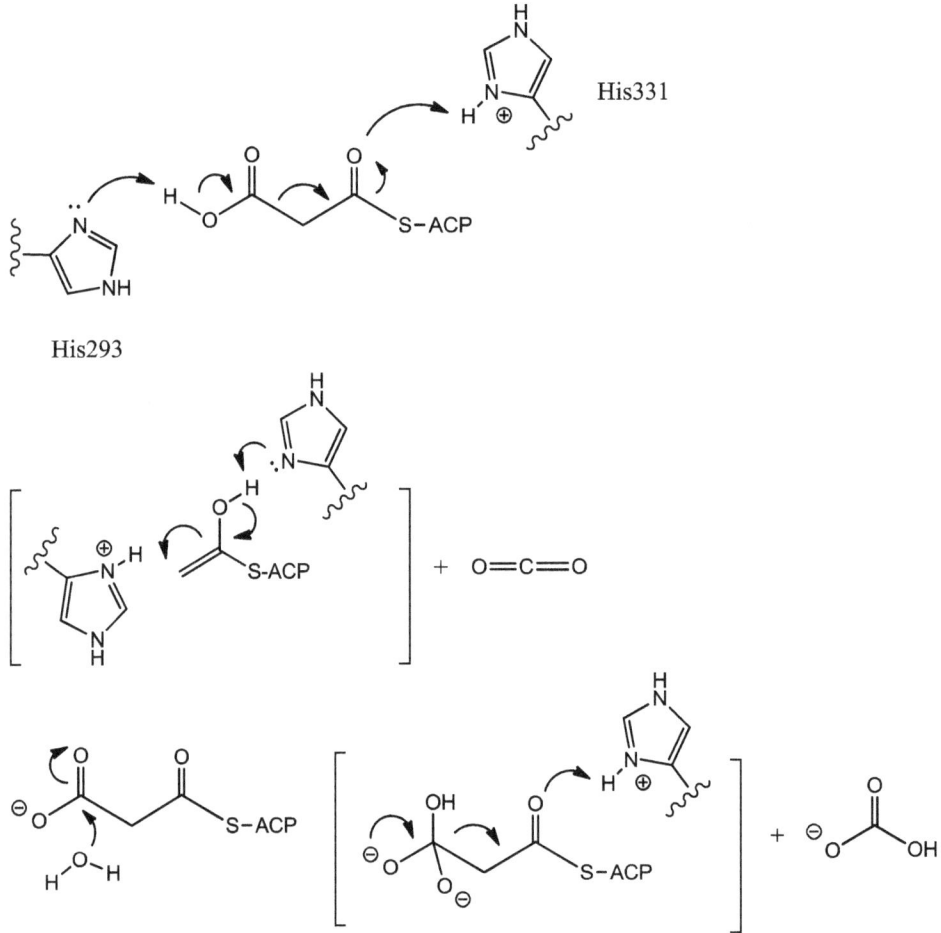

Fig. (106). Reaction mechanism for the malonyl decarboxylation process.

Human mitochondrial β-ketoacyl ACP synthase complexed with antifungal antibiotic fatty acid inhibitor cerulenin were analysed and the interaction identified between the residues and the inhibitor (Fig. **107**) [115].

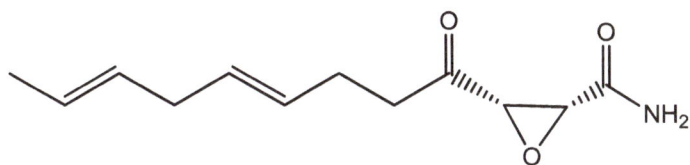

cerulenin

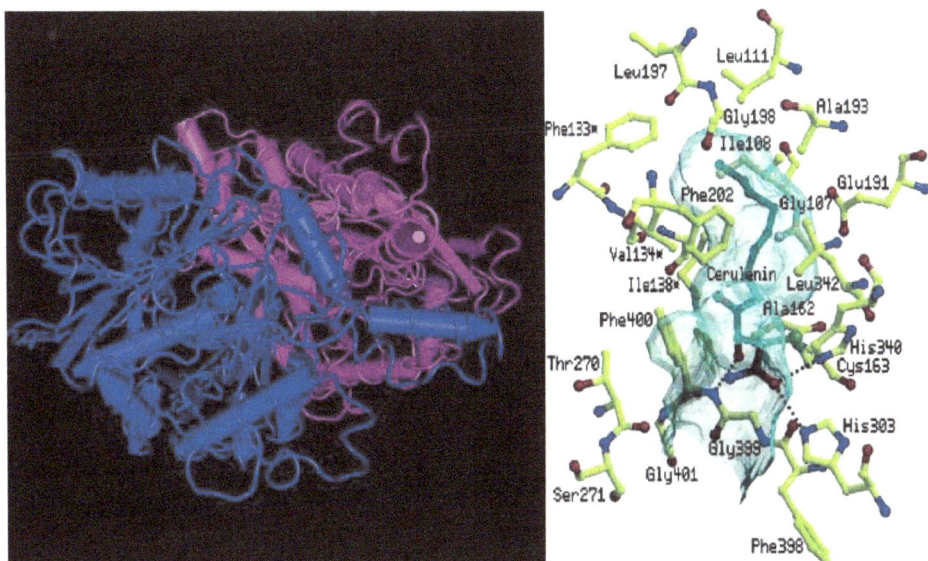

Fig. (107). Tridimensional structure and active site residues of human mitochondrial β-ketoacyl ACP synthase (PDB: 2IWY).

1.8. Ketoacyl Synthase Inhibitors

1,2,4-triazolone GSK2194069, and pyrimidine urea derivative GSK837149A (Fig. **108**) have been evaluated as a potent and specific inhibitors of the β-ketoacyl reductase (KR) in the human fatty acid synthase complex (FAS) [116, 117].

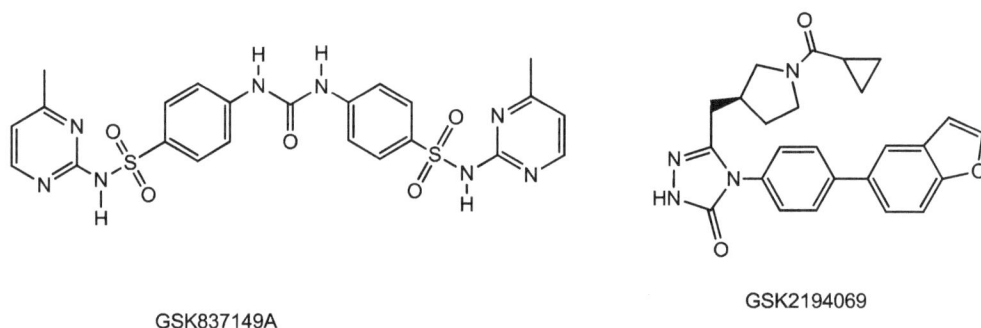

GSK837149A

GSK2194069

Fig. (108). Chemical structure of GSK2194069 and GSK837149A ketoacyl synthase inhibitors.

1.9. Stearoyl-CoA Desaturase (SCD)

The process for installing a double bond to the fatty acids stearoyl or palpalmitoyl-CoA is a crucial step in the fatty acid biosynthesis. This reaction is catalysed by the enzyme stearoyl-CoA desaturase which is present in all eukaryotes, giving place to unsaturated fatty acid oleoyl-CoA and palmitoleoyl-CoA, which are important precursors of triglycerides, cholesterol and membrane components.

For the introduction of the double bound specifically at the 9 position, furnishing monounsaturated fatty acids (MUFAs) the reaction requires oxygen, dimetal ion Zn^{2+} or Fe^{2+} and $NADPH+H^+$ in a complex known as cytochrome b5 reductase (Fig. **109**).

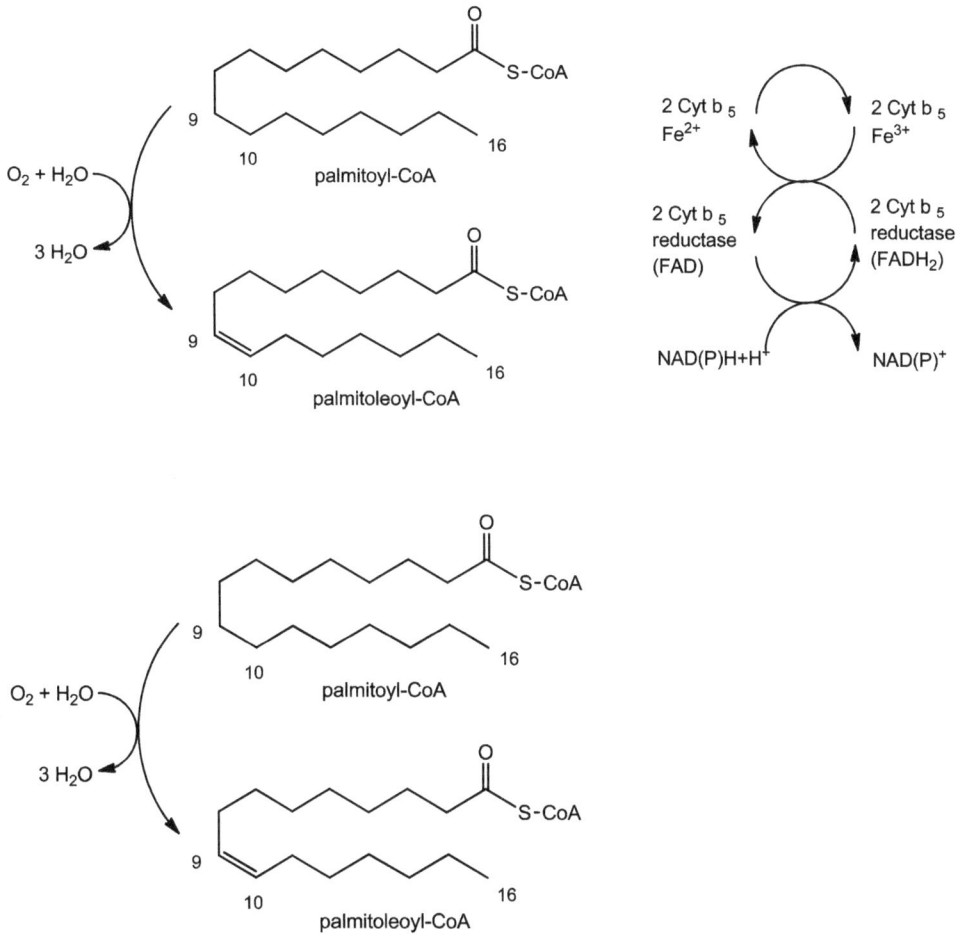

Fig. (109). Formation of monounsaturated fatty acids (MUFAs).

The crystal architecture of human stearoyl-CoA desaturase shows four transmembrane α-helices and a cytoplasmic cap domain. The stearoyl-CoA desaturase binding site presents a tunnel-like pocket for the acyl tail, the substrate-residues main interactions, and 2 zinc ions coordinated with histidines (Fig. **110**) [118].

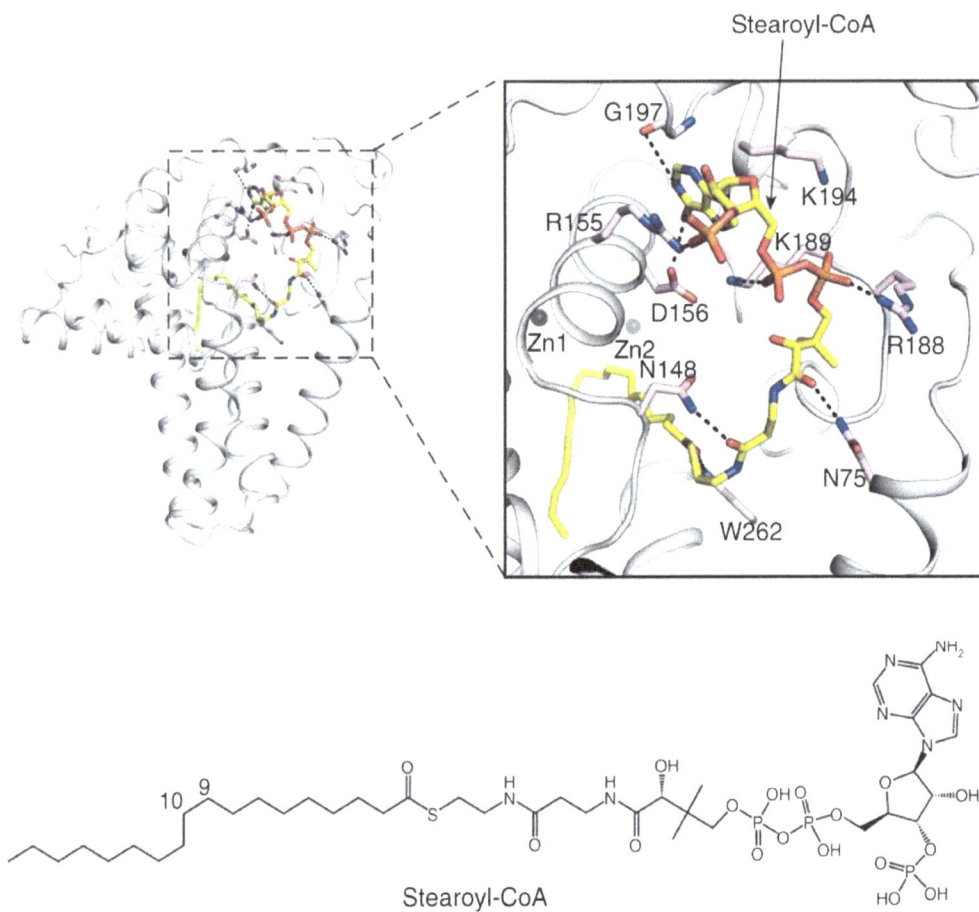

Fig. (110). Ribbon representation of human stearoyl-CoA desaturase and close- up view of the binding site region (PDB: 4ZYO).

1.10. Stearoyl-CoA Desaturase (SCD) Inhibitors

Inhibition studies on hepatitis C virus (HCV) demonstrates that the use of small molecule inhibitors on SCD-1 isoform disrupts the integrity of membranous HCV replication complexes provoking HCV RNS nuclease degradation [119].

It is known that the composition of the phospholipid bilayers modulates de curvature of the membranes and a higher proportion of oleic acid increase fluidity and negative curvature.

With this information in hand, small molecule inhibitors have been evaluated as inhibitors to prevent viral infections through the lipogenesis control. Notably macrocyclic NS3/4A protease inhibitor MK-4519 developed by Merck, was evaluated as an inhibitor of viral RNA replication, observing potent inhibition (EC_{50}=8.3 nM) (Fig. **111**) [120].

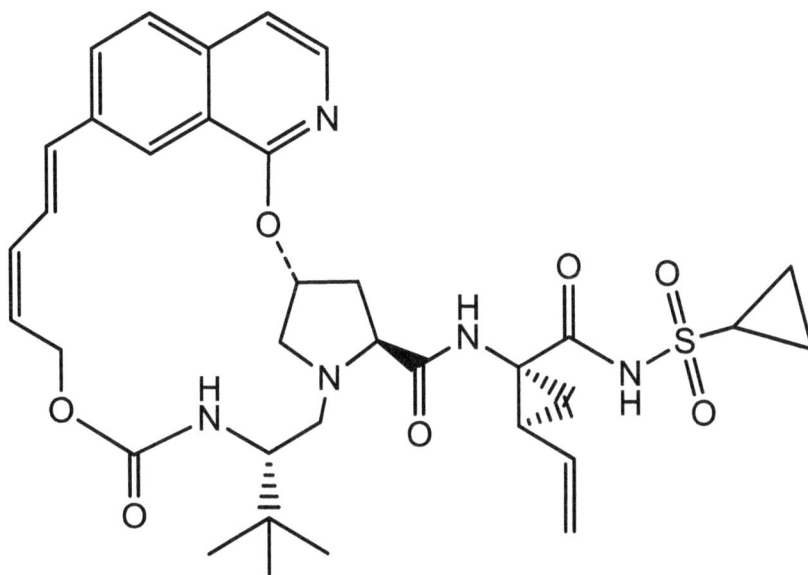

MK-4519

IC_{50} = 8.3 nM

Fig. (111). Chemical structure of macrocyclic NS3/4A protease inhibitor MK-4519.

It has been found that SCDs are overexpressed in many cancer types, increasing MUFAs levels, and consequently allowing cell viability, and proliferation. As a result of these findings, a series of small molecules have been tested as SCD inhibitors for cancer cells, showing some of them high potency against different cancer cells. Examples, MF-438, A-939572, LCF-369, CAY-10566 and natural product sterculic acid isolated from plants of *Sterculia foetida* (Fig. **112**) [121 - 122].

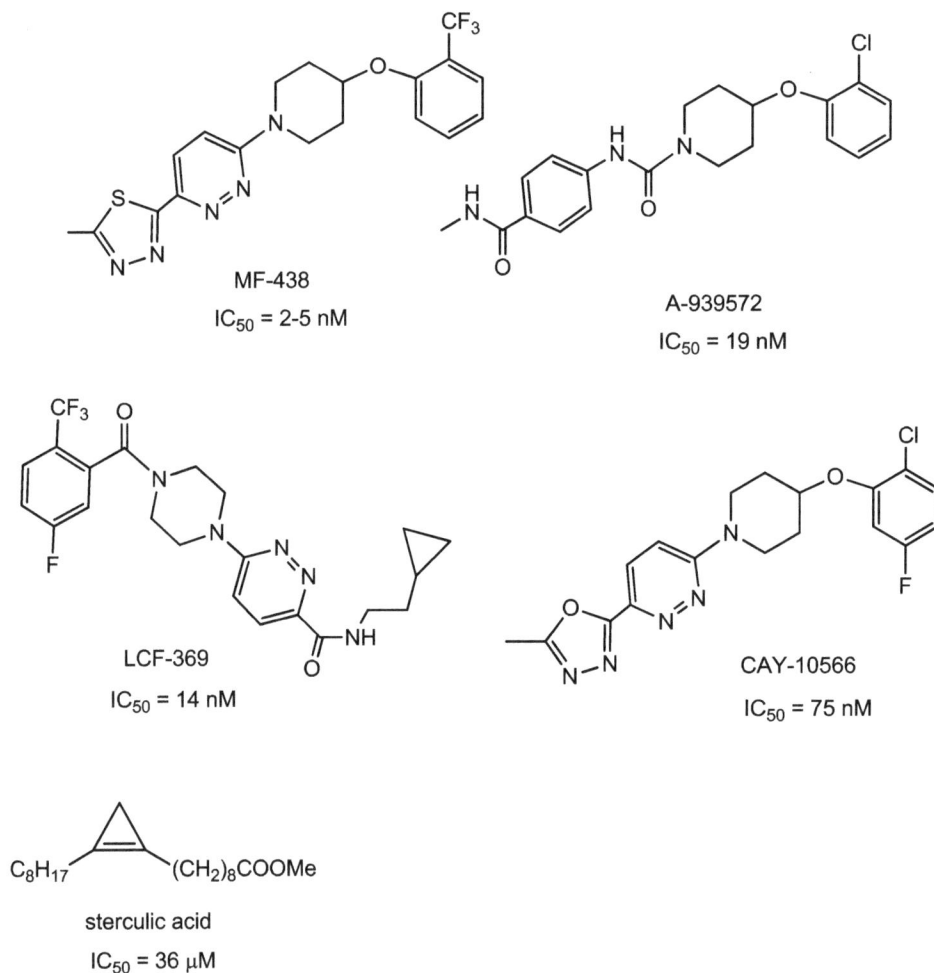

Fig. (112). Small molecule inhibitors of stearoyl-CoA desaturase against cancer cells.

2. SYSTHESIS OF TRIACYLGLYCEROL (TRIGLYCERIDES)

Triglycerides are important molecules with the main function of store calories as lipids and when needed to serve as energy supplier. However, the high levels are closely related with health problems associated with metabolic syndrome such as high blood pressure, obesity and diabetes. The triglycerides are formed by condensation of fatty acids with glycerol, but the process requires the incorporation stepwise of fatty acid units to fully esterified the glycerol molecule. Thus, the triglyceride synthesis initiates with glycerol as starting material converted to glycerol-3-phosphate by the enzyme glycerol kinase in the presence of ATP. The resulting glycerol-3-phosphate was esterified by the enzyme glycerol-3-phosphate acyl transferase and one equivalent with fatty acid-CoA to give lysophosphatidic acid. A second esterification reaction takes place through the acylglycerophosphate acyltransferase catalysis to produce phosphatidic acid. The next step involves the removal of the phosphate group by the enzyme phosphatidic acid phosphohydrolase to furnish diacylglycerol which is finally condensed by the enzyme diacylglycerol acyltransferase with the third fatty acid molecule to provide triacylglycerol as the end product (Fig. **113**).

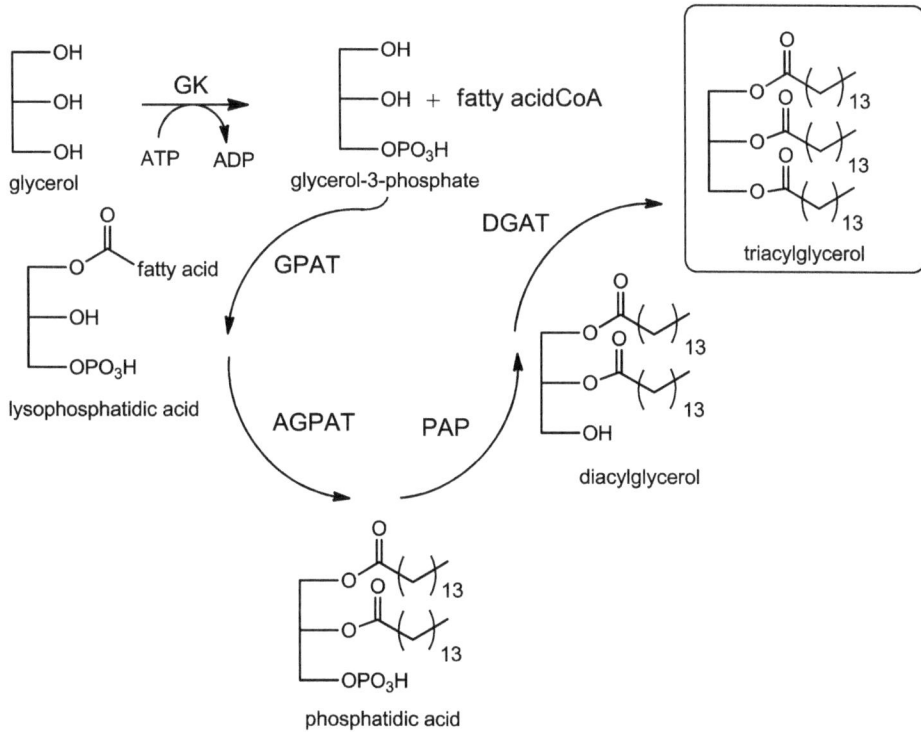

GK = glycerol kinase
GPAT = glycerol-3-phosphate acyltransferase
AGPAT = acylglycerophosphate acyltransferase
PAP = phosphatidic acid phosphohydrolase
DGAT = diacylglycerol acyltransferase

Fig. (113). The triacylglycerol cycle.

2.1. Glycerol Kinase (GLPK)

The conversion of glycerol into glycerol-3-phosphate (Fig. **114**) is considered the starting point in the triglyceride synthesis and it is accomplished by the enzyme glycerol kinase which also plays an important role in controlling the uptake into the cell [123]. Glycerol kinase is present in plants and animals and it can have reverse effects in humans when the activity is diminished [124].

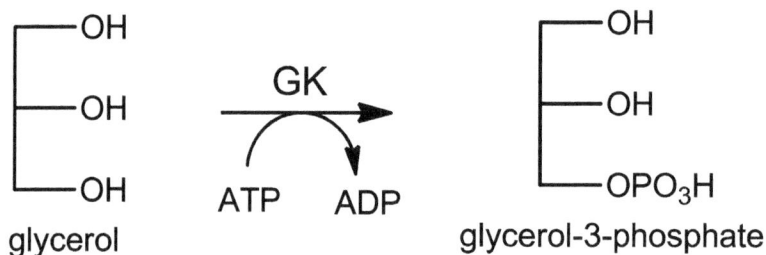

Fig. (**114**). The glycerol kinase reaction.

Different glycerol kinases complexed with glycerol and ethylene glycol have been reported from bacteria and parasites [125]. For instance, in trypanosomes GK can also catalyse the reverse reaction which is essential for the parasite survival and therefore it can be used as a target for the design of novel inhibitors. At the active site the catalytic glycerol establishes hydrogen bond interactions with conserved residues of domain I: Arg84, Glu85, Asp254. Also the active site ADP accepts hydrogen bonds from Thr12, and Gln255 and assume a Syn configuration with its ribose group (Fig. **115**) [126].

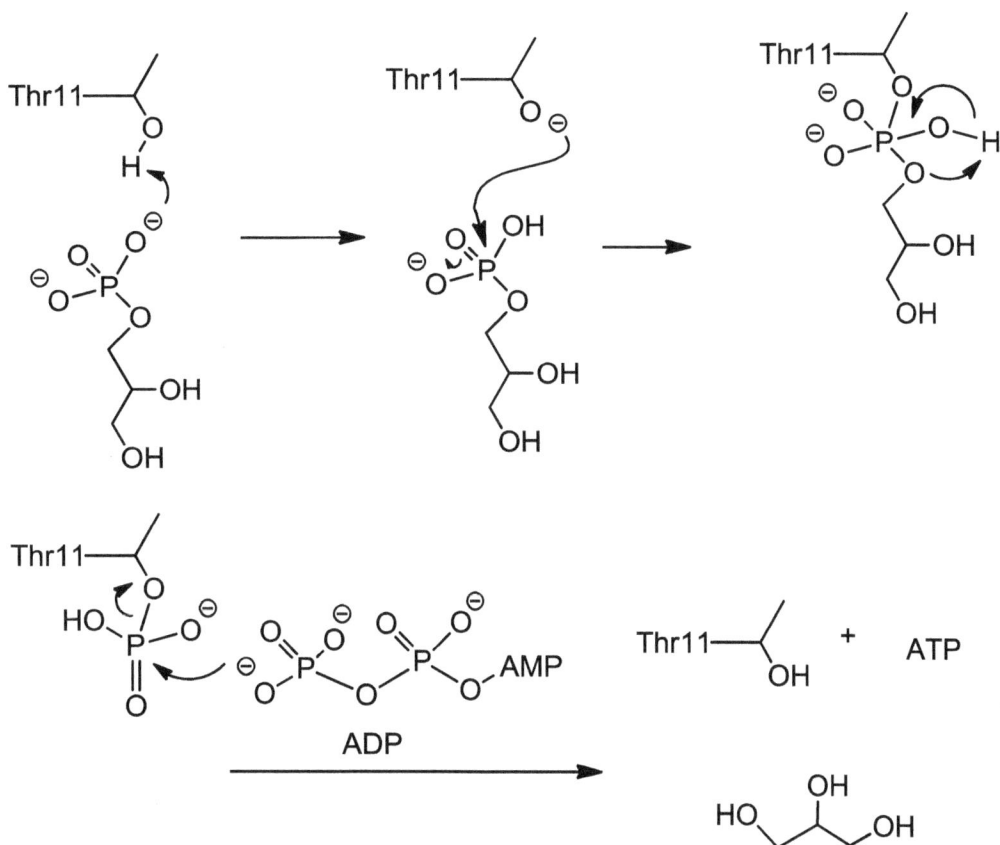

Fig. (115). Simplified mechanism representing the reverse process of glycerol dephosphorylation.

2.2. Glycerol Kinase Inhibitors

Racemic 2,3-dihydropropyl dichloroacetate was discovered as a potent inhibitor of glycerol kinase *in vitro* showing competitive inhibition with K_i value of 1.8×10^3 M (Fig. **116**) [127].

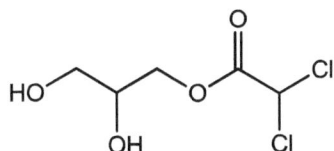

Fig. (116). Chemical structure of 2,3-dihydropropyl dichloroacetate.

2.3. Glycerol-3-phosphate Acyltransferase (GPAT)

This enzyme catalyses the first step in the triacylglycerol biosynthesis consisting in the attachment of glycerol-3-phosphate with fatty acid-ACP at the first position providing lysophosphatidic acid (Fig. **117**). In humans there are two isoforms one located in the mitochondrial outer membrane and the other in the endoplasmic reticulum [128]. On the other hand, the most important acyl donors found in bacterial glycerolipid synthesis are the acyl carrier protein (acyl-ACP), and acyl-CoA (from exogenous fatty acids), however also acyl-phosphate (acyl-PO_4) have also been identified, although exhibiting higher water instability than the thioester counterparts [129]. In bacteria, the transfer of the fatty acid to glycerol--phosphate uses acyl-phosphate and the enzyme involved is known as PlsY.

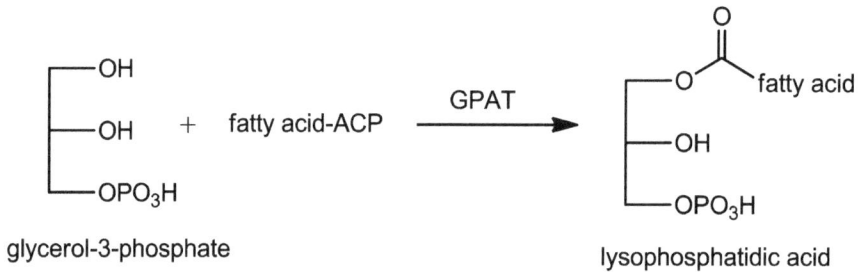

Fig. (**117**). The glycerol-3-phosphate acyltransferase reaction.

There are two distinct families identified for the acylation at 1-position of glycerol-3-phosphate, the PlsB acyltransferase found in *E. coli* with presence in many eukaryotes, and PlsY more widespread in bacteria.The PlsB acyltransferase uses either acyl-ACP or acyl-CoA as the donor and the PlsY requires acyl-PO_4 as acyl donor and it is the most abundant G3P acyltransferase in gram positive bacteria such as *S. pneumonia* and *S. aureus* [130].

In a stereo diagram of the glycerol-3-phosphate binding site from *Cucurbita moschata* (Fig. **118**), the phosphate group interacts with positively charged Arg235, Arg237, Lys193 and His194, and the C1-hydroxyl group is oriented with His139, Asp144 completing the catalytic triad [131].

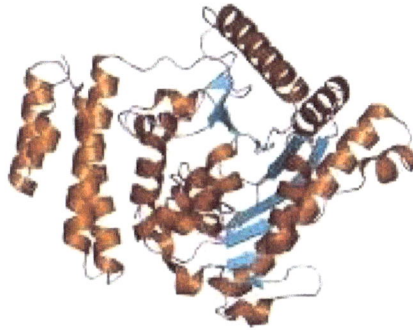

Fig. (118). Crystal structure of the glycerol-3-phosphate from *Cucurbita moschata* (PDB: 1K30).

More recently the membrane-integrated G3P acyltransferase PlsY was analysed, revealing a G3P pocket composed by a phosphate clamp (Ser35, Arg45, Lys104 and Asn180) and two side chains (Ser142 and His177), interacting with the 2-OH, and additional interactions between the carbonyl groups of Thr41, and Ala40 (Fig. **119**) [132].

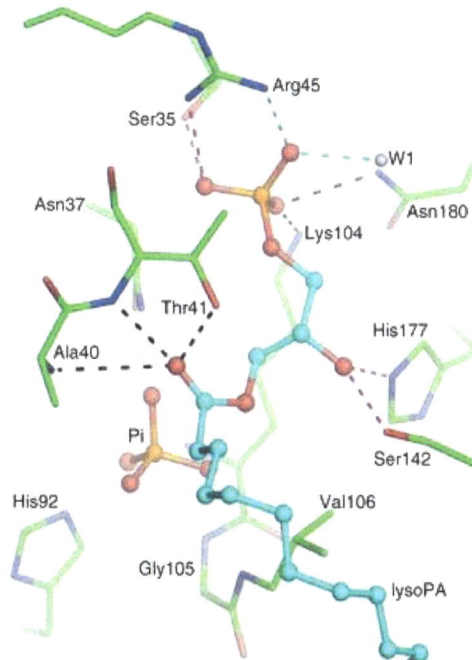

Fig. (119). Binding site diagram of the membrane-integrated G3P acyltransferase PlsY (PDB: 5XJ6).

2.4. Glycerol-3-phosphate Acyltransferase (GPAT) Inhibitors

Substituted sulphonamide derivatives having benzoic and phosphonic acids have been synthesized and evaluated as GPAT inhibitors displaying IC_{50} values in the mM range (Fig. **120**) [133].

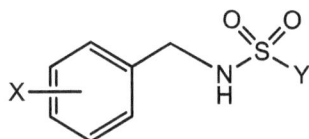

X = o-CO$_2$H, Y = C$_9$H$_{19}$ IC$_{50}$ = 66.5 μM

X = m-CO$_2$H, Y = C$_9$H$_{19}$ IC$_{50}$ = 83.5 μM

X = m-CH$_2$PO$_3$H$_2$, Y = C$_9$H$_{19}$ IC$_{50}$ = 62.8 μM

X = o-CH$_2$PO$_3$H$_2$, Y = C$_9$H$_{19}$ IC$_{50}$ = 81.1 μM

Fig. (120). Chemical structure of substituted sulphonamide derivatives.

Another small molecule GPAT inhibitor reported is the sulphonamide derivative 2-(nonylsulfonamide) benzoic acid (Fig. **121**) which shows a IC_{50} value of 24 μM and has been described as body weight reducer, increasing fatty acid oxidation and energy enhancer, besides other potential metabolic benefits [134].

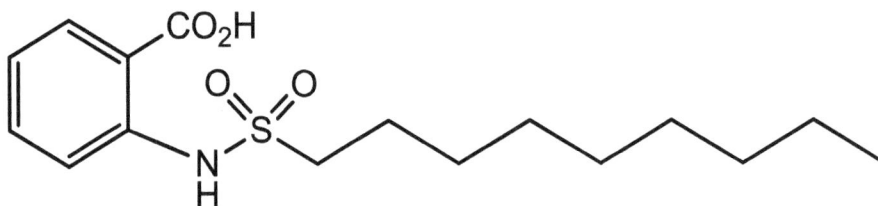

Fig. (121). Chemical structure of sulphonamide derivative 2-(nonylsulfonamide) benzoic acid (FSG67).

2.5. Acylglycerophosphate Acyltransferase (AGPAT or LPAAT)

This enzyme is involved in the synthesis of phosphatidic acid (PA) also known as 1,2-diacylglyceropho-3-poshate (diacyl-GP), and requires acyl-CoA as an acyl donor (Fig. **122**) [135].

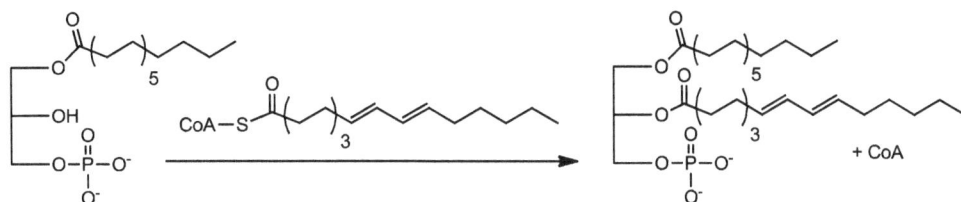

Fig. (122). The acylglycerophosphate acyltransferase reaction.

The AGPAT family has been identified with multiple names, among them AGPAT1 with LPAATα, and AGPAT8 with AGPAT9, LPAATθ, and GPAT3 because several groups independently cloned LPLATs. Human LPAAT1 (called AGAT1 or LPAATα) and human LPAAT2 (AGPAT2) were cloned based on their homologies to yeast orthologues [136].

Human AGPAT2 shows 35% amino acid identity to AGPAT1. The molecular masses of human and mouse AGPAT2 were both calculated to be 31 kDa. AGPAT2 was also located in the ER, but not in other subcellular organelles [135].

2.6. Acylglycerophosphate Acyltransferase Inhibitors

There have been 11 AGPAT described, however only few inhibitors are particularly for AGPAT2 while its involvement in ovarian cancer has been evaluated [137].

The amino triazine derivatives CT32228 inhibits LPAAT-β displaying an IC_{50} value of 50 nM, and the amino benzopyrrol CT32501 has a higher potency displaying IC_{50} value of 27 nM (Fig. **123**) [138, 139].

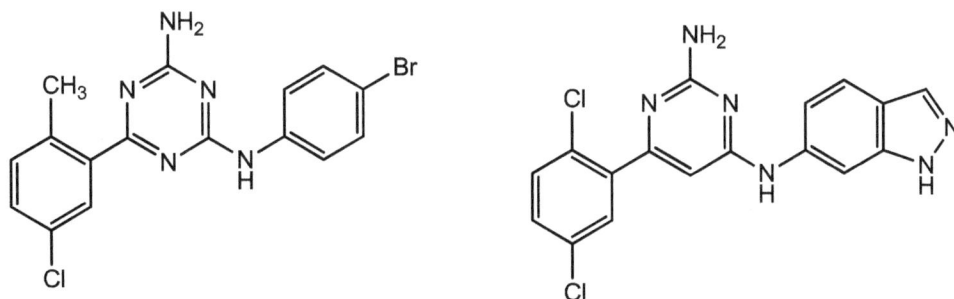

Fig. (123). The amino triazine derivative CT32228 and CT32501.

2.7. Phosphatidic Acid Phosphohydrolase (PAP)

This enzyme catalyses the conversion of phophatidic acid (PA) into diacylglycerol (DAG) in the presence of Mg^{2+} cation (Fig. **124**). It is a crucial enzyme in eukaryotic lipid homeostasis and also participates in signalling pathways that govern the expression of multiple genes associated with membrane and storage lipid [140].

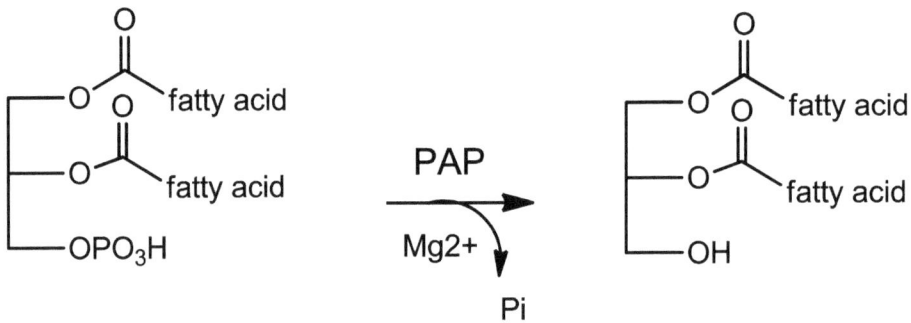

Fig. (124). The phosphatidic acid phosphohydrolase reaction.

The phosphatidic acid phosphatase (PAP) enzyme together with the diacylglycerol acyltransferase enzyme (DGAT) gained significance due to their potency for engineering microorganism oil production. Also PAP, and DGAT in combination with diacylglycerol kinase, CDP diacylglycerol synthase and phospholipase regulate the flux of materials between triacylglycerol (TAG) synthesis and phospholipids *via* the Kennedy pathway [141].

2.8. Phosphatidic Acid Phosphohydrolase Inhibitors

PAP enzyme activity is reversibly inhibited by fatty acids and their acyl-CoA esters. In rat liver containing PAP-1 and PAP-2, the 50% inhibition was observed at aprox. 0.6 mM and more than 80% at 1.2 mM with oleic acid (Fig. **125**) or by oleolylCoA. Palmitic acid inhibits PAP-1 with 30% inhibition at 1 mM [142].

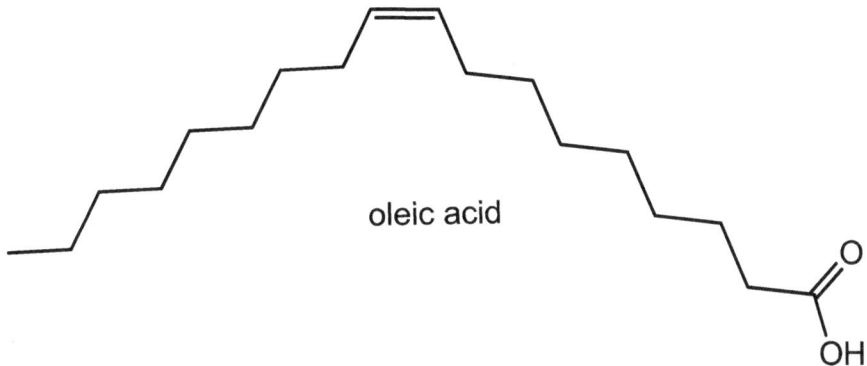

Fig. (125). Chemical structure of oleic acid.

2.9. Diacylglycerol Acyltransferase (DGAT)

The final step in the biosynthesis of triacylglycerol (TAG) depend on diacylglycerol acyltransferase (DGAT) and requires 1 equivalent of fatty acid acyl CoA and 1,2-diacylglycerol (Fig. **126**) which can also come from the Kennedy pathway and could be incorporated into phospholipids [143]. Despite its importance in the production of TAG limited amount of information is present in the literature [144].

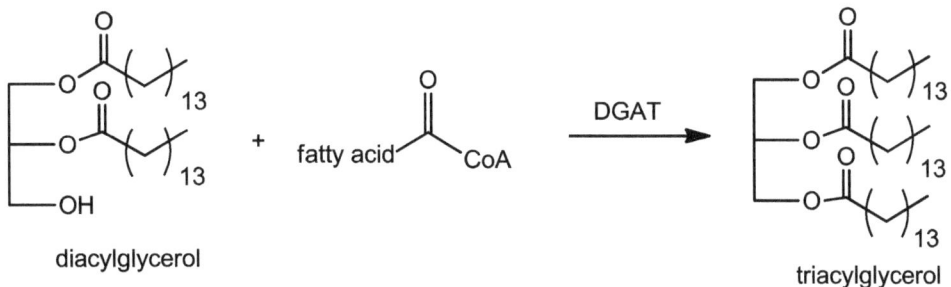

Fig. (126). The diacylglycerol acyltransferase reaction.

2.9. Diacylglycerol Acyltransferase Inhibitors

A variety of DGAT1 inhibitors (Fig. **127**), to treat metabolic disorders such as obesity, diabetes, dyslipidemia, hepatic steatosis, have been developed, some of which reach clinical trials although struggling with adverse effects. Notably, compound 4 exhibit high potency IC_{50} = 19 nM, while compound 5 has *in vivo* IC_{50} of 132 nM [145].

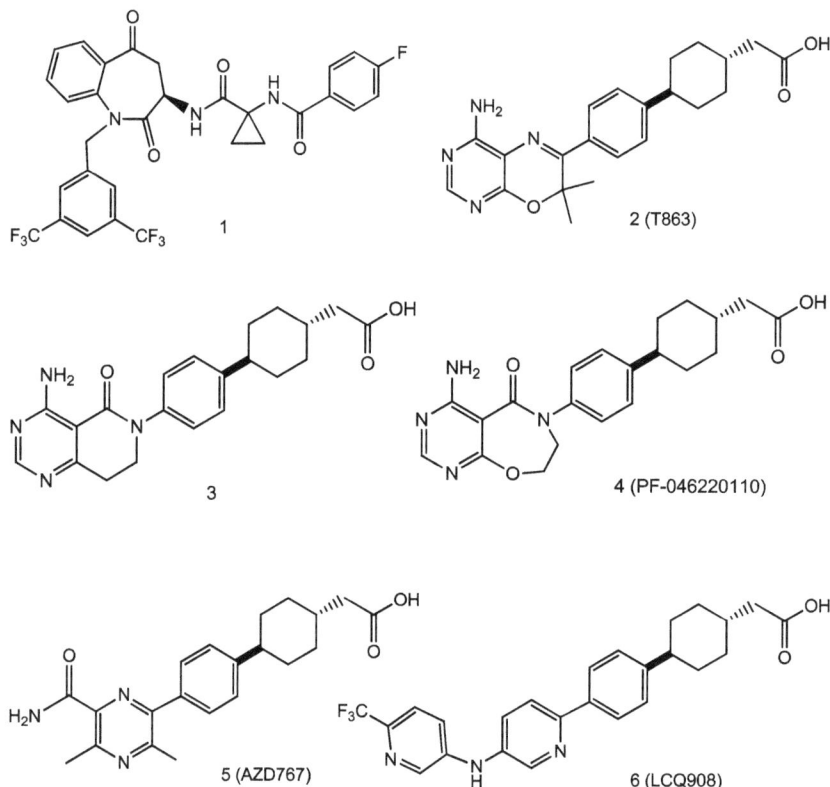

Fig. (127). Chemical structure of DGAT1 inhibitors.

2.10. The Kennedy Pathway

The phospholipid biosynthesis to produce phosphatidyl ethanolamine (PE) and phosphatidylcholine (PC) from ethanolamine or choline with diacylglycerol is commonly known as the Kennedy pathway.

The following full cycle represented in Fig. (**128**) summarizes the set of reactions involved, being identical for both phospholipids (PE and PC), it starts from the phosphorylation of the ethanolamine, catalysed by the enzyme ethanolamine kinase to yield phosphoethanolamine. Next, a coupling reaction with cytidine triphosphate mediated by the phosphoethanolamine cytidyltransferase, occurs to produce cytidine diphosphoethanolamine which finally, transfer the phosphoethanolamine group to the acylglycerol under the catalysis of ethanolaminephosphotransferase to finally produce phosphatidylethanolamine.

Fig. (128). The Kennedy pathway.

CHAPTER 4

Aminoacid Biosynthesis

Amino acids are biomolecules composed of an amino group and carboxylic acid as common features and different R substituents attached to a chiral carbon with L-configuration for the biologically active enantiomer in humans except for glycine devoid of chirality (Fig. **129**). Biosynthetically, the amino acids are derived from glycolysis, Krebs cycle or the pentose phosphate pathway.

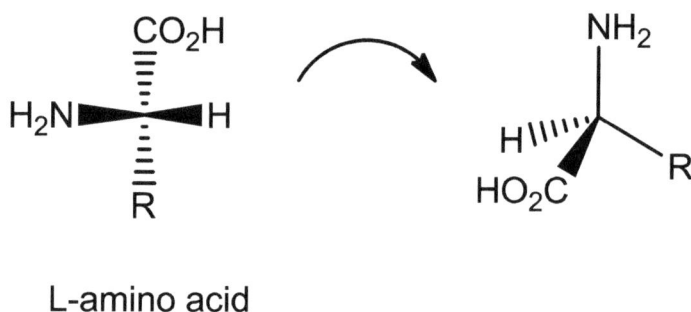

L-amino acid

Fig. (129). Tetrahedral projection of amino acids.

Based on additional functionalities present in the amino acids, they can be classified in aliphatic, aromatic, polar (hydroxyl and thiol groups), cationic, anionic, and heterocyclic groups (Fig. **130**).

1. GLYCINE BIOSYNTHESIS

Glycine is the simplest amino acid having hydrogen as R substituent, and therefore not presenting chirality. It is an amino acid with important implications in brain excitatory and inhibitory activities, and in the synthesis of other essential molecules such as muscle supplement creatine, antioxidant glutathione, and as an abundant component in the structural protein collagen. This amino acid is synthesized from amino acids serine, threonine or nutrient choline, and its detailed biosynthesis is described in the following sections (Fig. **131**).

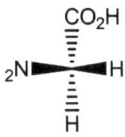

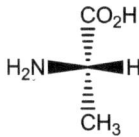

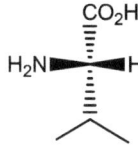

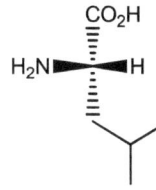

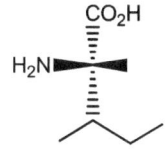

Glycine Gly/G	Alanine Ala/A	Valine Val/V	Leucine Leu/L	Isoleuci Ile/I

Aromatic

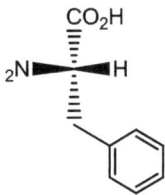

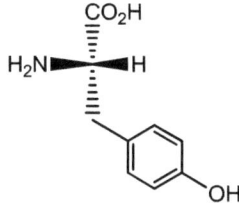

Phenylalanine   Tyrosine
Phe/F   Tyr/Y

Polar

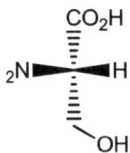

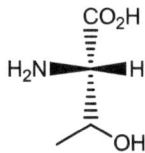

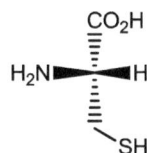

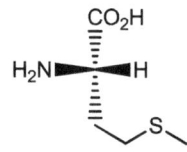

Serine   Threonine   Cysteine   Methionine
Ser/S   Thr/T   Cys/C   Met/M

Heterocyclic

Fig. 130 cont.....

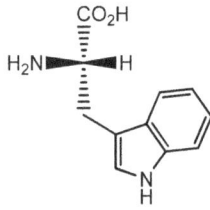

Tryptophan
Trp/W

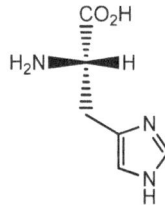

Histidine
His/H

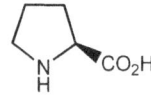

Proline
Pro/P

Cationic/anionic

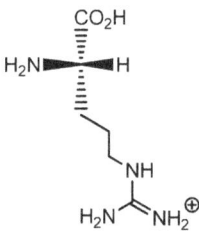

Arginine
Arg/R

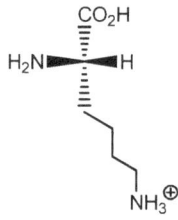

Lysine
Lys/K

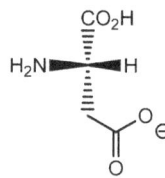

Aspartic acid
Asp/D

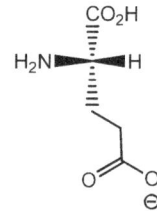

Glutamic acid
Glu/E

Amides

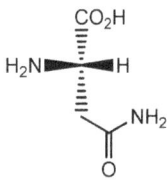

Asparagine
Asn/N

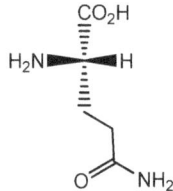

Glutamine
Gln/Q

Fig. (130). Classification and structure of amino acids.

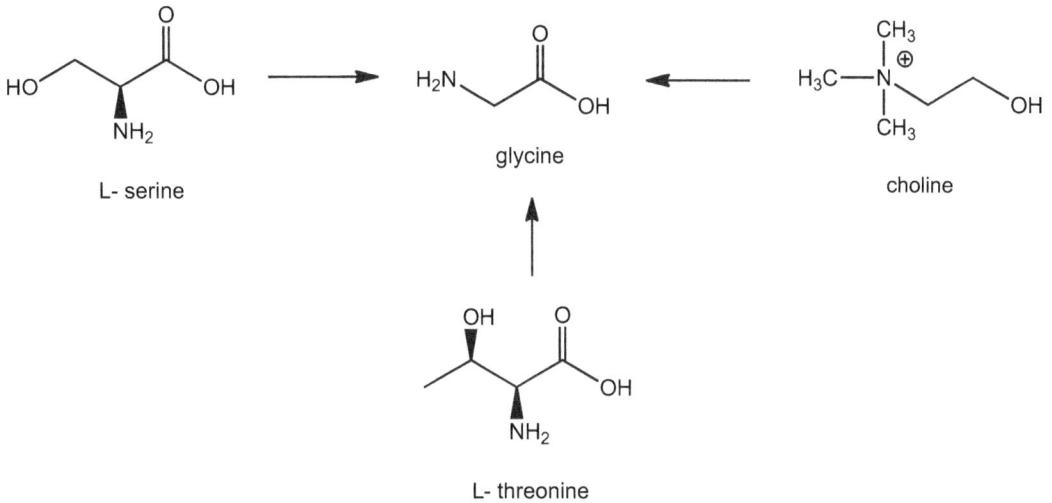

Fig. (131). General scheme of glycine biosynthetic precursors.

Glycine biosynthesis from serine as starting material is carried out by the catalysis of the enzyme serine hydroxyl methyl transferase (SHMT) with participation of tetrahydropteroylglutamate (H_4PteGlu), which is converted to 5,10-methylene tetrahydropteroylglutamate (5,10-MH_4PteGlu), and pyridoxal phosphate (B6) as cofactors (Fig. **132**).

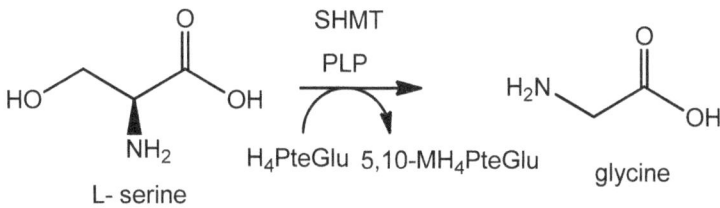

Fig. 132 cont.....

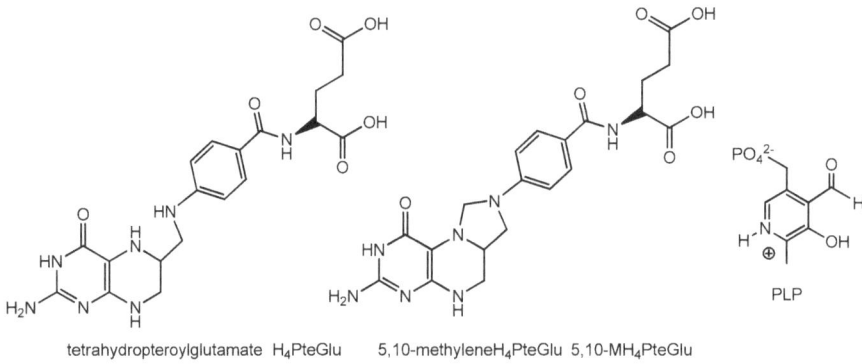

tetrahydropteroylglutamate H₄PteGlu   5,10-methyleneH₄PteGlu 5,10-MH₄PteGlu

Fig. (132). General scheme of glycine biosynthesis from serine in the presence of tetrahydrofolate and pyridoxal phosphate cofactors.

An early mechanism suggested that the conversion of L-serine to glycine involved formaldehyde formation, however further studies with mutants determined that instead of formaldehyde formation, a direct nucleophile displacement of the serine hydroxyl by N^5 of tetrahydropteroylglutamate (H₄PteGlu) occurs, giving place to glycine and 5,10-methylene-H₄PteGlu. The reaction mechanism satisfying this transformation proposes a retroaldol from N^5 of THF making an attack on C3 serine resulting in C3-C2 cleavage of serine (Fig. **133**) [141].

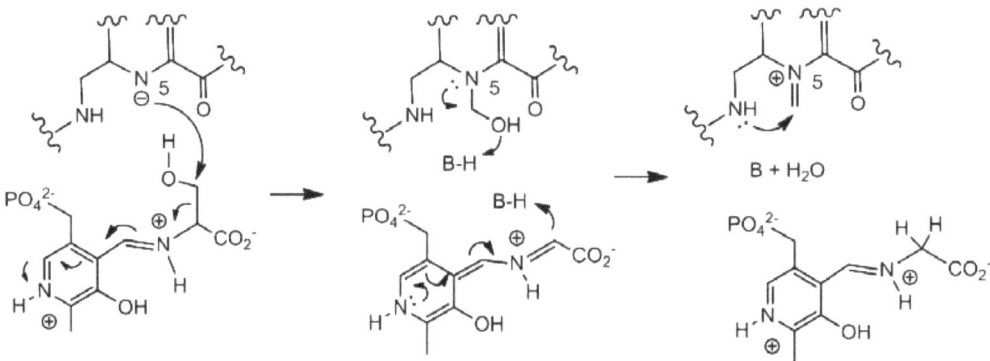

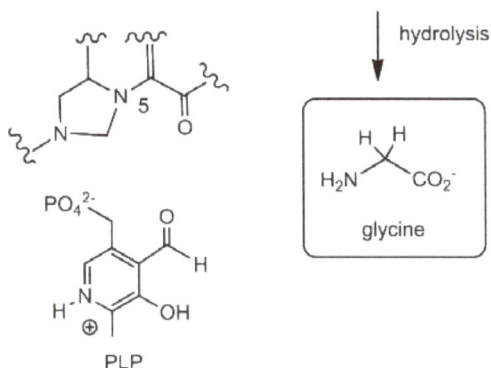

Fig. (133). Proposed reaction mechanism for glycine formation involving PLP and H$_4$PteGlu.

1.1. Serine Hydroxyl Methyl Transferase (SHMT)

The crystal structure of SHMT from *Methanocaidococcus jannaschii* in the apo form (*mj*SHMT without cofactor PLP) and holo form (*mj*SHMT with PLP) was determined, showing the catalytic domain in purple, the C-terminal in green, the N-terminal α-helix in yellow, and cofactor PLP in red spheres (Fig. **134**) [142, 143].

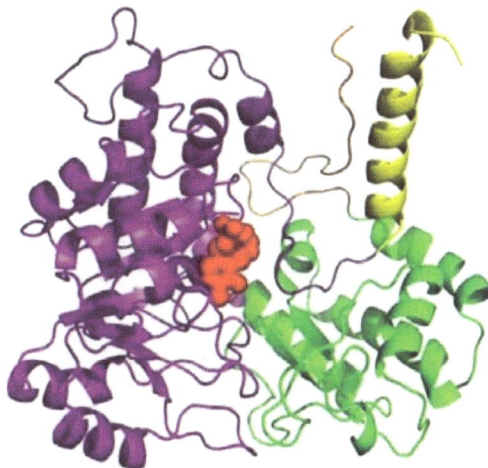

Fig. (134). The ribbon structure of G monomer SHMT from *Methanocaidococcus jannaschii* G monomer showing the catalytic domain in purple, the C-terminal in green, and PLP as spheres in red (PDB: 4BHD).

Native serine hydroxymethyltransferase from *Bacillus stearothermophilus* complexed with serine has been described, allowing the identification of the residues and the cofactors participating. The SHMT stereoview shows the aldimide and THF interacting with serine and glycine (Fig. **135**) [144]. The crystal structure of *Escherichia coli* SHMT (eSHMT) in complex with glycine was determined, showing evidence on the participation of pyridoxal 5'-phosphate and tetrahydropteroylglutamate (5-formyl-H4PteGlu) as the one-carbon carrier. Other crucial interactions are observed such as ionic interaction between the pyridinium nitrogen with Asp200, hydrogen bond of the C3 hydroxyl group with the hydroxyl group of Ser175, and hydrogen bonds of phosphate with Tyr550 and Ser99 [145].

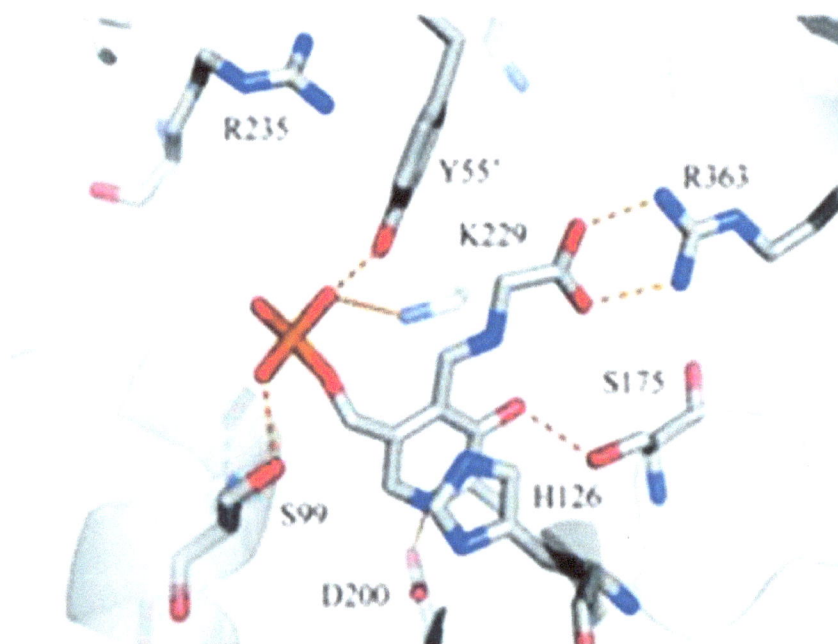

Fig. (135). Stereoview shows the aldimide and THF interacting with serine and glycine (PDB: 1DFO).

1.2. Serine Hydroxymethyl Transferase Inhibitors

A natural product containing pyrazolopyran ring isolated from plant SHIN-1 and optimized by chemical modification SHIN-2 and 3 were reported as inhibitors (Fig. **136**) of plant and plasmodia SHMT although for human SHMT, the inhibition was less significant [146].

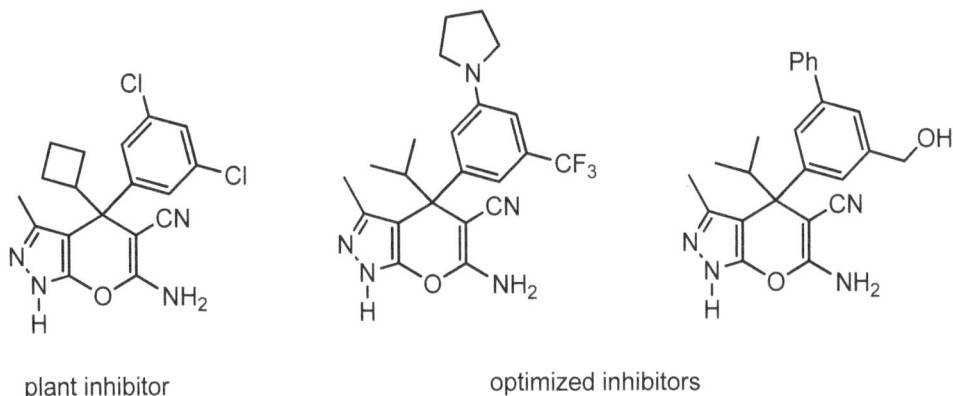

plant inhibitor   optimized inhibitors

Fig. (136). Chemical structure of pyrazolopyran analogues as SHMT inhibitors.

2. ALANINE BIOSYNTHESIS

L-alanine is obtained through a transamination process in which L-valine behaves as an amino group donor and pyruvate as the acceptor (Fig. **137**). The overall reaction requires pyridoxal and is catalysed by pyridoxamine-pyruvate aminotransferase (PPAT).

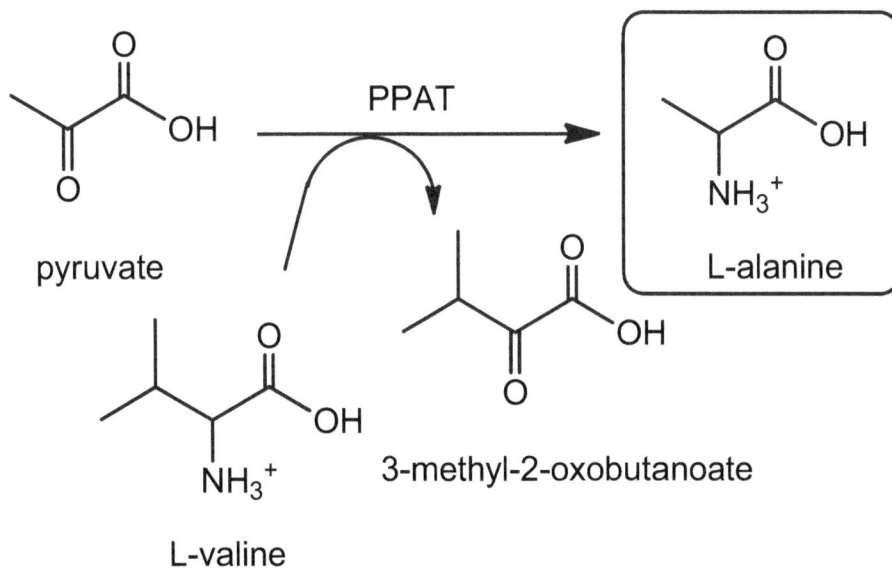

PPAT= pyridoxamine-pyruvate aminotransferase

Fig. (137). Biosynthesis of alanine from pyruvate.

The proposed mechanism starts with a nucleophilic addition of the pyroxamide to the α-keto position of pyruvate to generate the imine. Next, a lysine residue takes a hydrogen-producing resonance within the pyridoxal ring evolving to a quinonoid intermediate. The electrons move back to the imine pushed by the nitrogen electron pair to get an external aldimide which follows imine hydrolysis to produce the target aminoacid alanine and restore the pyridoxal cofactor [147]. The transamination process is usually composed of the intermediates ketamine, quinonoid, external and internal aldimide in equilibrium states which eventually lead to the transamination products, as shown in Fig. (138).

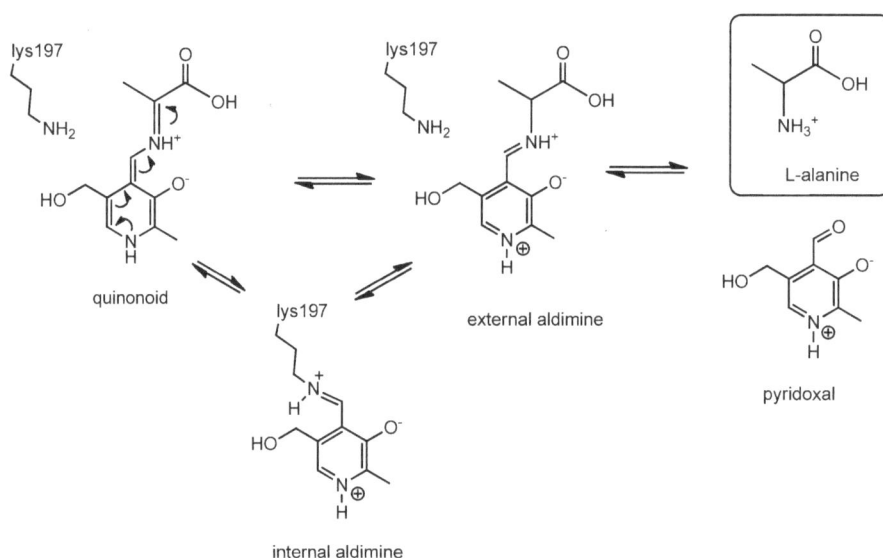

Fig. (138). The main intermediates formed during the transamination process.

2.1. Pyridoxamine-pyruvate Aminotransferase (PPAT)

The crystal structure of pyridoxamine-pyruvate aminotransferase from *Mesorhizobium loti* was described as a ribbon diagram of an asymmetric dimer in complex with PL in red, conformed by subunit A and B, with N- and C- terminal domains (Fig. **139**) [148].

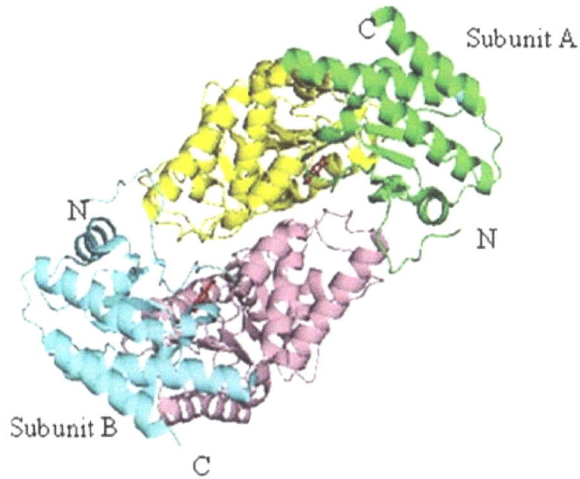

Fig. (139). Ribbon diagram of PPAT from *Mesorhizobium loti* (PDB: 2Z9X).

2.2. Pyridoxamine-pyruvate Aminotransferase Inhibitors

The amino acid analogue β-chloro-L-alanine is a nonspecific transaminase reversible inhibitor, displaying its inhibitory effects on *Escherichia coli* K-12 transaminase (Fig. **140**) [149].

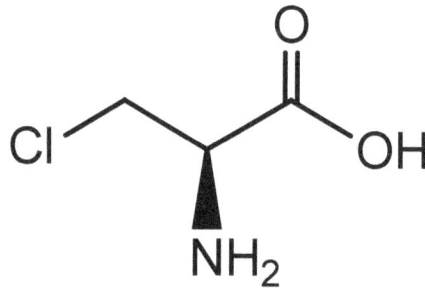

Fig. (140). Chemical structure of transaminase reversible inhibitor β-chloro-L-alanine.

3. VALINE BIOSYNTHESIS

Branched-chain amino acids (valine, leucine and isoleucine) are not synthesized by mammals, and their biosynthesis comes from pyruvate although the latter is related to the threonine pathway. This group of branched chain amino acids are currently used to build muscle and to improve mental focus [150].

Thus, valine biosynthesis starts with pyruvate that is converted to 2-acetolactate by the enzyme acetohydroxyacid synthase (AHAS) catalysing the condensation of two pyruvate molecules to produce (S)-2-acetolactate [151]. Further keto formation promotes methyl migration giving place to 3-hydrohy-2-keto isovalerate by the catalysis of enzyme acetolactate mutase, and subsequent ketone reduction resulting in 2,3-dihydroxy isovalerate. The next step involving dehydration and the enol tautomerism to the keto form gives 2-ketoisovalerate, which is converted to the target valine transaminase by conjugation with L-glutamate (Fig. **141**).

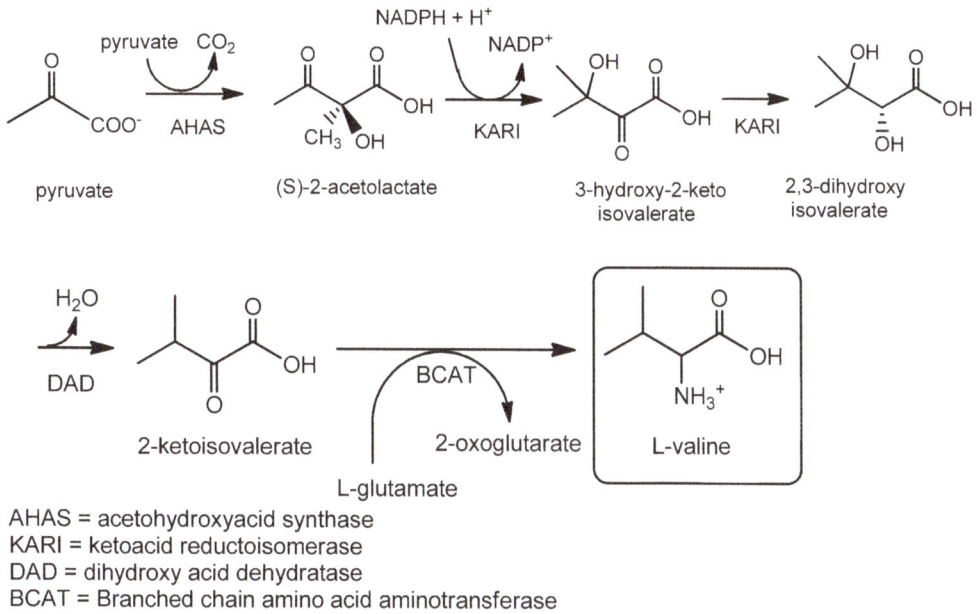

Fig. (141). Biosynthesis of L-valine from pyruvate.

AHAS = acetohydroxyacid synthase
KARI = ketoacid reductoisomerase
DAD = dihydroxy acid dehydratase
BCAT = Branched chain amino acid aminotransferase

3.1. Acetohydroxyacid Synthase (AHAS)

It is a thiamine diphosphate (ThDP) dependent enzyme responsible for the condensation of two pyruvate molecules to produce acetolactate and carbon dioxide (Fig. **142**) It is a key enzyme in the branched-chain amino acids (BCAAs) biosynthesis.

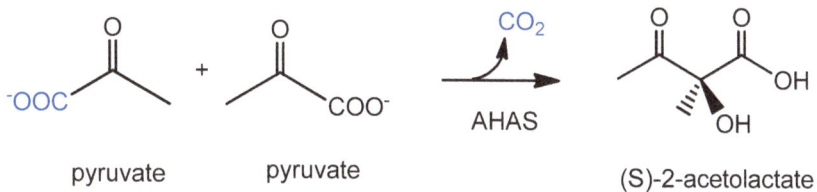

Fig. (142). The acetohydroxyacid synthase reaction.

Consistent mechanisms, which explain the transformation, establish an initial attack of a reactive ylide to pyruvate molecule to form the E-lactylThDP complex. A decarboxylation is produced generating the hydroxyethylthiamine diphosphate (HEThDP) complex. Next, the enamine formed attacks the second pyruvate molecule giving place to the additional product E-AHAThDP which is finally cleaved to release the acetolactate molecule (Fig. **143**) [152].

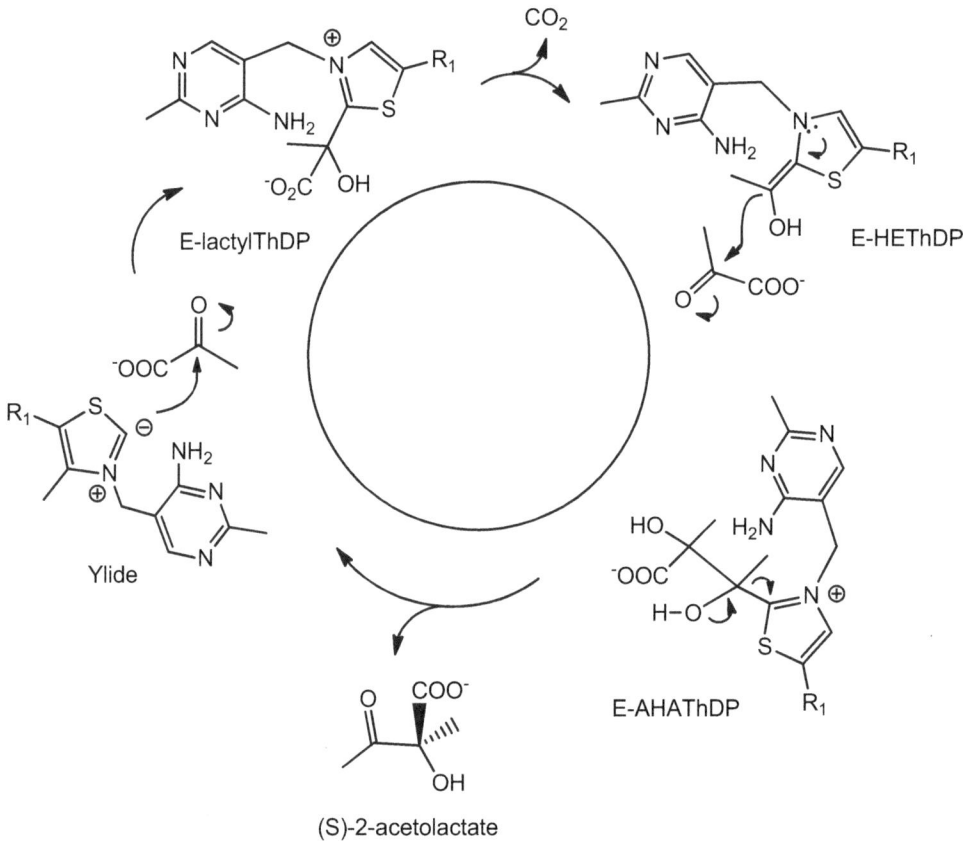

Fig. (143). Proposed mechanism for acetolactate formation.

The tertiary structure of acetohydroxyacid synthase (AHAS) from yeast was identified as a dimer, showing the α-, β-, and γ-domains, and in ball and stick representing the thiaminediphosphate (ThDP) and FAD regions as shown in Fig. (**144**) [153].

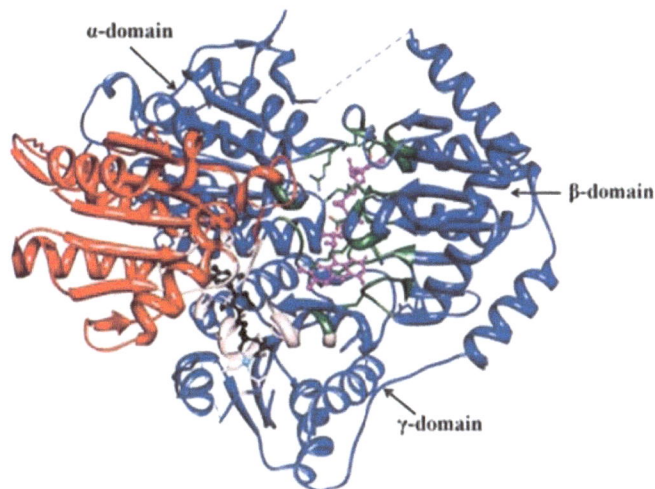

Fig. (144). Tertiary structure of acetohydroxyacid synthase from yeast (PDB: 1JSC).

3.2. Acetohydroxyacid Synthase Inhibitors

The compound KHG20612 (3-phenyldisulfanyl-[1,2,4]triazole-1-carboxylic acid phenylamide, and its analogue KHG20614 (Fig. **145**) were identified as potent *M. tuberculosis* AHAS, and also against *H. influenza*, *E. coli*, and *S. sonnei* [154].

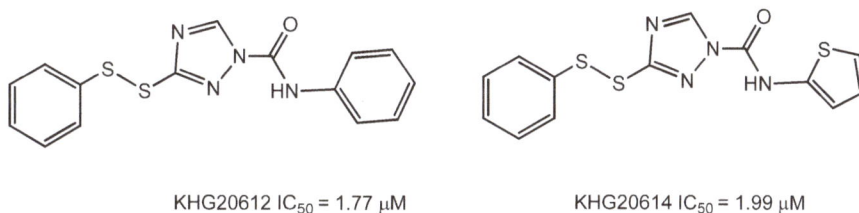

KHG20612 IC_{50} = 1.77 µM   KHG20614 IC_{50} = 1.99 µM

Fig. (145). Chemical structure and inhibition constants of 1,2,4 triazol derivatives KHG20612 and KHG20614.

3.3. Ketol-acid Reductoisomerase (KARI)

These enzymes catalyse the second step in the BCAA pathway corresponding to the transformation of (2S)-acetolactate to (2R)-2,3-dihydroxy-3-isovalerate, or when 2-aceto-2-hydroxybutyrate is the substrate, converting to (2R, 3R)-2,3-dihydroxy-3-methylvalerate (Fig. **146**).

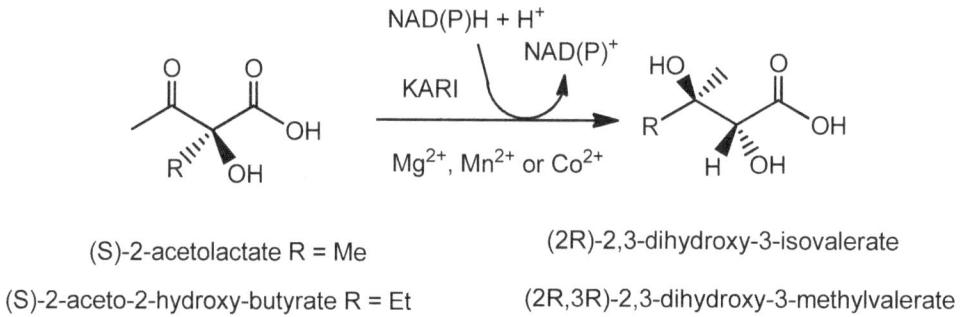

Fig. (146). The ketol-Acid Reductoisomerase reaction.

A simplified mechanism of KARI as a bifunctional enzyme postulates a concerted ketone formation and alkyl transposition to produce 3-hydroxy-2-keto isovalerate (R = Me) and subsequently a hydride reduction promoted by NADPH to provide 2,3-dihydroxy-isovalerate (Fig. **147**) [155].

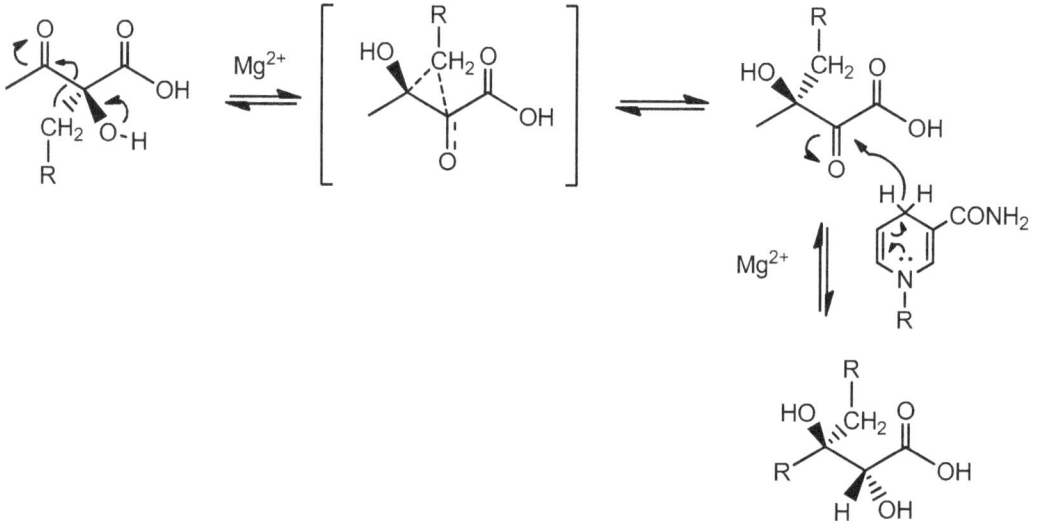

Fig. (147). Reaction mechanism for the conversion of (2S)-acetolactate to (2R)-2,3-dihydroxy-3-isovalerate.

The crystal structure of KARI enzyme from different microorganism and plants has been determined, among them, rice, *E. coli*, *M. tuberculosis*, *S. aureus*, *S. solfataricus* have been determined. The structure of KARI from archaea

Sulfolobus solfataricum (Sso-KARI) using cryogenic electron microscopy (cryo-EM) was solved, observing dodecameric architecture with the active site localized at the interface within the dimer. Also the cryo-EM of Sso-KARI:2Mg^{2+} complexed with NADPH with an analogue CPD was elucidated (Fig. **148**) [156].

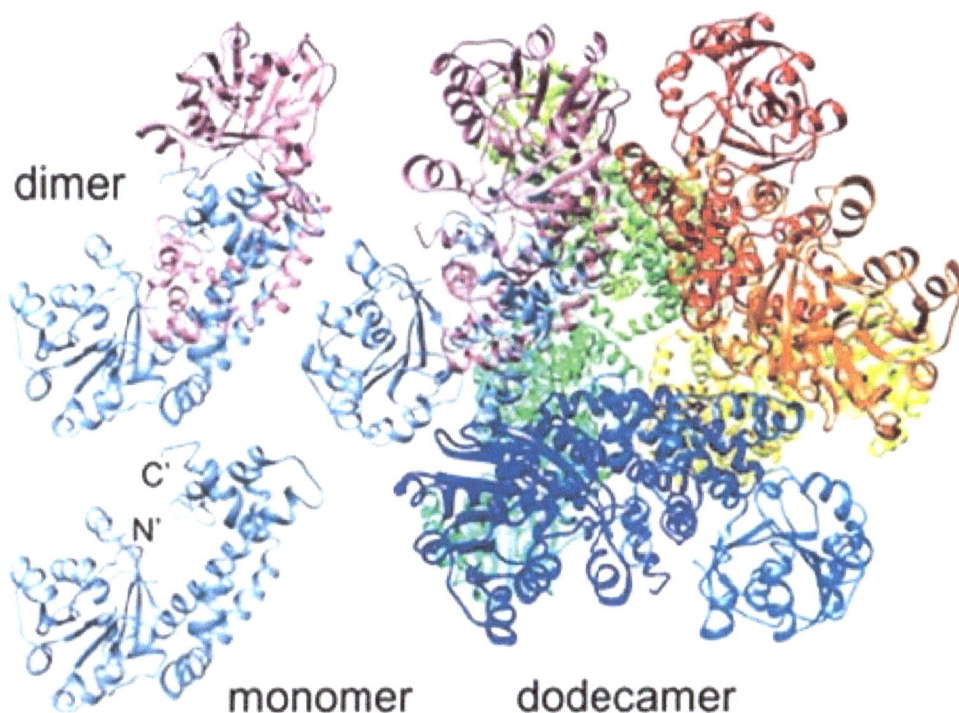

Fig. (148). Cryo-EM structure of KARI from *Sulfolobus solfataricus* (PDB: 6JCV).

3.4 Ketol-Acid Reductoisomerase Inhibitors

The herbicides N-isopropyloxalyl hydroxamate (IpOHA) and cyclopropane-1,--dicarboxylate (CPD) shown in Fig. (**149**), were tested as KARI inhibitors displaying potent activity against *Staphylococcus aureus* at micro molar and nanomolar scales [157].

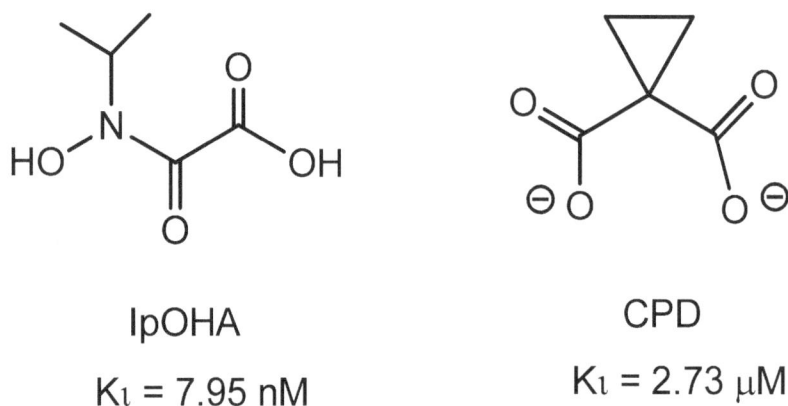

Fig. (149). Chemical structure and inhibition constants of N-isopropyloxalyl hydroxamate (IpOHA) and cyclopropane-1,1-dicarboxylate (CPD).

3.5 Dihydroxy Acid Dehydratase (DAD)

These enzymes catalyse the conversion of 2,3-dihydroxy isovalerate to 2-ketoisovalerate as shown in Fig. (150).

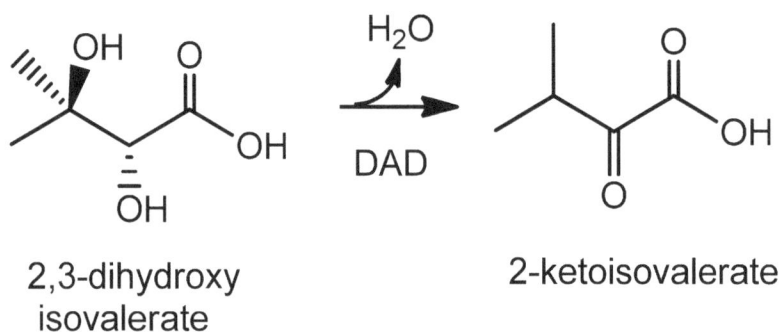

Fig. (150). The dihydroxy isovalerate dehydratase reaction.

A simplified mechanism implies the dehydration type E1 assisted by the F_2S_2 complex producing an enol intermediate which tautomerizes to the α-keto form producing 2-keto isovalerate (Fig. **151**) [158].

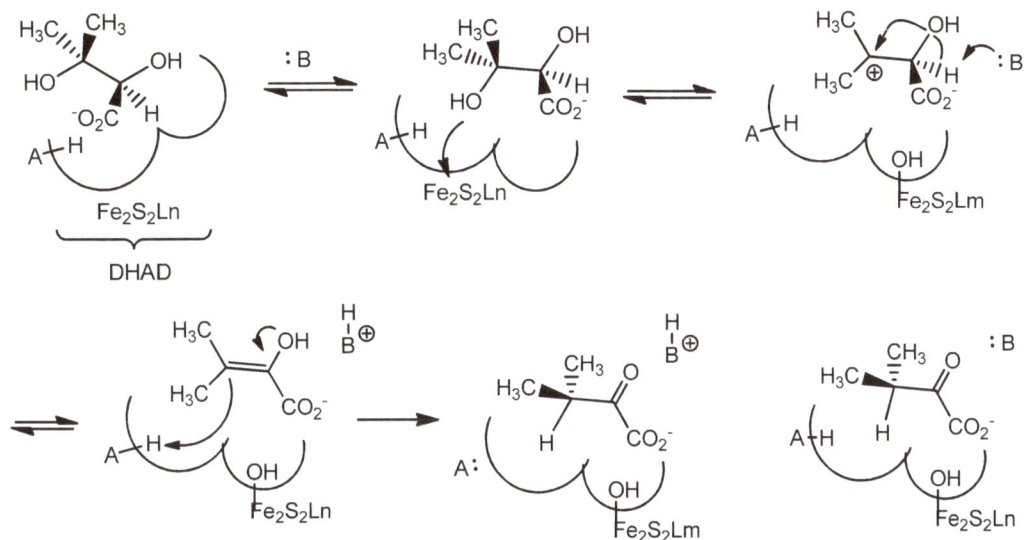

Fig. (151). Proposed mechanism for the conversion of 2,3-dihydroxy isovalerate to 2-ketoisovalerate.

The crystal structure of dimeric dihydroxy isovalerate dehydratase from *Arabidopsis thaliana* (holo-AthDHAD) and detailed closeup view showing cofactor 2F2-2S cluster and Mg^{2+} ion at the active site (Fig. **152**) [159].

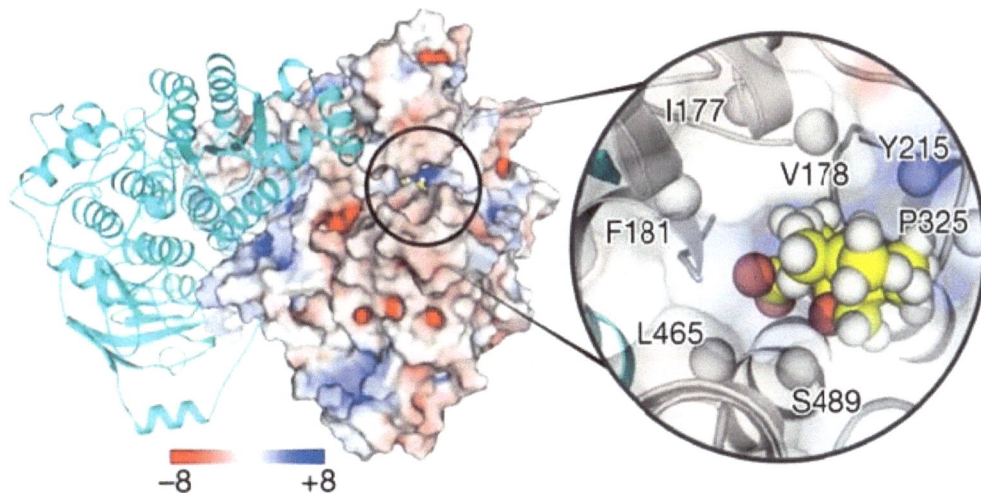

Fig. (152). Crystal structure of dihydroxy isovalerate dehydratase from *Arabidopsis thaliana* (PDB: 5ZE4).

3.6. Dihydroxy Acid Dehydratase Inhibitors

Aspterric acid (Fig. **153**) is a sesquiterpene plant growth regulator isolated from *Aspergillus terreus* with inhibitory activity toward *Arabidopsis thaliana*, *Zea mays*, and *Solanum lycopersicum*.

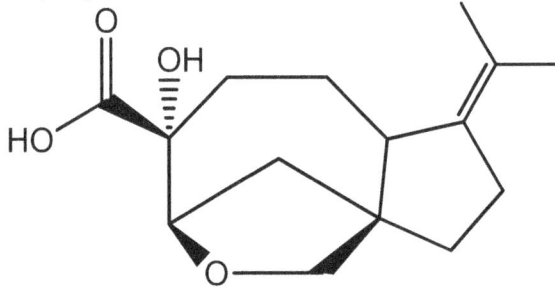

Fig. (153). Chemial structure of aspterric acid.

Inhibition assays on DAD from *A. terreus* (*Ate*DAD) and *Arabidopsis thaliana* (*Ath*DAD) as target enzymes were carried out, observing inhibitory concentration IC_{50} values of 0.31 mM and 0.50 mM, respectively [159].

3.7. Branched Chain Amino Acid Aminotransferase (BCAT)

The final step in the branched chain amino acids (BCAAS) leading to the formation of L-valine, L-leucine, and L-isoleucine is performed by the enzyme branched chain amino acid aminotransferase (BCAT), dependent of pyridoxal 5'-phosphate cofactor (PLP). The reaction consists of the amino group transfer from branched chain α-keto acids 2-ketoisovalerate, 2-ketoisocaproate and 2-keto-3-methylvalerate to yield branched chain amino L-valine, L-leucine and L-isoleucine, respectively (Fig. **154**) [160].

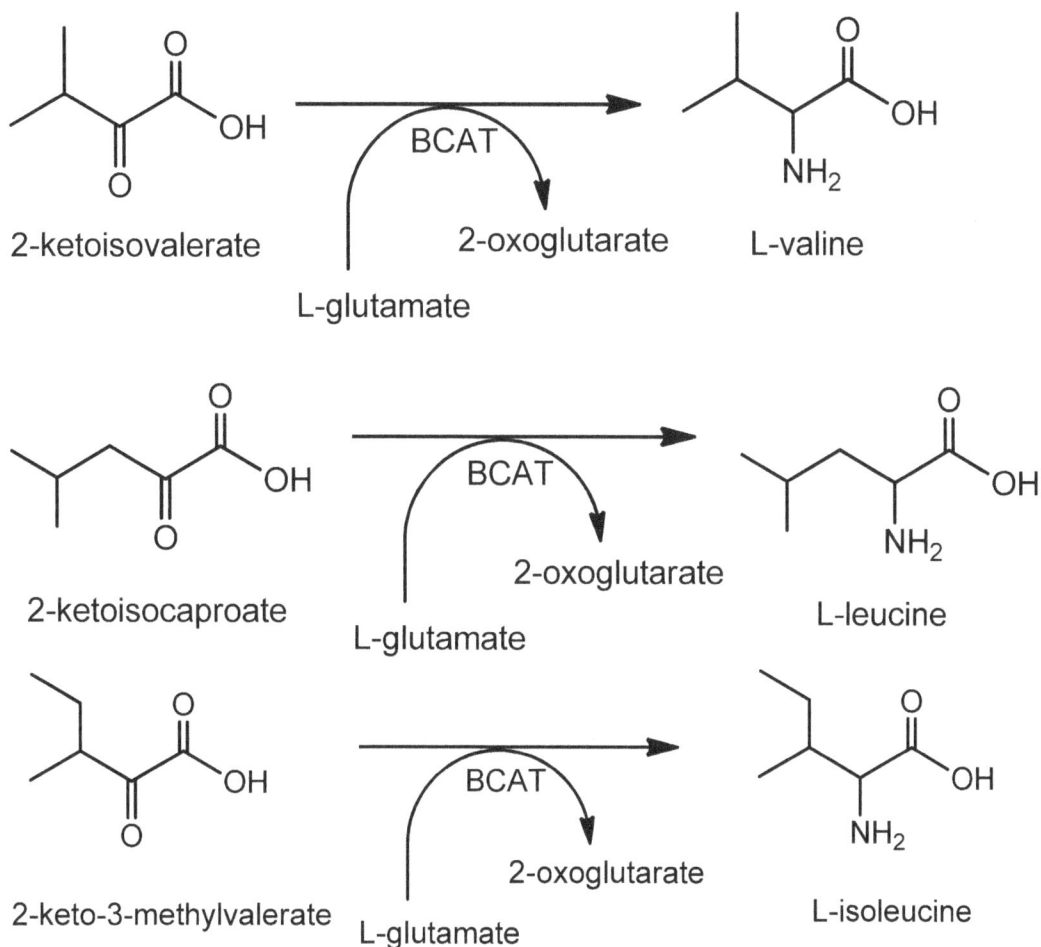

Fig. (154). The branched chain transaminase reaction.

The structure of branched chain amino acid transaminase from *M. tuberculosis* was determined as a homodimer with N-terminus in blue, and C-terminus in red, with each monomer composed of two domains. At the active site, the pyridoxal monophosphate is located between the two domains surrounded by residues forming and extensive hydrogen bond interactions network (Fig. **155**) [160].

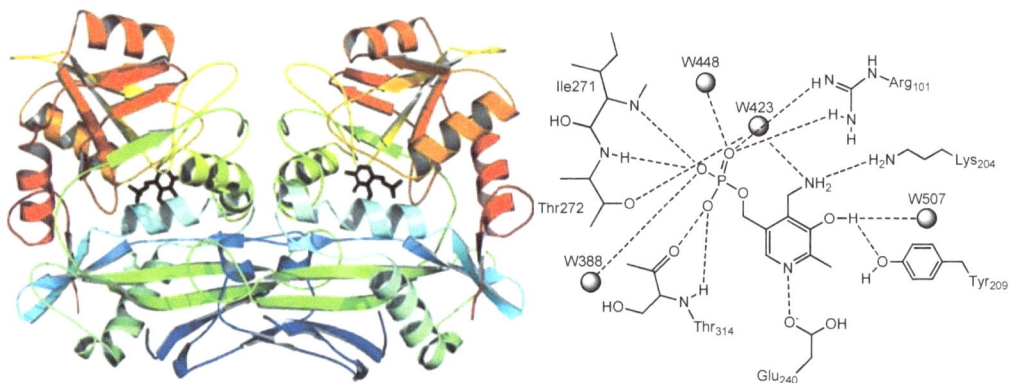

Fig. (155). Ribbon representation of branched chain amino acid transaminase (*Mt*BCAT) from *M. tuberculosis* (PDB: 3HT5) and hydrogen bond interactions with pyridoxal cofactor.

3.8. Branched Chain Amino Acid Aminotransferase Inhibitors

Cycloserine (DCS) is an isoxazolone isolated from *Streptomyces* species with inhibitory capacity against *Staphylococcus aureus* acting as an irreversible inhibitor of bacterial D-amino acid transaminase (Fig. **156**) [161]. Based on these findings, DCS and its enantiomer LCS were evaluated as inhibitors against *M. tuberculosis*, observing that LCS was more efficient inhibitor in about 10 fold than DCS, with a *Ki* of 88 μM and MIC value of 0.3 μg/mL for the L-isomer and 2.3 μ/mL for the D-isomer [162].

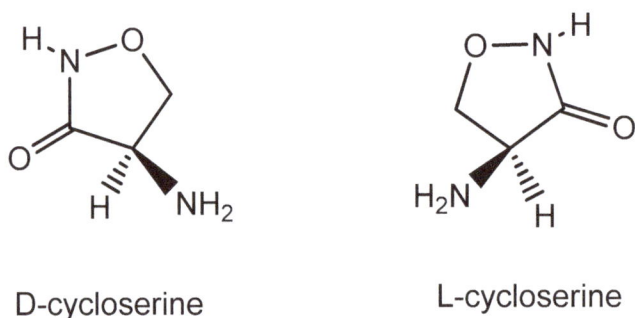

D-cycloserine

L-cycloserine

Fig. (156). Chemical structure of cycloserine.

4. LEUCINE BIOSYNTHESIS

Leucine is a branched chain amino acid (BCAA) closely associated with muscle performance along with valine and isoleucine. Its biosynthesis starts from 2-ketoisovalerate, which is a common starting material for the preparation of L-leucine and L-valine. For the case of L-valine, the reaction follows the course of a transamination reaction using as coupling system L-glutamate, providing L-valine and α-ketoglutarate. In leucine biosynthesis, the precursor 2-ketoisovalerate is condensed with acetyl CoA under the catalysis of α-isopropylmalate synthase to generate α-isopropylmalate. The next step consists of isomerization catalysed by isopropylmalate isomerase to afford β-isopropylmalate which is decarboxylated and oxidized in the presence of NAD^+ by the enzyme isopropylmalate dehydrogenase to α-ketocaproate, and finally a transamination reaction occurs with the coupling system in which L-glutamate to α-ketoglutarate is catalysed by branched chain amino acid aminotransferase (BCAT) to furnish leucine (Fig. **157**) [163].

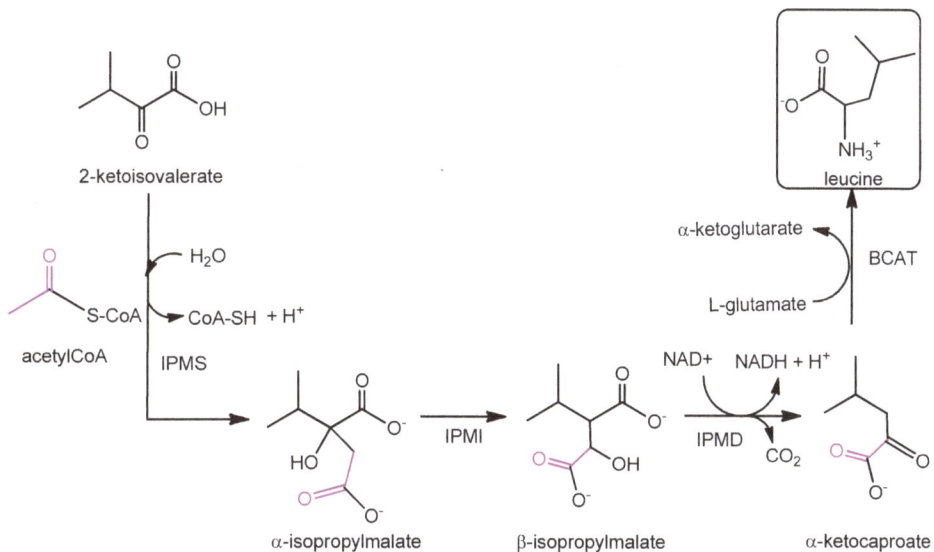

IPMS = a-isopropylmalate synthase
IPMI = isopropylmalate isomerase
IPMD = isopropylmalate dehydrogenase
BCAT = Branched chain amino acid aminotransferase

Fig. (157). Biosynthesis of L-Leucine and L-valine from 2-ketoisovalerate.

4.1. α-Isopropylmalate Synthase (IPMS)

α-Isopropylmalate synthase is involved in the initial step of the L-leucine biosynthesis requiring α-ketoisovalerate and acetyl-CoA to produce (S)--isopropylmalate.The reaction is considered an adol type condensation and uses a variety of divalent cations including Mg^{2+}, Mn^{2+}, Co^{2+}, Ni^{2+}, and Zn^{2+} which binds effectively to the active site (Fig. **158**).

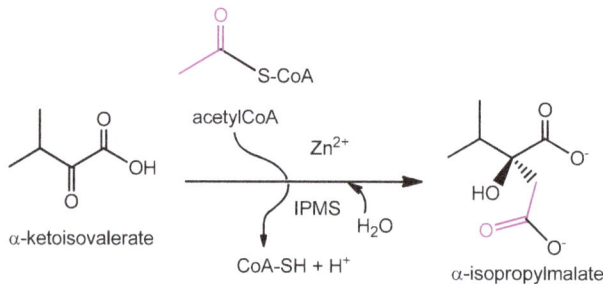

Fig. (158). The α-isopropylmalate synthase reaction.

The crystal structure of α-IPMS from *M. tuberculosis* (*Mt*IPMS) and subdomain II of *Leptospira biflexa* (*Lb*IPMS) showing the regulatory domain, leucine binding site, the catalytic domain and two active sites are represented in Fig. (**Fig. 159**) [164, 165].

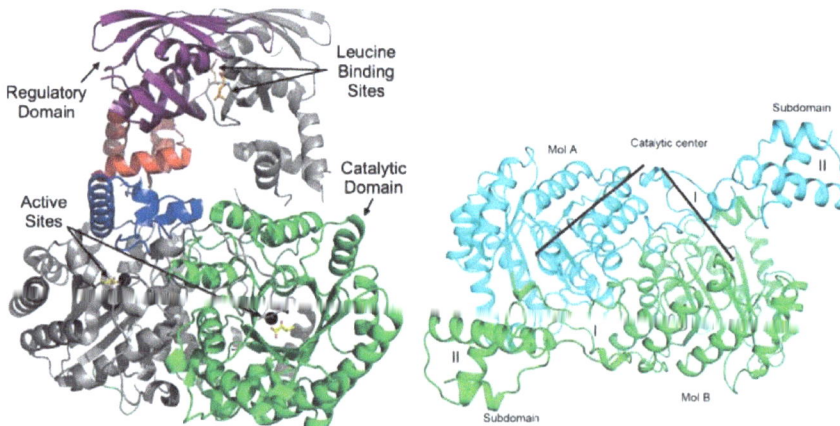

Fig. (159). The crystal structure of α-IPMS from *M. tuberculosis* (PDB: 3RMJ) and subdomain II of *Leptospira biflexa* (PDB: 4OV9).

At the binding site it is possible to see the substrate α-ketoisovalerate (α-Kiv) coordinated to Zn^{2+}, and the residues represented as green sticks (Fig. **160**).

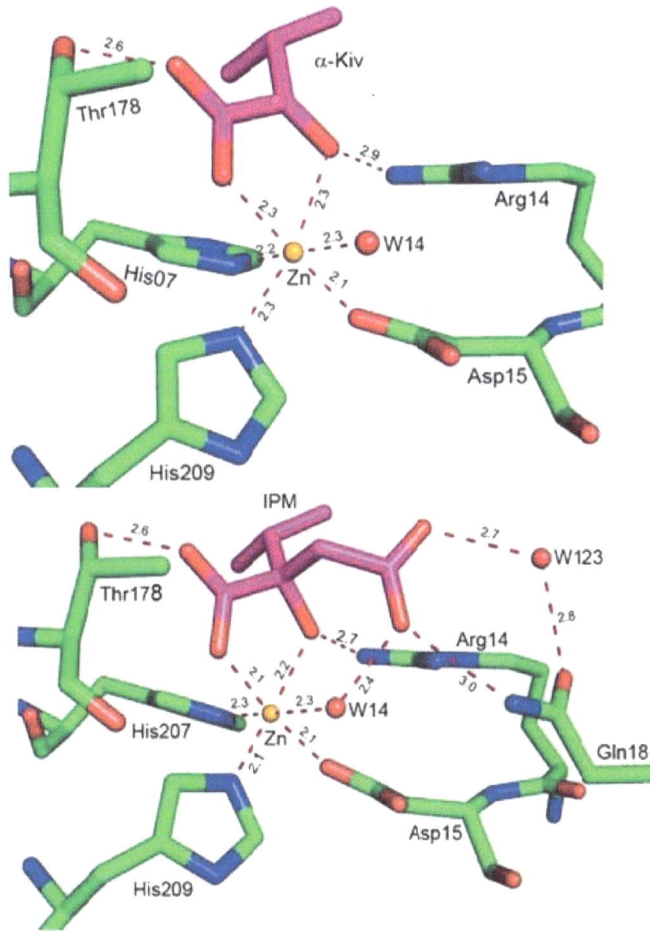

Fig. (160). Ball and stick representation showing the α-ketoisovalerate substrate coordinated to Zn^{2+} and water molecules.

The proposed mechanism for the biosynthesis of (S)-α-isopropylmalate initiate with an enolization process requiring a residue acting as base, presumably Arg80, Glu218, Glu317, Arg318, His379, Tyr410. The next step involves a nucleophilic attack of a water molecule to the thioester position, passing through the formation of a tetrahedral intermediate, which by following a nucleophilic substitution step in which the negative charge returns to provide the ester group, with the simultaneous release of CoA (Fig. **161**) [166].

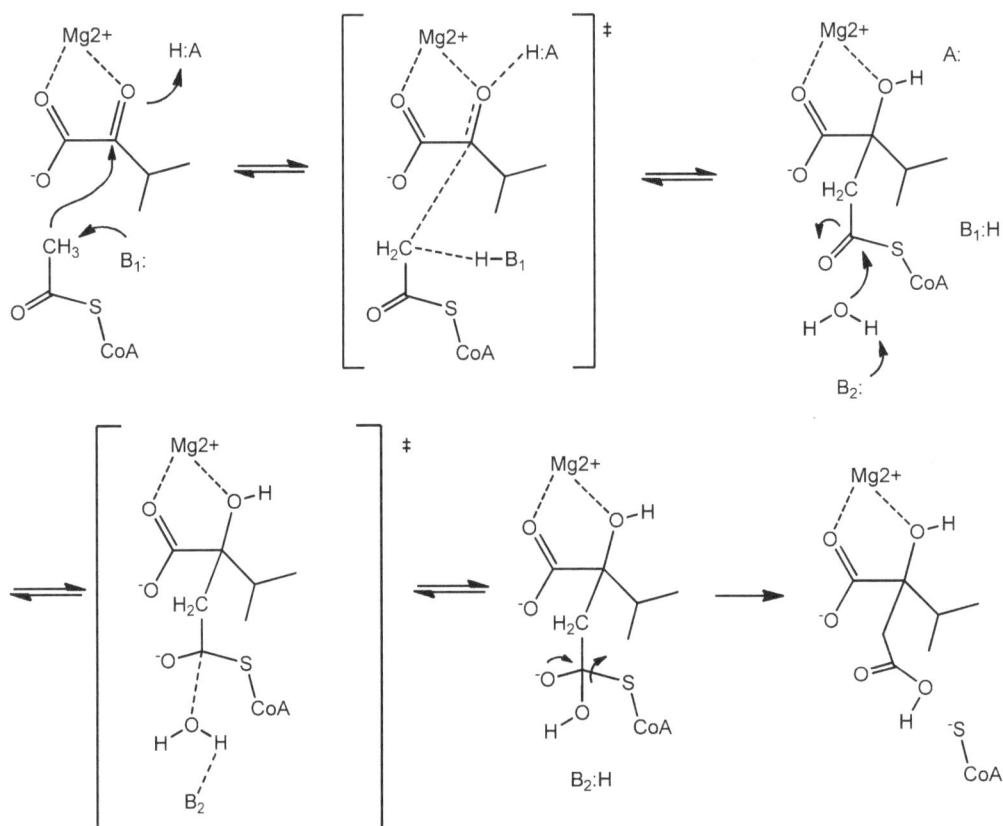

Fig. (161). Proposed mechanism for the biosynthesis of (S)-α-isopropylmalate.

4.2. α-Isopropylmalate Synthase Inhibition

Inhibition studies of α-IPMS as a target enzyme haven't been described, even though *in silico* identification of candidates for inhibition were analysed. By exploring chemical data from NCI, DrugBank and CheMBL data banks some candidates were proposed as potential inhibitors, such as D-glucitol biphosphate derivatives CHEMBL404748 and CHEMBL1235112 (Fig. **162**) [167].

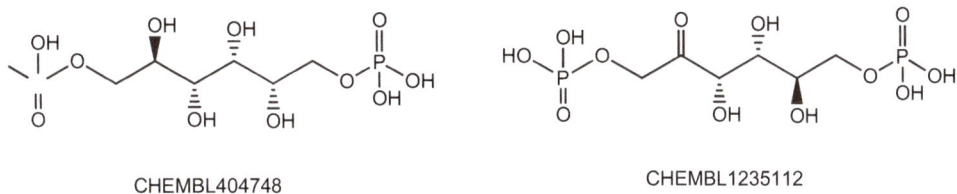

CHEMBL404748   CHEMBL1235112

Fig. (162). Chemical structure of α-isopropylmalate synthase potential inhibitors.

4.3. α-Isopropylmalate Isomerase (IPMI)

This enzyme belongs to the aconitase family of proteins and catalyses the second step in the leucine biosynthesis consisting in the conversion of (2S)--isopropylmalate into (2R,3S)-3-isopropylmalate (Fig. **163**) [168].

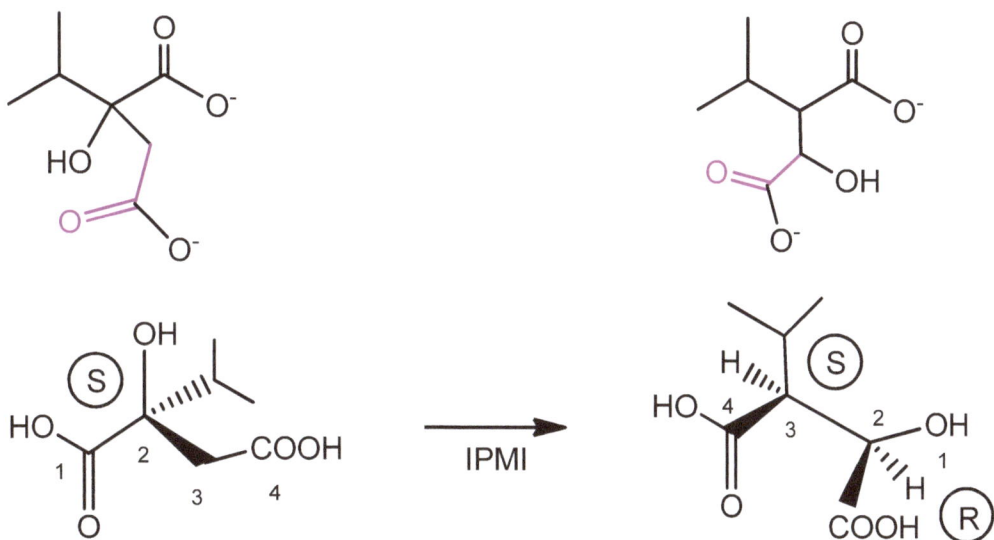

Fig. (163). The α-isopropylmalate isomerase reaction.

IPM isomerases are present in all prokaryotic and several eukaryotic species, and contain a large (LeuC) and small (LeuD) subunits. Until recently the large subunit has been solved by X-Ray spectroscopy and several report describes the small unit resolution. Thus, the IPMI large subunit from *Methanococcus jannaschii* was determined as monomer and dimer, the former containing three domains, and the active site interacting with MPD 2-methylpentane-2,4-diol and DMPD 2,4-dimethylpentane-2,4-diol (Fig. **164**) [169].

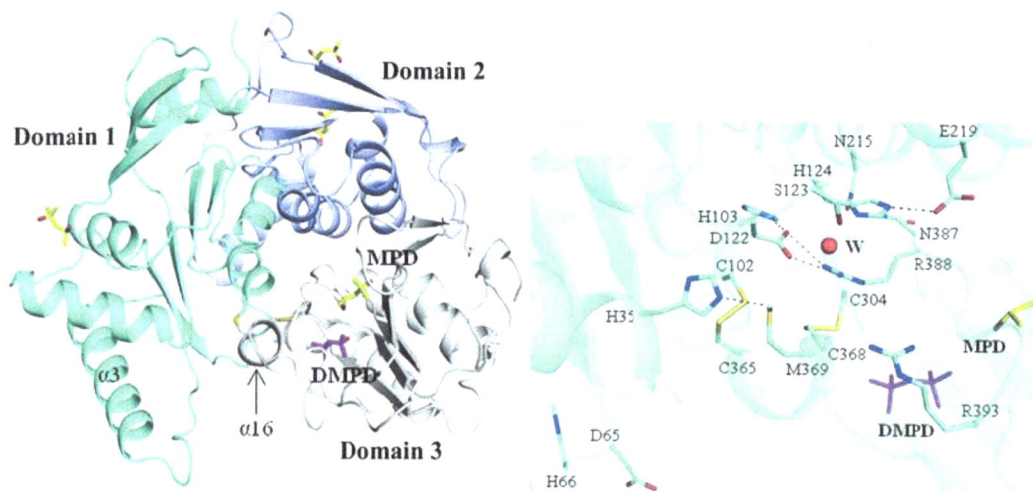

Fig. (164). Crystal structure of IPMI large subunit from *Methanococcus jannaschii* (PDB: 4KP1).

The mechanism of IPMI is homologous to aconitase, and isocitrate dehydrogenase since the catalytic reaction is considered almost identical. The basic steps are dehydration reaction to produce the double bond as a result of β-elimination, and re-hydration by the water attack at the less hindered position to yield the corresponding isomer (Fig. **165**) [170].

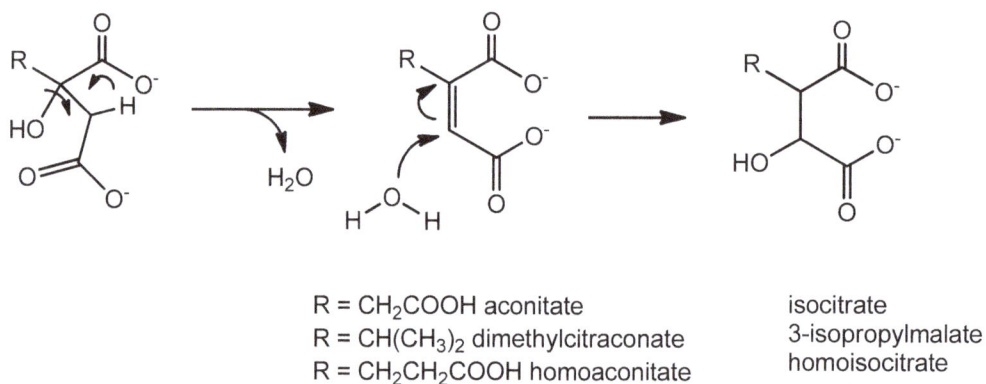

R = CH$_2$COOH aconitate
R = CH(CH$_3$)$_2$ dimethylcitraconate
R = CH$_2$CH$_2$COOH homoaconitate

isocitrate
3-isopropylmalate
homoisocitrate

Fig. (165). The basic steps for the isomerization reaction.

An extended mechanism involving [4Fe-4S] cluster proposes a base assisted dehydration of IPM to produce cis-alkene and rehydration at the opposite position, leading to the formation of 3-isopropylmalate (Fig. **166**) [171].

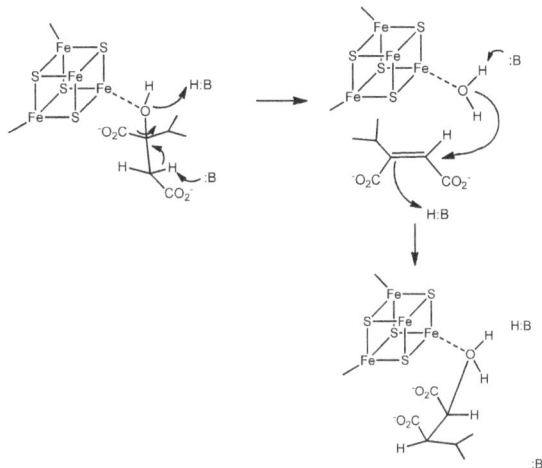

Fig. (166). Isomerization mechanism involving [4Fe-4S] cluster.

4.4. α-Isopropylmalate Isomerase Inhibitors

Nitronate compounds (Fig. **167**) with structural similarity with dimethylcitraconate were synthesized and tested as IPMI inhibitors. The results show that cyclic nitronate structures were equally potent on carrot cell growth [172].

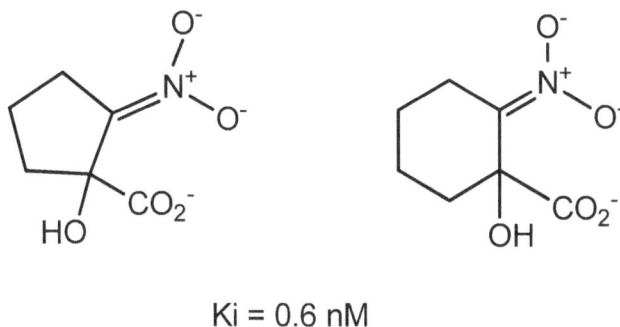

$K_i = 0.6$ nM

Fig. (167). Chemical structure and inhibition constants of nitronate compounds.

4.5. α-Isopropylmalate Dehydrogenase (IPMD)

The penultimate step in the leucine biosynthesis is carried out by the enzyme α-isopropylmalate dehydrogenase converting β-isopropylmalate to α-ketocaproate (Fig. **168**).

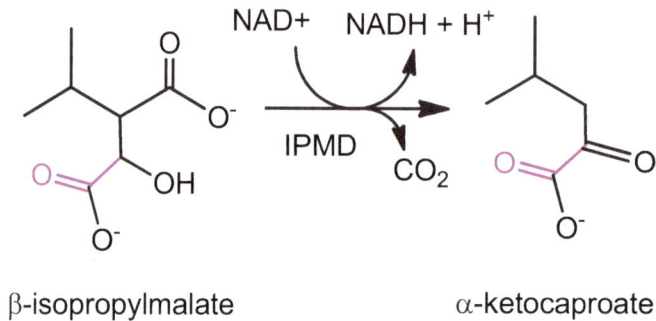

Fig. (168). The α-isopropylmalate dehydrogenase reaction.

A proposed mechanism of IPMD from *Arabidopsis thaliana* establish the interactions of β-isopropylmalate with Tyr181, Lys232, Asn234, Asp264, Asp288, Asp292 and NAD+ cofactor, and the basic steps involving decarboxylation, enol and keto formation (Fig. **169**) [173].

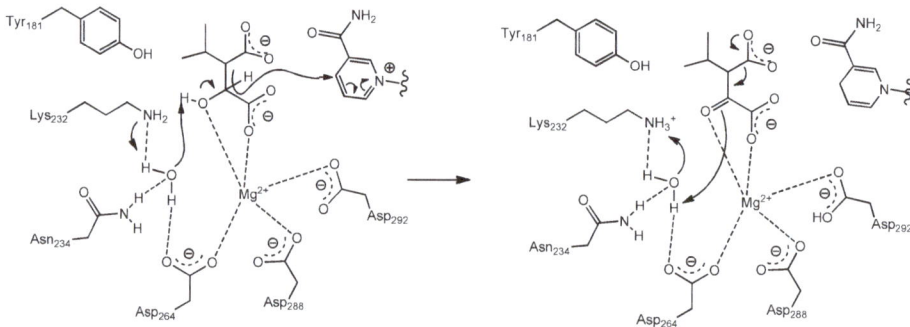

Fig. 169 cont.....

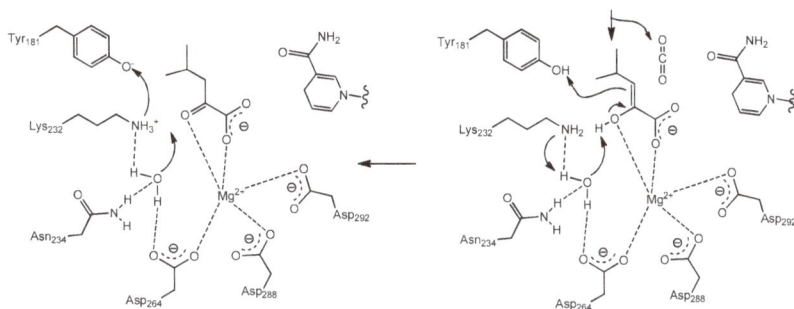

Fig. (169). Proposed mechanism for the biosynthesis of α-ketocaproate.

The IPMD dimer from *Arabidopsis thaliana* (*At*IPMD2) is composed by α-helices and β-strands as well as N- and C-terminal positions. Also the region of β-isopropylmalate, NAD^+ and Mg^{2+} are indicated as spheres and at the binding site the interaction established with the residues (Fig. **170**) [173].

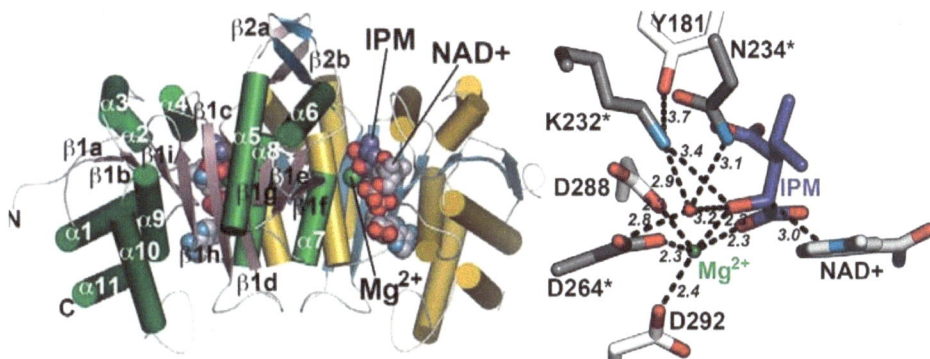

Fig. (170). Ribbon representation of IPMD dimer from *Arabidopsis thaliana* and the binding site region (PDB: 3R8W).

4.6. α-Isopropylmalate Dehydrogenase Inhibition

Herbicide compound O-isobutenyl oxalylhydroxamate (O-IbOHA) shown in Fig. (**171**) has been described as potent IPMD inhibitor with an apparent inhibition constant K_i of 5 nM, behaving as competitive inhibitor as observed in the Lineweaver-Burk plot [174].

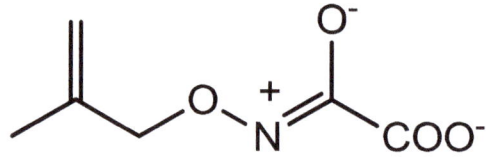

Fig. (171). Chemical structure of O-isobutenyl oxalylhydroxamate.

5. ISOLEUCINE BIOSYNTHESIS

Isoleucine is with valine and leucine a branched chain amino acid with important implications in the muscle protein synthesis, considered stronger than valine but weaker that leucine in terms of increasing glucose uptake to the muscle. Its biosynthesis requires L-threonine as starting material, which is converted to 2-ketobutyrate (or 2-oxobutyrate) by the catalysis of threonine ammonia lyase. The next step involves a transfer of acetaldehyde from pyruvate to 2-ketobutyrate to form 2-acetoxyhydroxybutyrate under the catalysis of acetolactate synthase. The following step is catalysed by acetohydroxyacid reductoisomerase (KARI) promoting an alkyl migration from the acetohydroxyacid intermediate and further reduction to give 2,3-dihydroxy-3-methylvalerate [175]. Further dehydration allows the enol formation, which tautomerize to the keto form, providing 2-ketomethylvalerate, and final conversion to isoleucine by transamination reaction catalysed by isoleucine aminotransferase (Fig. **172**).

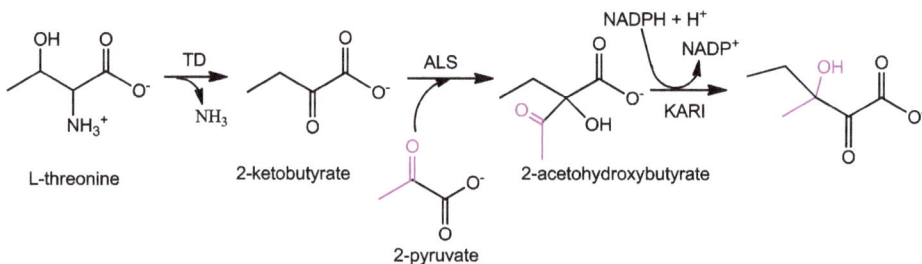

Fig. 172 cont.....

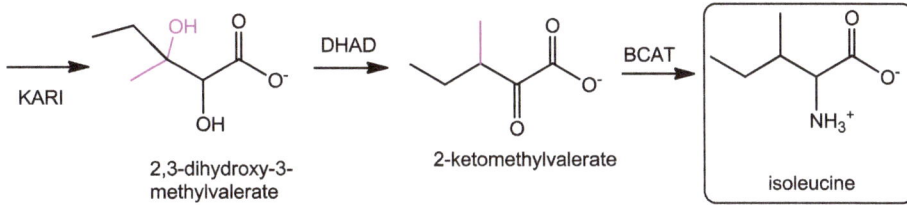

KARI

2,3-dihydroxy-3-
methylvalerate

DHAD

2-ketomethylvalerate

BCAT

isoleucine

TD = threonine ammonia lyase
ALS = acetolactate synthase
KARI = acetohydroxyacid reductoisomerase
DHAD = dihydroxyacid dehydratase
BCAT = Branched chain amino acid aminotransferase

Fig. (172). Isoleucine biosynthesis from L-threonine.

6. PHENYLALANINE BIOSYNTHESIS

Phenylalanine is with tyrosine a benzene containing amino acid found in animal food products such as meat, eggs and milk, and it is the precursor of adrenaline needed in vital processes related to brain functioning. Its deficiency is related to skin defects such as vitiligo, or degenerative disease such as Parkinson.

The aromatic amino acids phenylalanine, tyrosine and tryptophan are produced from carbohydrates through the shikimic acid pathway. Plants and microorganisms produce these aromatic compounds through *de novo* synthesis *Via* shikimate pathway, by using phosphoenolpyruvate from glycolysis and erythrose 4-phosphate [176], excepting mammals, which did not develop the pathway and therefore are not able to produce aromatic amino acids from their own sources [177].

The last steps in the shikimic acid pathway directs the process to the production of phenylalanine involving prephenate aminotransferase to convert prephenate to arogenate and arogenate dehydratase to transform arogenate into phenylalanine (Fig. **173**). Moreover, the shikimic pathway is highly conserved in plant, bacteria and fungi and provides besides the aromatic amino acid mentioned other important metabolites such as alkaloids, flavonoids and antibiotics [178].

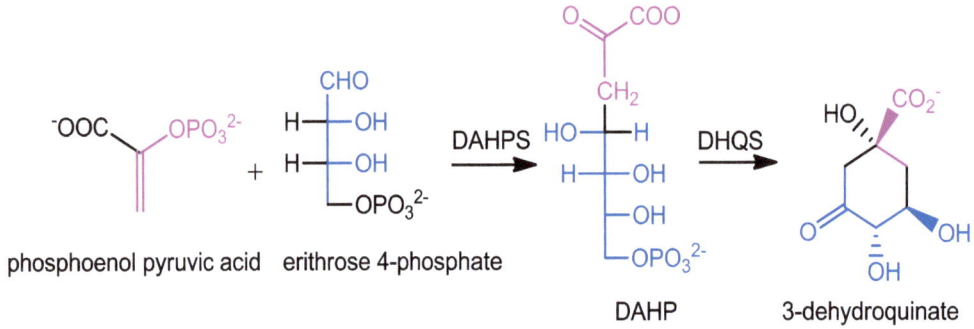

phosphoenol pyruvic acid   erithrose 4-phosphate   DAHP   3-dehydroquinate

DAHP = 2-keto-3-deoxy-D-arabinoheptulosonic acid 7-phosphate

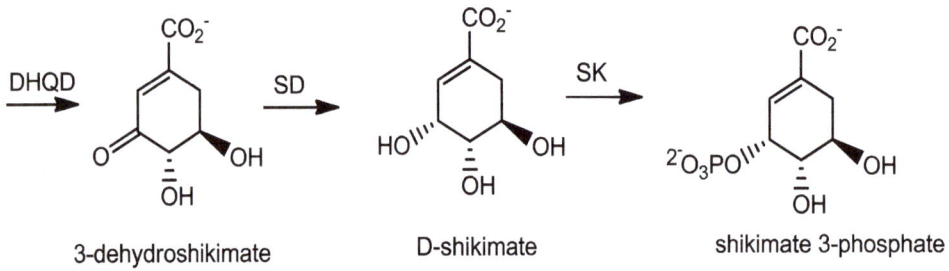

3-dehydroshikimate   D-shikimate   shikimate 3-phosphate

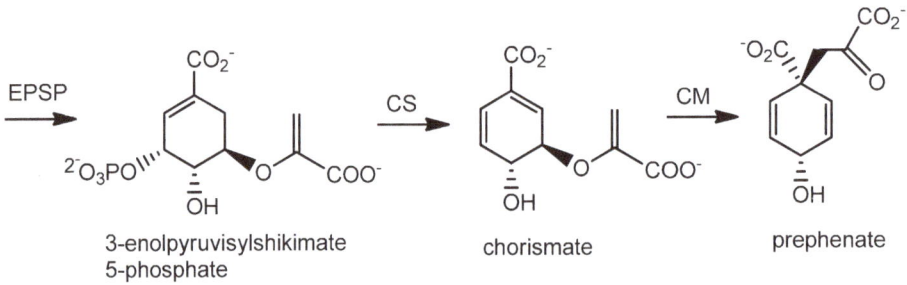

3-enolpyruvisylshikimate   chorismate   prephenate
5-phosphate

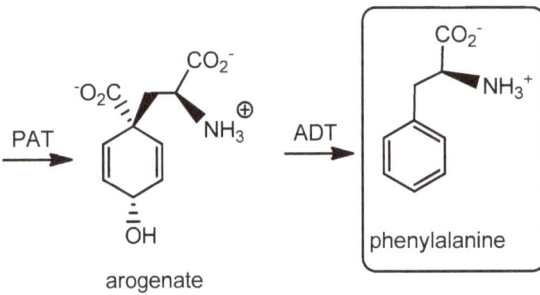

arogenate   phenylalanine

Fig. 173 cont.....

DAHPS = deoxyarabinoheptulosonatephosphate synthase

DHQS = dehydroquinate synthase

DHQD = dehydroquinate dehydratase

SD = shikimate dehydrogenase

SK = shikimate kinase

EPSP = enolpyruvisylshikimatephosphate synthase

CS = chorismate synthase

CM = chorismate mutase

PAT = prephenate aminotransferase

ADT = arogenate dehydratase

Fig. (173). The phenylalanine biosynthetic pathway.

6.1. 3-Deoxy-D-Arabinoheptulosonate 7-Phosphate Synthase (DAHPS)

It is the first enzyme in the shikimic acid pathway, consisting in the condensation of phosphoenol pyruvic acid (PEP) and erithrose 4-phosphate (E4P) to give 3-deoxy-D-arabino-heptulosonate 7-phosphate (DAHP) and inorganic phosphate (Fig. **174**).

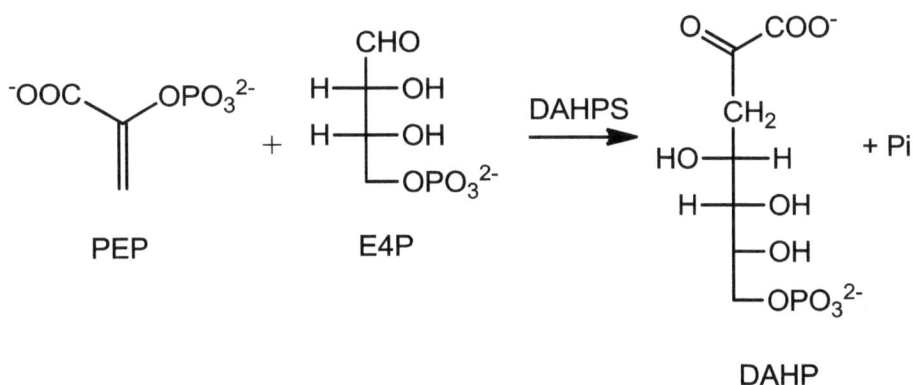

Fig. (174). The 3-deoxy-D-arabino-heptulosonate 7-phosphate formation.

The crystal structure of DAHPS was elucidated as a tetramer with identical subunits forming a buried interspace containing residues interacting in monomers B-D, and A-C. At the active site the residues interacts with manganese ion, as well as with the PEP substrate and water molecules (Fig. **175**) [179].

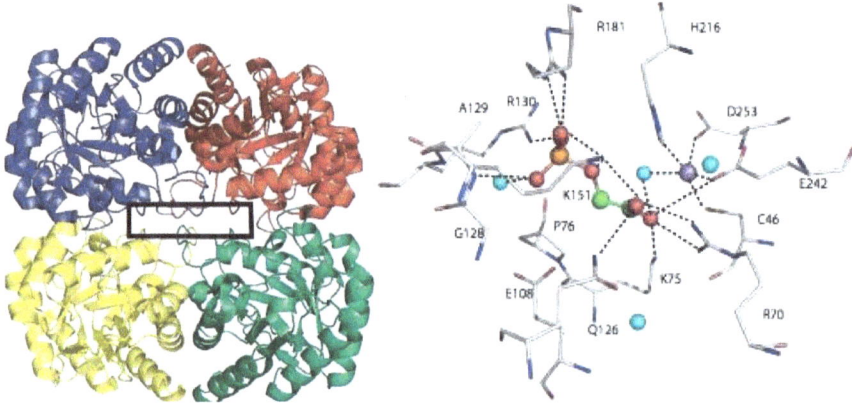

Fig. (175). The crystal structure and binding site of DAH7PS from *Aeropyrum pernix* in complex with Mn2+ and PEP (PDB: 1VS1).

A reaction mechanism proposes a double bond addition from PEP to the aldehyde to furnish a methylene and carbocation specie. Next a water molecule is added and the resulting hydroxyl group forming the keto group, which expel the phosphate group as a leaving group, providing 3-deoxy-D-arabino-heptulosonate 7-phosphate (DAHP) and inorganic phosphate (Fig. **176**) [180].

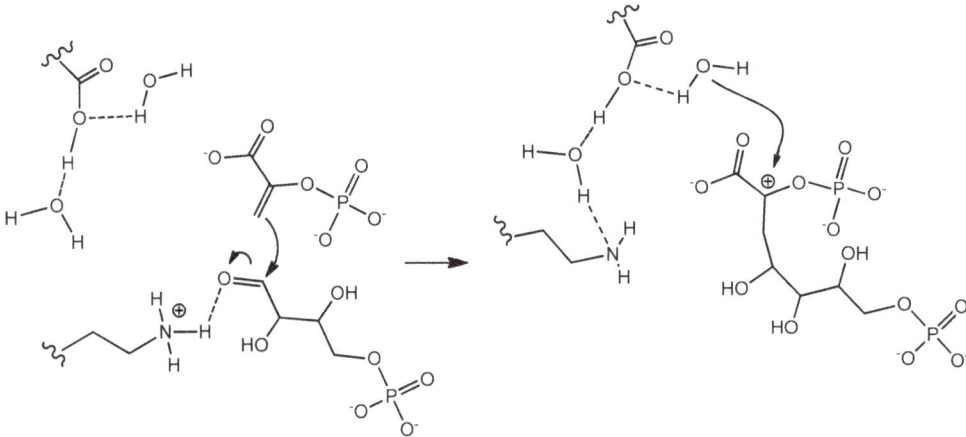

Fig. 176 cont.....

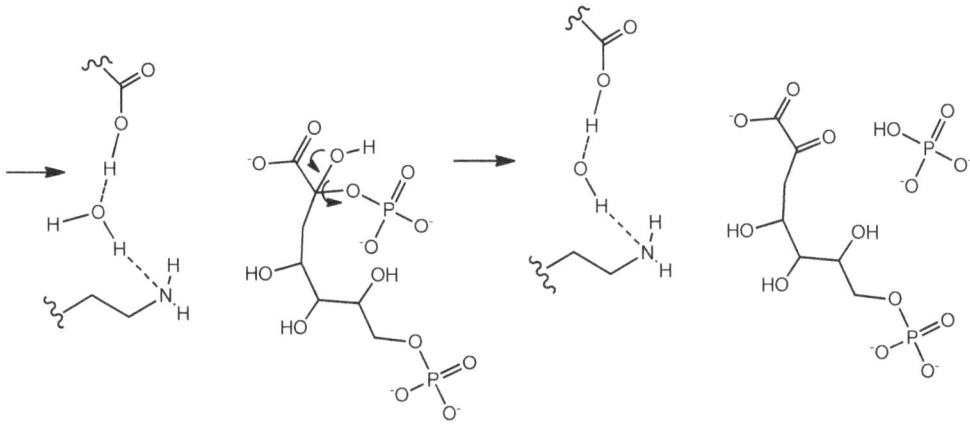

Fig. (176). Simplified reaction mechanism for the formation of 3-deoxy-D-arabino heptulosonate 7-phosphate.

6.2. Dehydroquinate Synthase (DHQS)

The process to convert 3-deoxy-D-arabino-heptulosonate 7-phosphate (DAHP) into the carbocycle 3-dehydroquinate (DHQ) is mediated by the enzyme dehydroquinate synthase (DHQS), requiring NAD^+ as cofactor (Fig. **177**).

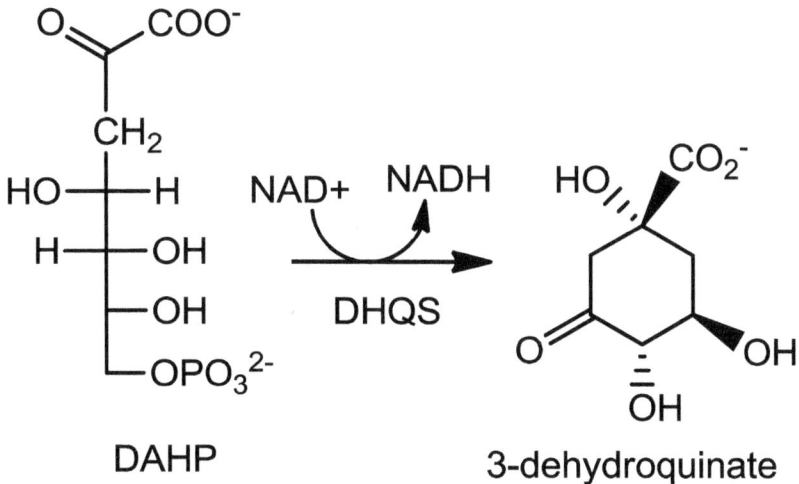

Fig. (177). The dehydroquinate synthase reaction.

The mechanisms proposed involves step series, starting with the initial cyclization to the chair conformation followed by C-5 alcohol oxidation, β-elimination of inorganic phosphate, reduction of ketone, ring opening and aldol type reaction to yield 3-dehydroquinate (DHQ) as shown in Fig. (**178**) [181].

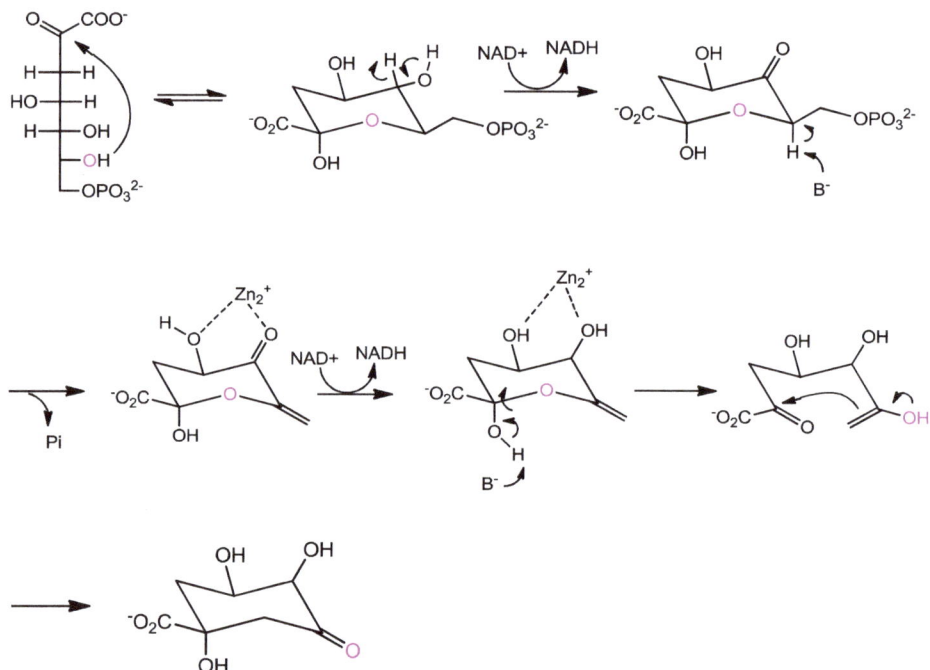

Fig. (178). The proposed mechanism for 3-dehydroquinate formation.

Dehydroquinate synthase is a dimer highly conserved with NAD^+ and three phosphate groups attached to each monomer, presenting three coordination sites (Glu198, His261, His278) to interact with the metal ion, which can be either Co^{2+} or Zn^{2+}. At the catalytic site two phosphate molecules, and glycine residue are near the NAD^+, in what it seem to be critical in the binding and orientation of the substrate (Fig. **179**) [182].

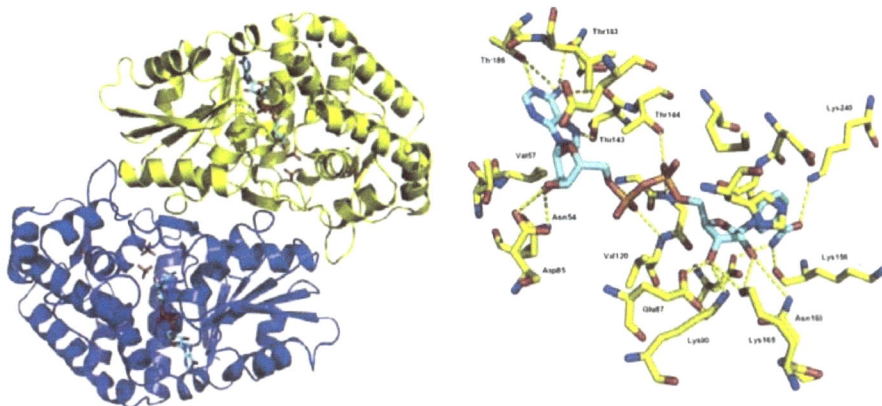

Fig. (179). Ribbon representation and binding site of dehydroquinate synthase from *Actinidia chinesis* in complex with NAD (PDB: 3ZOK).

6.3. Dehydroquinate Synthase Inhibitors

The design of inhibitors for DHQS became of interest because their potential use as herbicides. Thus, different small molecules have been assayed as herbicides, among them phosphonate 3-deoxy-D-arabino-heptulosonic acid 7-phosphonate (Fig. **180**), displays potent competitive inhibition of DHQS with K_i = 1.1 μM [183].

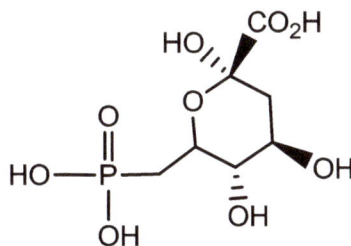

Ki = 1.1 μM

Fig. (180). Chemical structure of DHQS inhibitor.

Benzoic acids and catechols have been synthesized and evaluated as DHQS inhibitors in the presence of Co^{2+} and Zn^{2+}, obtaining inhibitory constants with significant variations, being the most potent the phosphate benzoic catechols (Fig. **181**) [184].

	Ki µM (Zn^{2+}, Co^{2+})	
R = CH$_2$PO$_3$H$_2$	0.35	21
R = CH$_2$CH$_2$PO$_3$H$_2$	1.7	5
R = H	65	44

	Ki µM (Zn^{2+}, Co^{2+})	
R = CH$_2$CH$_2$PO$_3$H$_2$	90	200
R = H	630	880

Fig. (181). Chemical structure and DHQS inhibitory constants of benzoic acid and catechol derivatives.

6.4. Dehydroquinate Dehydratase (DHQD)

The third step in the shikimate pathway is catalysed by the enzyme DHQD, involving the conversion of 3-dehydroquinate into 3-dehydroshikimate as shown in Fig. (**182**). Dehydroquinate dehydratase has been found as two subtype families, the type I and type II not related to each other without structural homology, and following different reaction mechanism. The first catalyse through the formation of an imine intermediate, and the second through an enolate formation [185].

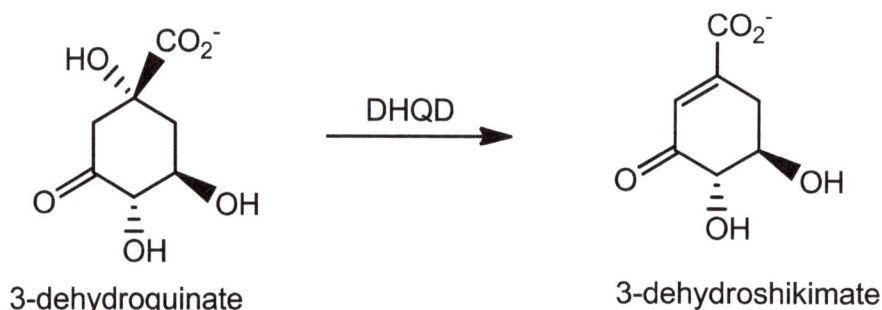

Fig. (182). The dehydroquinate dehydratase reaction.

The proposed mechanism for DHQD type I started with a base Schiff formation between the ketone and residue K170, followed by proton withdrawal assisted by histidine143 resulting in formation of enamine. A conjugated dehydration promoted by the enamine take place to generate an intra cyclic double bond and iminium neutralized by K170, which undergoes an imine hydrolysis to yield 3-dehydroshikimate (Fig. **183**) [185].

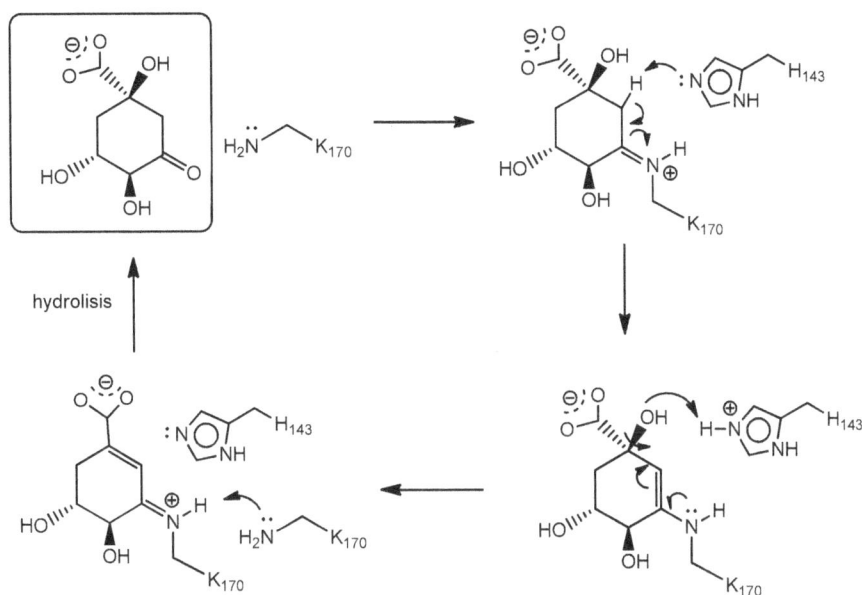

Fig. (183). The proposed mechanism for DHQD type I.

DHQD type II has been found in *Streptomyces coelicolor* and the reaction mechanism starts with a proton withdrawal, to form an enolate, which moves forward to C-1 position to expel the hydroxyl group as illustrated in Fig. (**184**) [186].

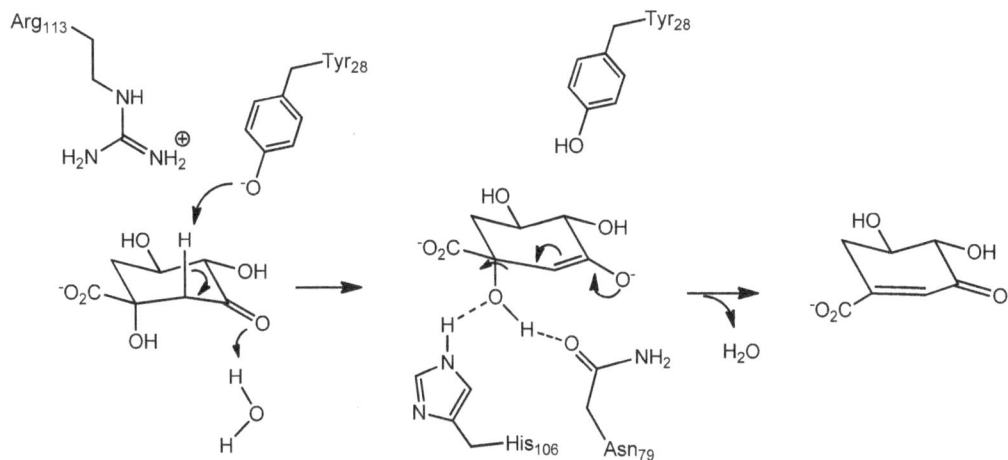

Fig. (184). Proposed mechanism for DHQD type II.

The crystal structure of DHQD type II from *S. coelicolor* and from *M. Tuberculosis* is dodecameric or actually a tetramer of trimmers. At the active site Asn79 and His106 establish hydrogen bond contact with C-1 hydroxyl group, which favors the dehydration process (Fig. **185**) [186, 187].

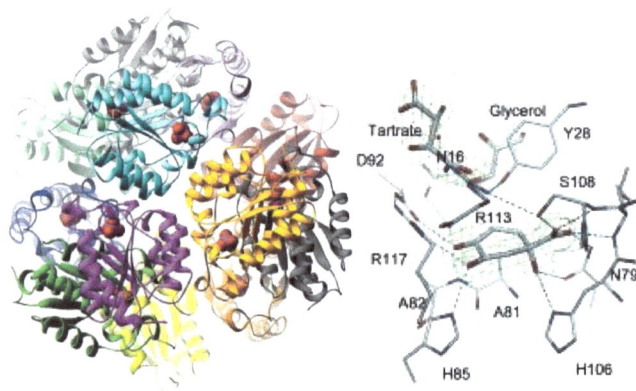

Fig. (185). The crystal structure and binding site of DHQD type II from *Streptomyces coelicolor* (PDB: 1GTZ).

6.5. Dehydroquinate Dehydratase Inhibitors

The development of effective dehydroquinate dehydratase type II inhibitors for treating drug-resistant *Mycobacterium tuberculosis* co-infected with VIH virus is gaining attention [188]. Thus, the carbocycles mimics of enol intermediate were described as DQDH type II inhibitors, showing different degrees of potency (Fig. **186**) [189, 190].

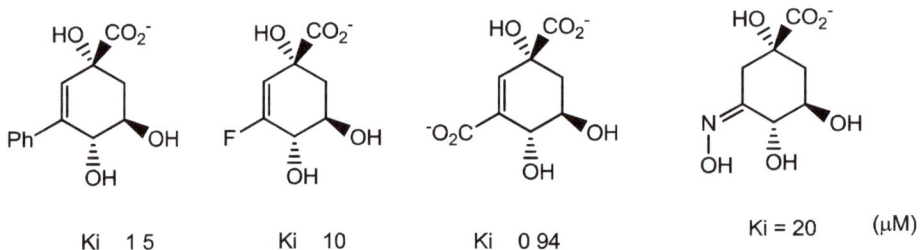

Ki 1 5   Ki 10   Ki 0 94   Ki = 20   (μM)

Fig. 186 cont.....

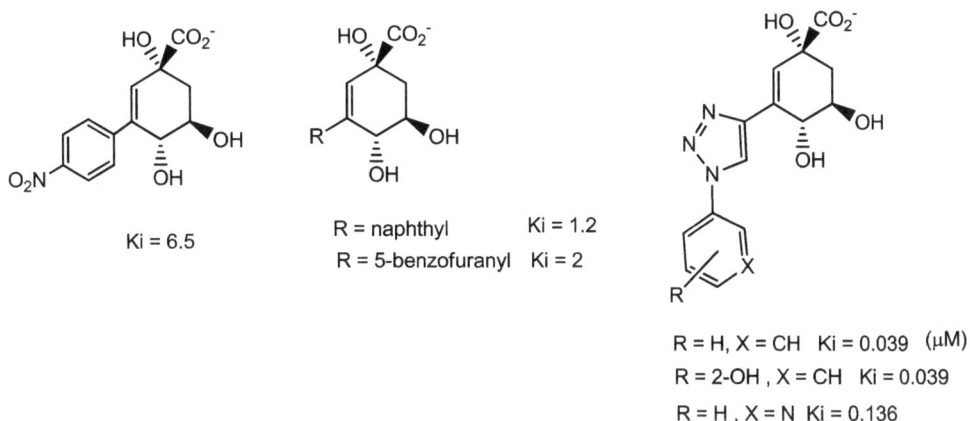

R = naphthyl   Ki = 1.2
R = 5-benzofuranyl Ki = 2

Ki = 6.5

R = H, X = CH Ki = 0.039 (µM)
R = 2-OH , X = CH Ki = 0.039
R = H , X = N Ki = 0.136

Fig. (186). Chemical structure and inhibition constants of dehydroquinate dehydratase inhibitors.

6.6. Shikimate Dehydrogenase (SD)

The reduction step of 3-dehydroshikimate to provide shikimate (Fig. **187**) is catalysed by the NADPH dependent enzyme shikimate dehydrogenase (SD).

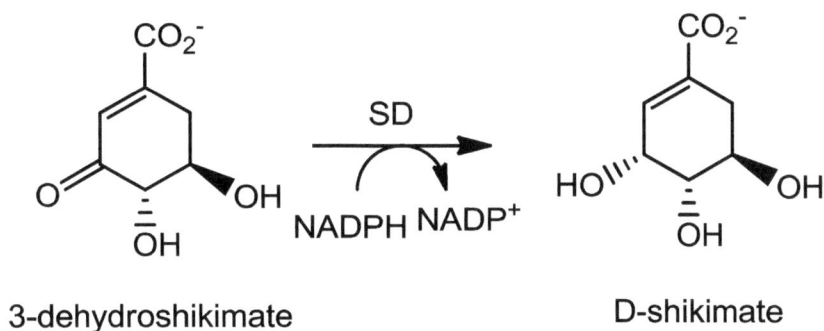

3-dehydroshikimate   D-shikimate

Fig. (187). The shikimate dehydrogenase reaction.

A simplified mechanism considers a hydride addition to the ketone in a cis mode with the vicinal hydroxyl group to yield D-shikimate (Fig. **188**) [191].

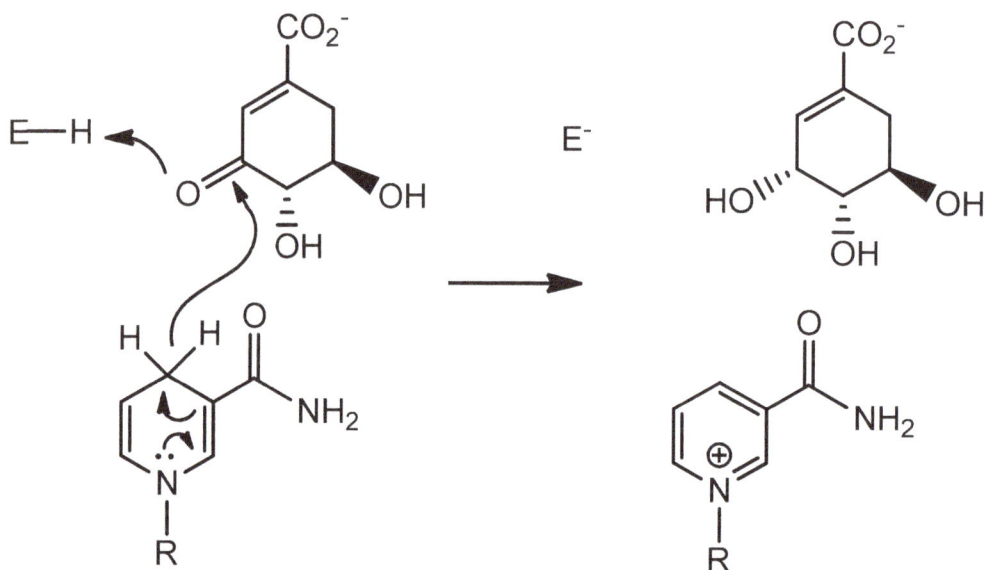

Fig. (188). Simplified reaction mechanism of D-shikimate formation.

The crystal structure of SD from *Haemophilus influenza* in complex with NADPH shows two domains, the catalytic and the NADPH domain. The catalytic domain adopting a twisted α/β structure, and in the NADPH domain an unusual binding mode were only adenine, ribose, 2'-phosphate and 5'-diphosphate of NADPH interact with the protein (Fig. **189**) [192].

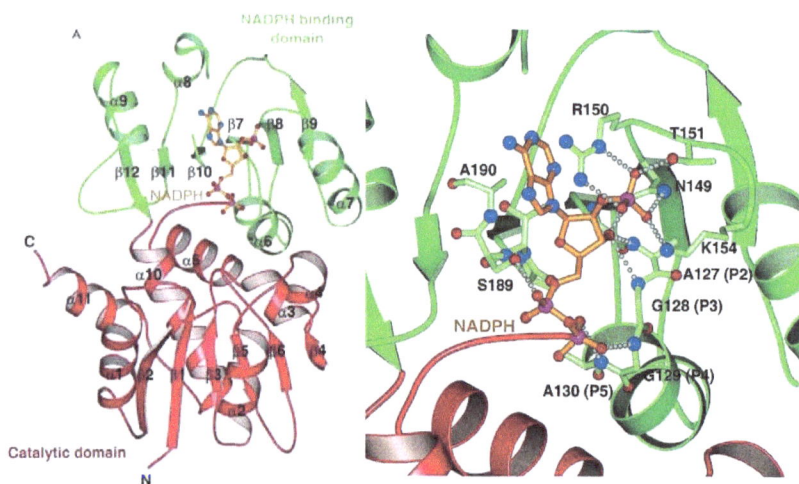

Fig. (189). The crystal structure of SD from *Haemophilus influenza* in complex with NADPH (PDB: 1P74).

6.7. Shikimate Dehydrogenase Inhibitors

Shikimic acid derivatives has been synthesized and evaluated as inhibitors, such as the dimeric diamides with different size linker (Fig. **190**), although with not significant inhibition since the constant values falls in the range of K_i 400-458 μM [193].

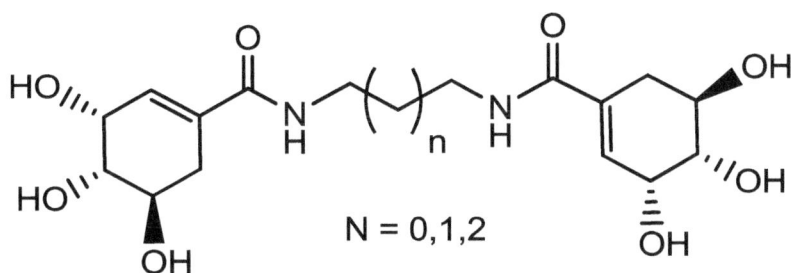

Fig. (190). Chemical structure of dimeric diamides.

Other inhibitors exhibiting significant structural variations were found through virtual screening and analysed as *Staphylococcus aureus* SD inhibitors (Fig. **191**) [194].

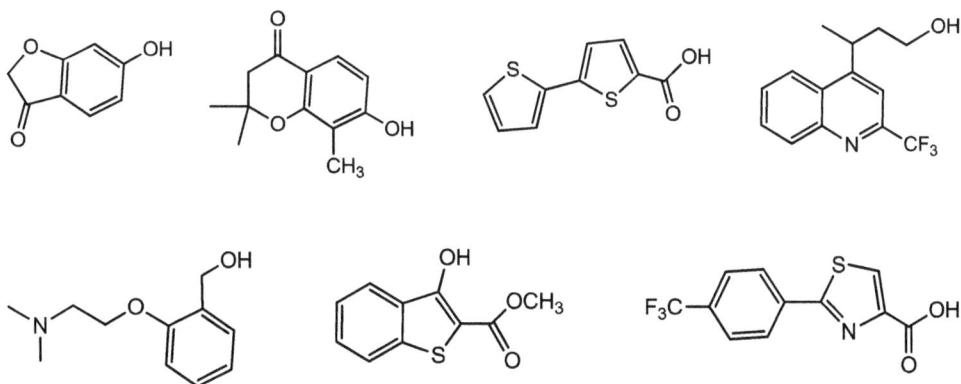

Fig. 191 cont.....

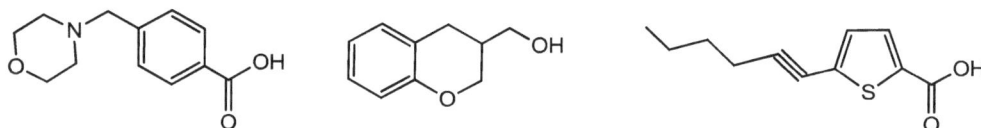

Fig. (191). Chemical structure of SD inhibitors.

6.8. Shikimate Kinase (SK)

Also known as shikimate 3-phosphotransferase catalyses the fifth step the in shikimate pathway, catalysing the reaction from D-shikimate to shikimate 3-phosphate (Fig. **192**).

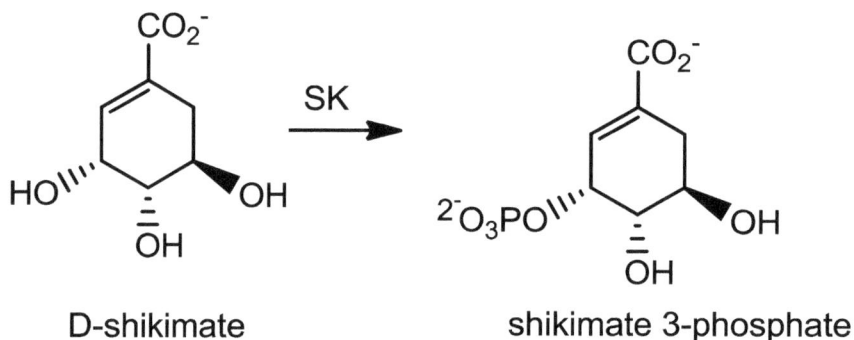

Fig. (192). The shikimate kinase reaction.

The crystal structure of shikimate kinase from *Mycobacterium tuberculosis* (*Mt*SK) complexed with MgADP reveals a six coordination for Mg^{2+}, and MtSK-MgADP-shikimate a distorted six-coordinated, with a conserved pocket with residues Asp34, Arg58, Gly80 and Arg136. Substantial evidence on the conformational arrangements and orientations suggest that shikimate adopts the same orientation in all complexes with *Mt*SK (Fig. **193**) [195, 196].

Fig. (193). The crystal structure and binding site of shikimate kinase from *Mycobacterium tuberculosis* (*Mt*SK) complexed with MgADP (PDB: 2IYR).

6.9. Shikimate Kinase Inhibitors

Inhibition of SK is receiving substantial attention due the need of designing effective drugs to confront the multidrug-resistant *Mycobacterium tuberculosis* (MDR-MT), which remains as a challenging health problem. As an effort for finding effective candidates, high-throughput virtual screening investigations have been implemented, using software programs which are able to perform flexible docking simulations with the active site of *Mt*SK. As a result of these studies the compounds shown in Fig. (**Fig. 194**) [197] were found as promising candidates, although further experimental inhibition test will be required [198 - 200].

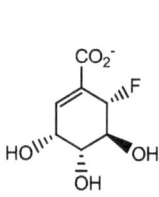

6-S-fluoroshikimate

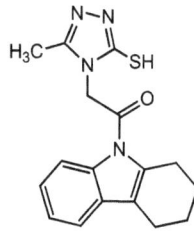

asxe1

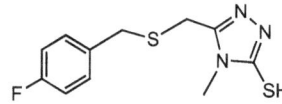

MW1

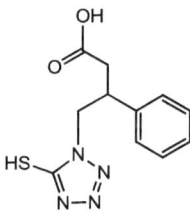

diver1

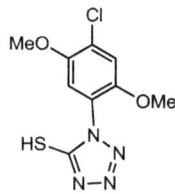

spch1

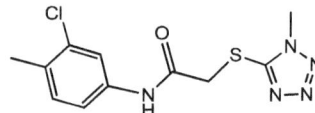

asxc2

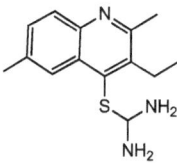

asxc1

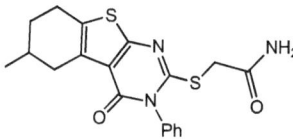

spca1

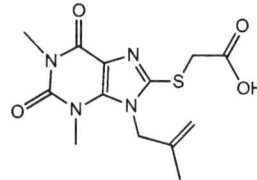

spca2

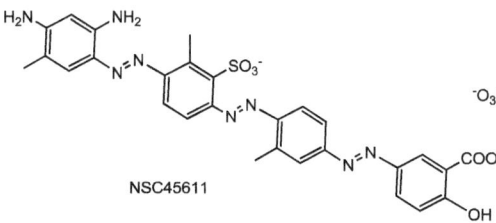

NSC45611

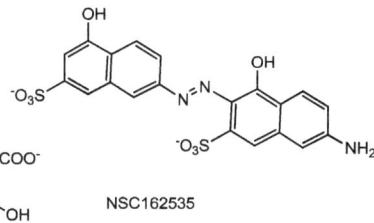

NSC162535

Fig. 194 cont.....

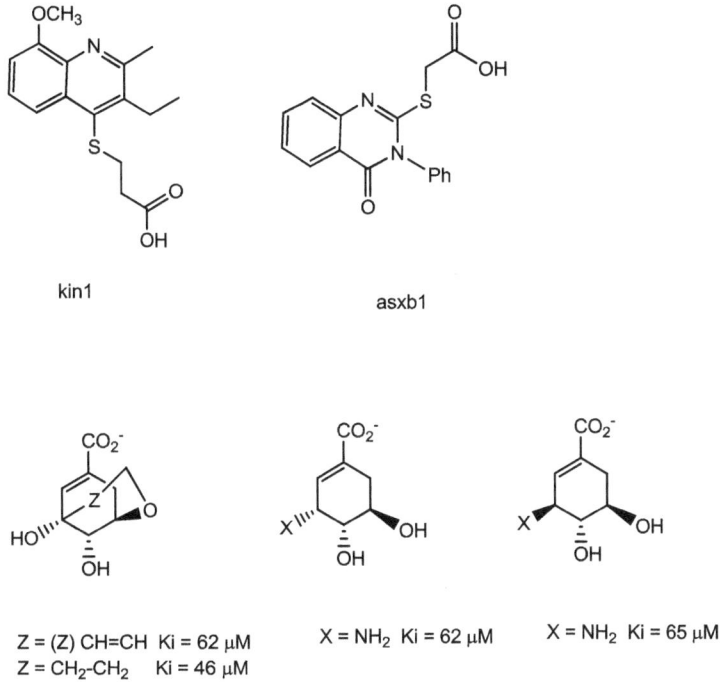

kin1   asxb1

Z = (Z) CH=CH Ki = 62 μM   X = NH$_2$ Ki = 62 μM   X = NH$_2$ Ki = 65 μM
Z = CH$_2$-CH$_2$ Ki = 46 μM

Fig. (194). Chemical structure of shikimate kinase inhibitors.

6.10. Enolpyruvisylshikimate Phosphate Synthase (EPSP)

This enzyme participates in the sixth step of the shikimate pathway involving shikimate-3-phosphate with phosphoenolpyruvate (PEP) to give 3-enolpyruvisylshikimate-5-phosphate (EPSP) and phosphate (Fig. **195**) [201].

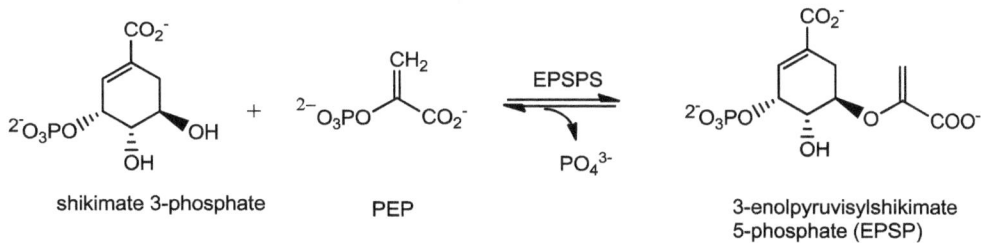

shikimate 3-phosphate   PEP   3-enolpyruvisylshikimate
                                                                5-phosphate (EPSP)

Fig. (195). The enolpyruvisylshikimate phosphate synthase reaction.

The mechanism proposes an unusual hydroxyl attack to the double bond to produce a phospho hemiacetal intermediate, which undergoes β-elimination of the phosphate group (Fig. **196**) [202].

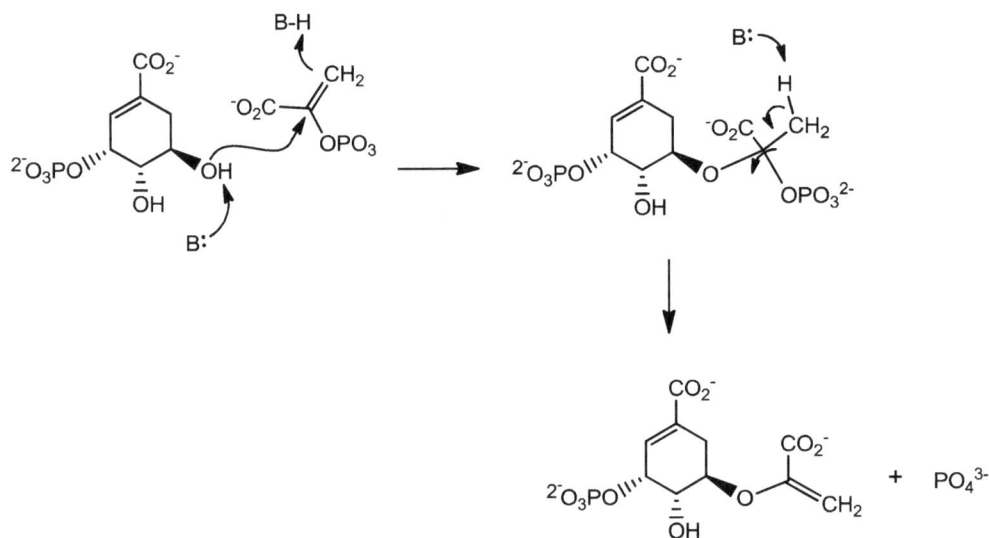

Fig. (196). Proposed mechanism for 3-enolpyruvisylshikimate-5-phosphate (EPSP) formation.

6.11. Enolpyruvisylshikimate Phosphate Synthase Inhibitors

EPSPS enzyme has been identified as an attractive target enzyme due his participation in crucial processes such as microbial survival, parasite growth and production of essential metabolites in plants and fungi. The well-known herbicide glyphosate inhibits EPSPS and consequently interfere with the biosynthesis of aromatic compounds in plants, bacteria, parasites and fungi, without affecting supposedly animal metabolism (Fig. **197**) [203].

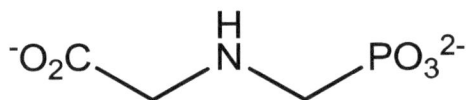

Fig. (197). Chemical structure of herbicide glyphosate.

Other EPSPS inhibitors described are phosphonate enantiomers of shikimate 3-phosphate (Fig. **198**), assayed on EPSPS from *Petunia hybrida*, observing that

(R)-diastereomer is the most potent with K_i value = 16 nM, in comparison with the (S)-diastereomer having a K_i value of 750 nM. Additionally it was found that the (R)-phosphonate induces enzyme's flexibility, which can explain the increased inhibition [204].

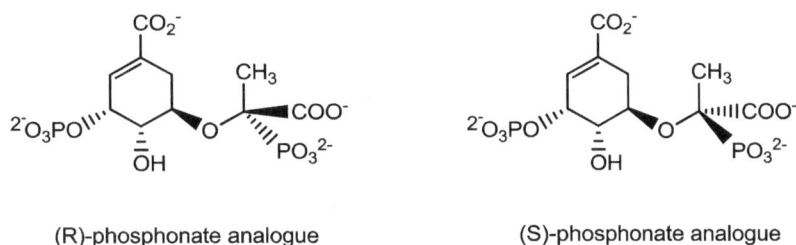

(R)-phosphonate analogue　　　　　　　　(S)-phosphonate analogue

Fig. (198). Chemical structure of shikimate 3-phosphate enantiomers.

6.12. Chorismate Synthase (CS)

The last step in the shikimate pathway leading to the aromatic metabolites formation is mediated by the enzyme chrosimate synthase which catalyse the conversion of 3-enolpyruvisylshikimate 5-phosphate to chorismate (Fig. **199**) a key intermediate in a number of biosynthetic processes, including the aromatic amino acids phenylalanine, tyrosine and triptophane.

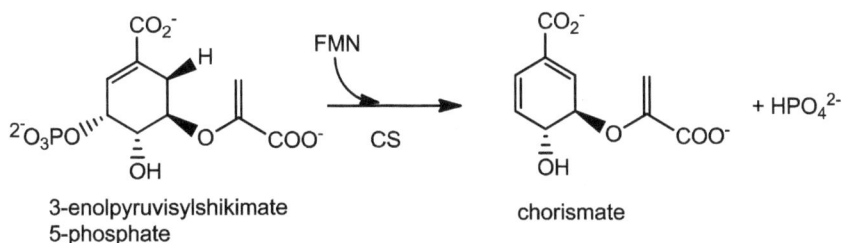

3-enolpyruvisylshikimate 5-phosphate　　　　　　　chorismate

Fig. (199). The chorismate synthase reaction.

The mechanism involves a 1,4-trans elimination assisted by reduced flavin mononucleotide (FMN) and two histidine molecules. Based on site directed mutagenesis, a proposed mechanism involves an electron transfer from flavin to the 3-enolpyruvisylshikimate-5-phosphate (EPSP) promoting the phosphate exit as a leaving group. Next a free radical tautomerism will result in the formation of an unpaired electron installed at N(5). Final free radical coupling reaction gives place to the double bond formation (Fig. **200**) [205].

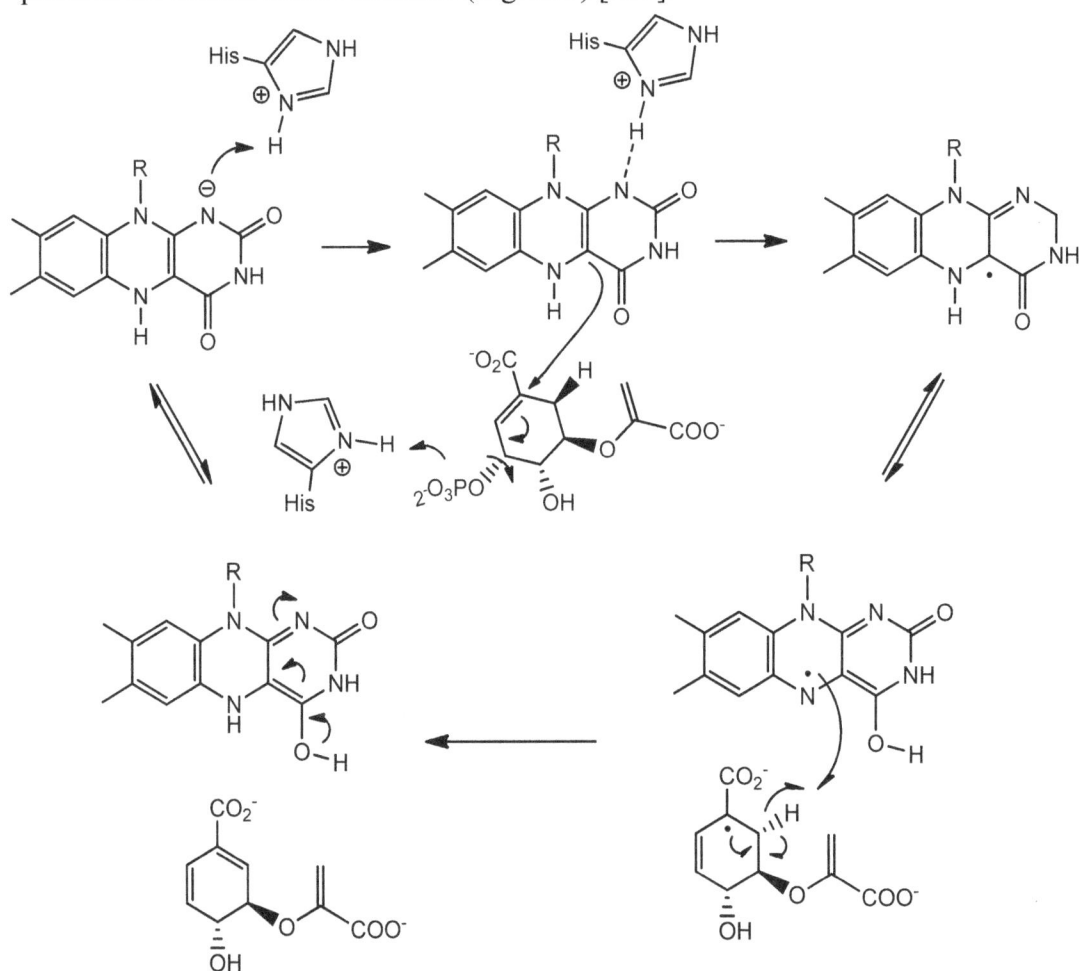

Fig. (200). Proposed mechanism for chorismate formation.

The chorismate synthase structure from *Aquifex aeolicus* was established as a tetrameric structure, also identified as dimer of dimers with a core interface of two ionic networks, providing stabilization to the oligomeric state (Fig. **201**) [206]. The structural analysis of chorismate synthase from *Plasmodium falciparum* has been described, showing an EPSP substrate interacting with a number of positively charged residues such as Arg123, Lys457, and Pro333 [207].

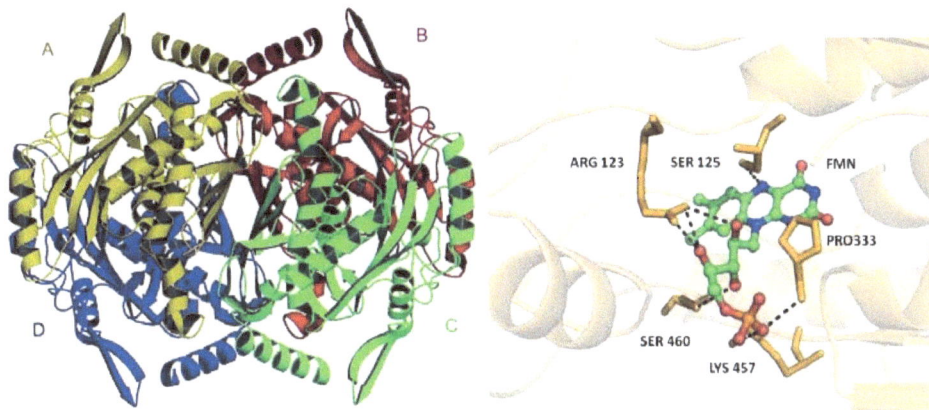

Fig. (201). Ribbon representation and bindig site of chorismate synthase from *Aquifex aeolicus* (PDB: 1Q1L).

6.13. Chorismate Synthase Inhibitors

Fluorinated shikimate isomer 6(S)-fluoroshikimate is converted to 6-fluoro-EPSP, inhibiting growth of *Escherichia coli* chorismate synthase *in vitro* (Fig. **202**) [208].

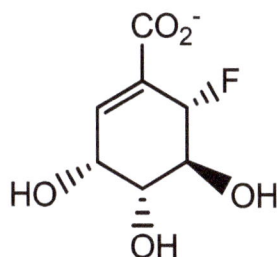

6(S)-fluoroshikimate

Fig. (202). Chemical structure of fluorinated shikimate isomer 6(S)-fluoroshikimate.

Another report describes the synthesis of dihydroxybenzofuranone analogues, being the most potent against chorismate synthase from *Pseudomona pneumoniae* the *para*-pentoxy analogue displaying an IC_{50} value of 0.22 µM, followed by the non substituted analogue with IC_{50} value of 3.5 µM (Fig. **203**) [209].

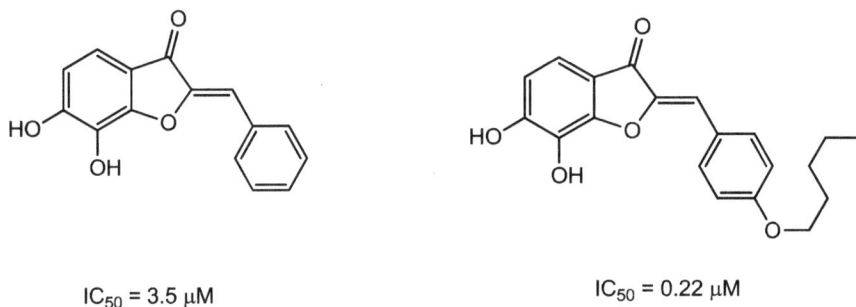

IC$_{50}$ = 3.5 μM   IC$_{50}$ = 0.22 μM

Fig. (203). Chemical structure and IC$_{50}$ values of dihydroxybenzofuranone analogues.

6.14. Chorismate Mutase (CM)

Chorismate mutase catalyses the conversion of chorismate to prephenate (Fig. **204**) involving Claisen rearrangement defined as a [3,3]-sigmatropic rearrangement, which requires allyl vinyl ether, being transformed to unsaturated carbonyl product.

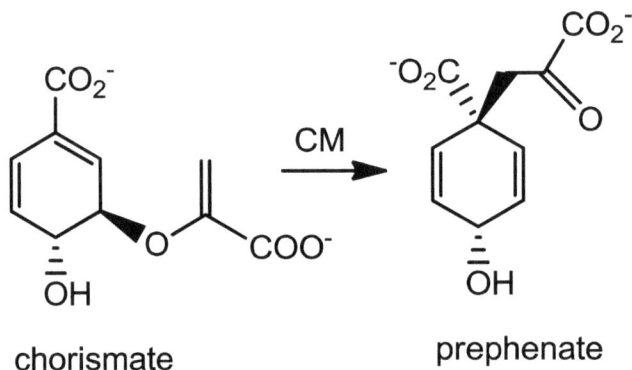

chorismate   prephenate

Fig. (204). The chorismate mutase reaction.

In order to have a better understanding on this transformation a conformational analysis has been described, considering two non-stable forms, defined as pseudodiequatorial conformation in equilibrium with a pseudodiaxial conformation, which finally result in a more stable chair-like transition state as shown in Fig. (**205**) [210].

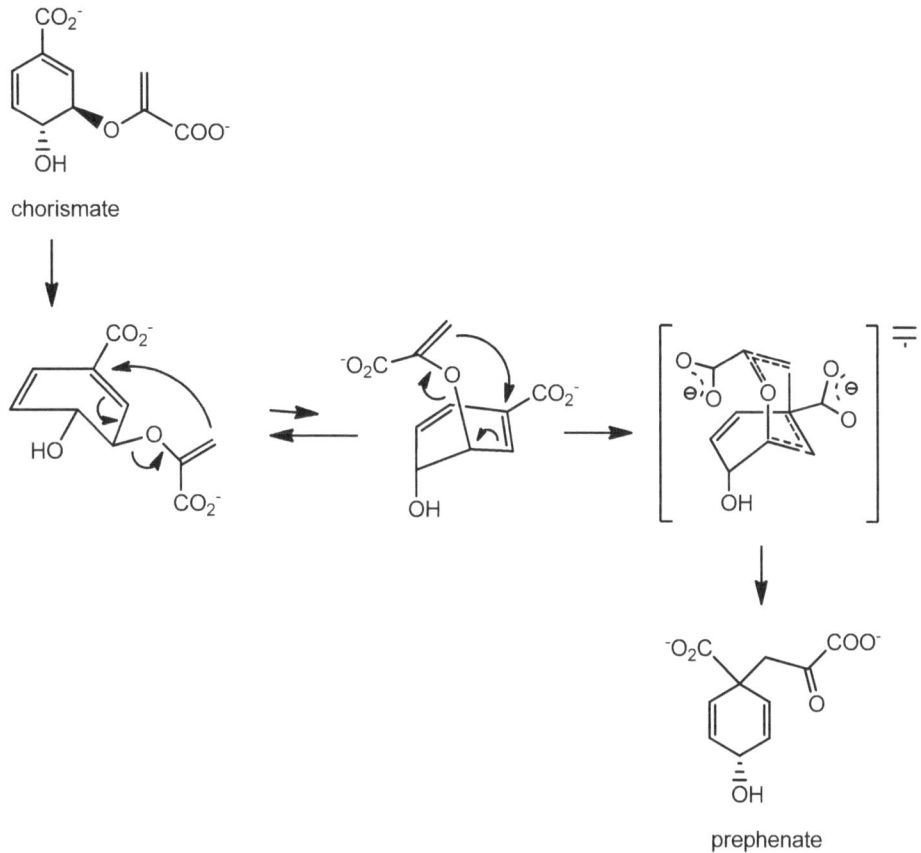

Fig. (205). Conformational analysis describing the formation of prephenate from chorismate.

The molecular analysis on the chorismate mutase from *Mycobacterium tuberculosis* (*Mt*CM), *Saccharomyces cerevisiae* (ScCM) and *Escherichia coli* (*Ec*CM) allowed to recognize the residues interacting with the transition state analogues (TSA) observing conserved interactions of Arg, Glu and Lys with the substrate (Fig. **206**) [211].

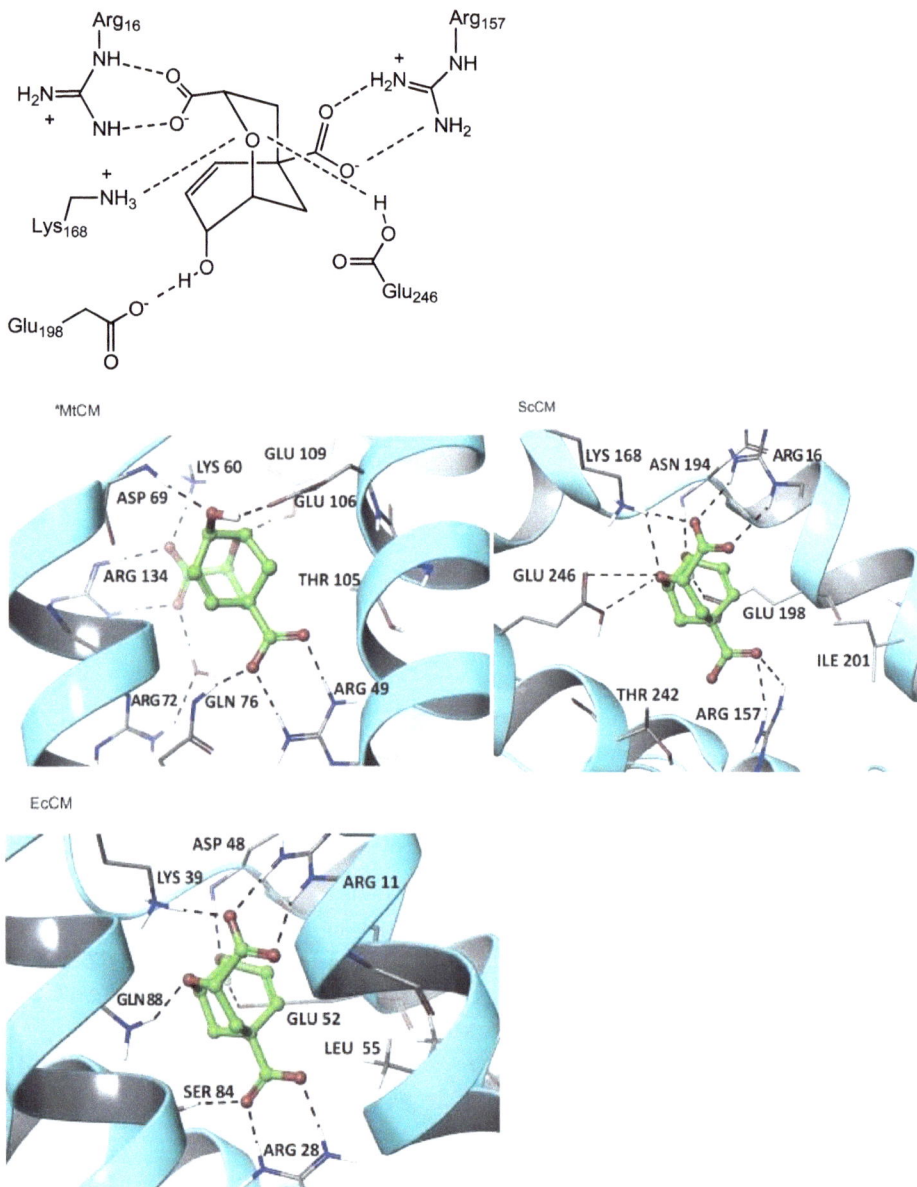

Fig. (206). Binding site of chorismate mutase from *Mycobacterium tuberculosis* (PDB: 2FP1), *Saccharomyces cerevisiae* (PDB: 1R52) and *Escherichia coli* (PDB: 1ECM).

6.15. Chorismate Mutase Inhibitors

Series of aza bicyclic analogues were prepared and tested against *E. coli* chorismate mutase, however none of them were as potent as their bicyclic pyrane analogues **1** and **2** showing inhibition values at nanomolar scale (Fig. **207**) [212].

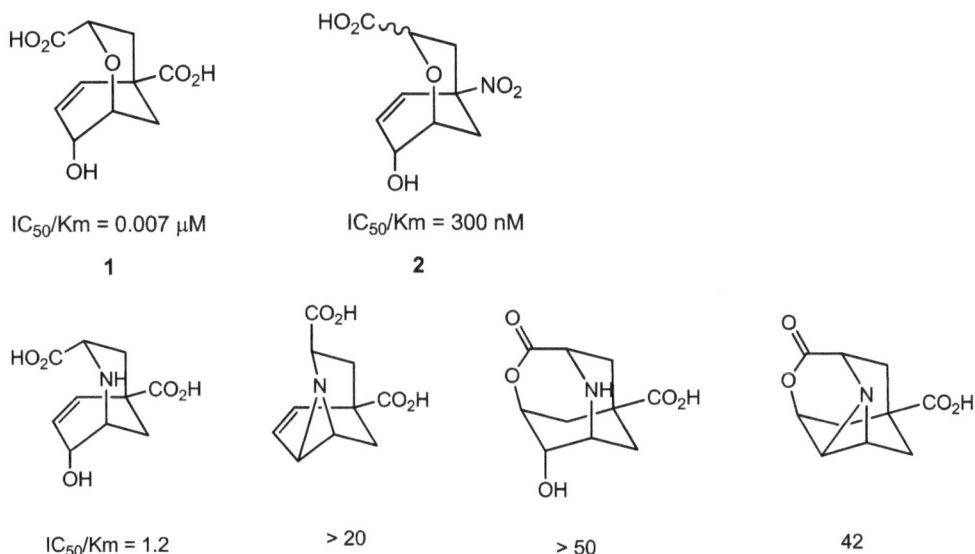

Fig. (207). Chemical structure of oxo and aza bicyclic analogues and their inhibitory values against chorismate mutase from *E. coli*.

Other inhibitors evaluated against chorismate mutase from *Mycobacterium tuberculosis* include sulphonamide, indole, chromone, a variety of heterocycles and carvacrol a potent CM inhibitor present in popular herb oregano (Fig. **208**) [213].

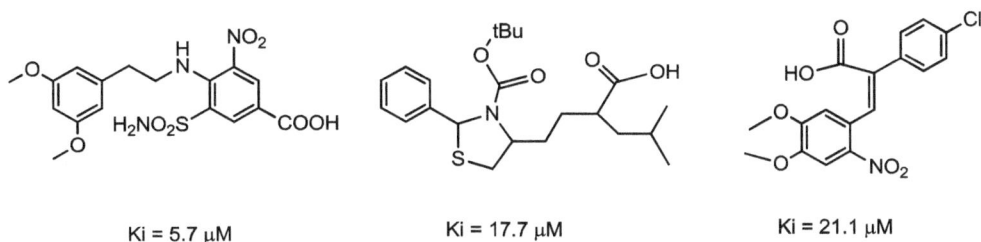

Fig. 208 cont.....

IC$_{50}$ = 17.02 µM   IC$_{50}$ = 15.63 µM   IC$_{50}$ = 15.13 µM

IC$_{50}$ = 23.88 µM   IC$_{50}$ = 19.80 µM   IC$_{50}$ = 14.76 µM

IC$_{50}$ = 1.06 µM   IC$_{50}$ = 19.74 µM

IC$_{50}$ = 1.01 µM

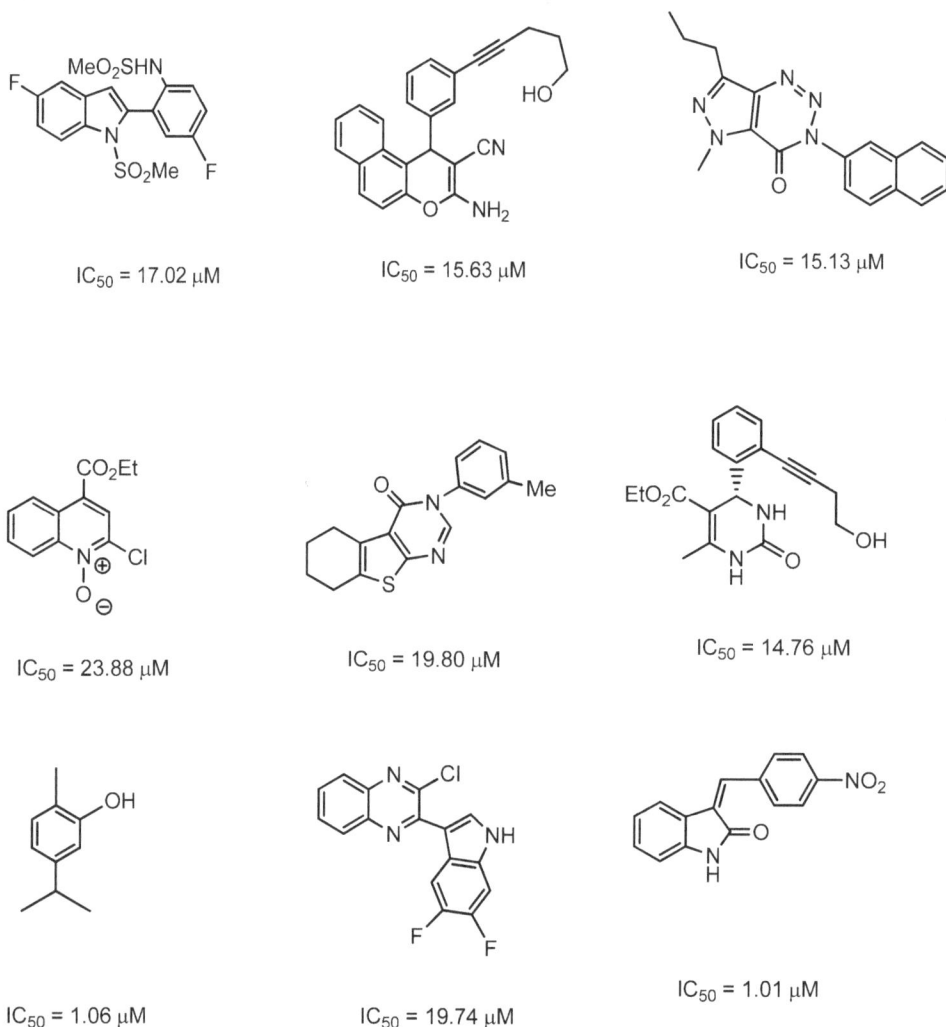

Fig. (208). Chemical structure and IC$_{50}$ values of miscellaneous compounds exhibiting chorismate mutase inhibition.

6.16. Prephenate aminotransferase (PAT)

The step involving the conversion of prephenate to arogenate is carried out by the pyridoxal dependent prephenate aminotransferase (PAT) as shown in Fig. (209).

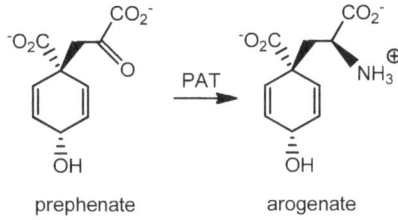

Fig. (209). The prephenate aminotransferase reaction.

Crystallographic studies and molecular docking on prephenate aminotransferase from *Arabidopsis thaliana* were conducted providing a wealth of evidence on the catalytic site summarized in the initial attachment of Lys306 residue with pyridoxal phosphate (PLP) to produce an internal aldimide. The resulting Schiff base undergoes a transamination with asparagine to form the conjugate PLP-Asp, which after hydrolysis generate the external aldimide PLP-Asp. A second hydrolysis takes place producing an internal aldimide, which reacts with prephenate forming the complex PLP-prephenate. Further transamination reaction between lys306 and PLP-prephenate results in the formation of arogenate and the aldimide Lys306-PLP (Fig. **210**) [214].

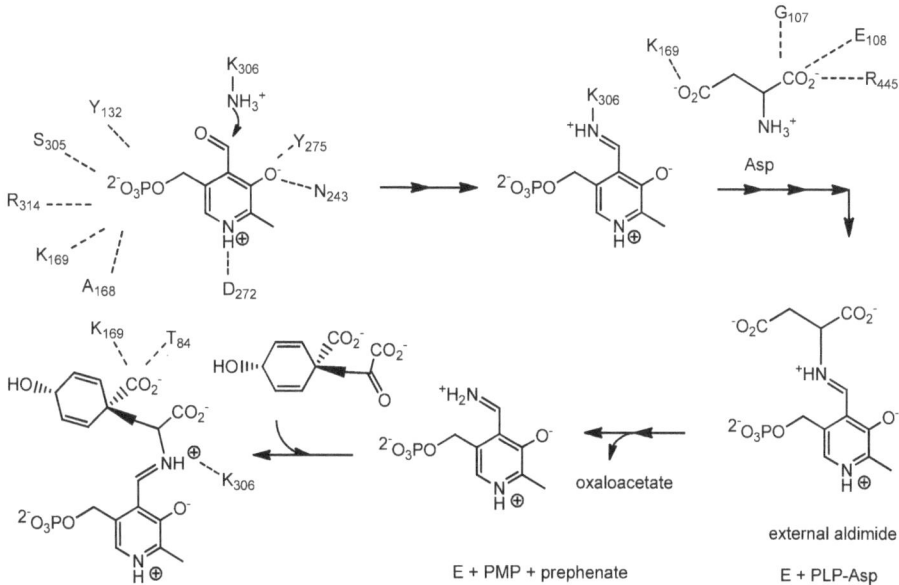

Fig. 210 cont.....

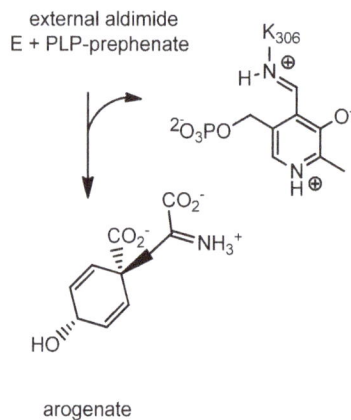

Fig. (210). Diagram of interactions and intermediates formed during the arogenate biosynthesis from prephenate.

Prephenate aminotransferase from *Arabidopsis thaliana* (*At*PAT) crystallizes as a homodimer and each monomer is made of 15 α-helices an 9 β-strands divided between two structural domains (Fig. **211**) [214].

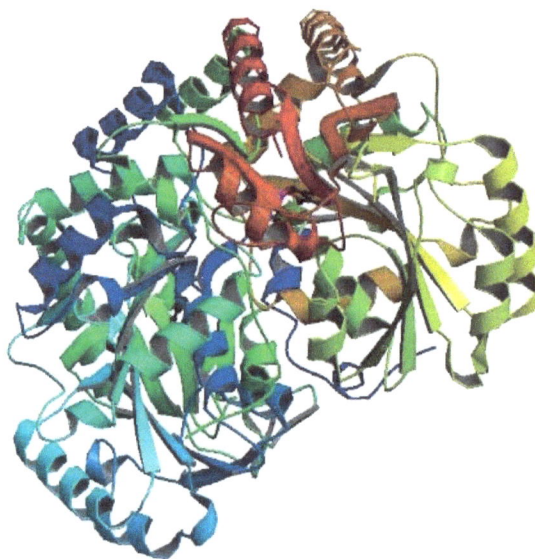

Fig. (211). Ribbon structure of prephenate aminotransferase from *Arabidopsis thaliana* (PDB: 5WMH).

6.17. Prephenate Aminotransferase Inhibitors

Only amino acid cysteine has been described to reduce PAT activity of plants and microorganism, showing IC_{50} values of 1.3 µM [215].

6.18. Arogenate Dehydratase (ADT)

The conversion of arogenate to L-phenylalanine is catalysed by the enzyme arogenate dehydratase (ADT), involving decarboxylative-dehydration steps (Fig. **212**). However in *Ecoli* and yeast the arogenate route is absent [216] while in plants L-phenylanaline might be prepared also from phenylpyruvate by prephenate dehydratase (PDT) [217].

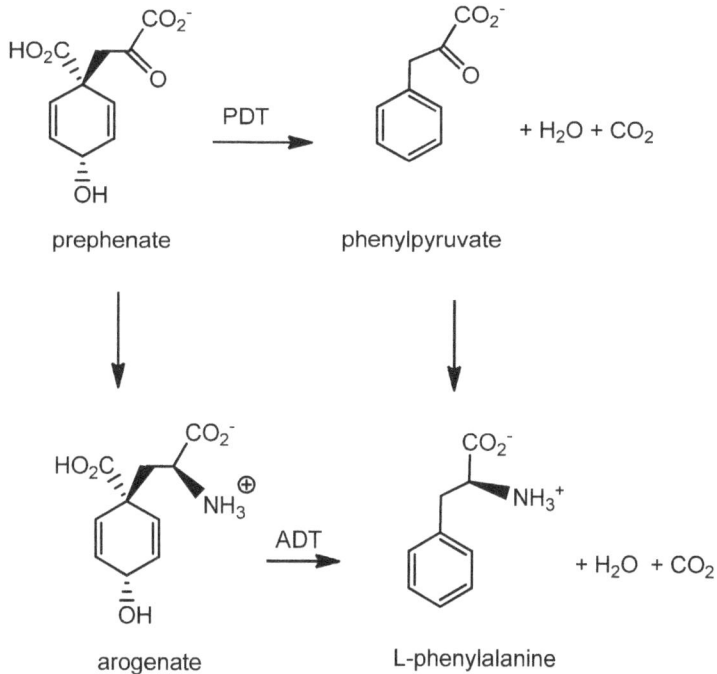

Fig. 212 cont.....

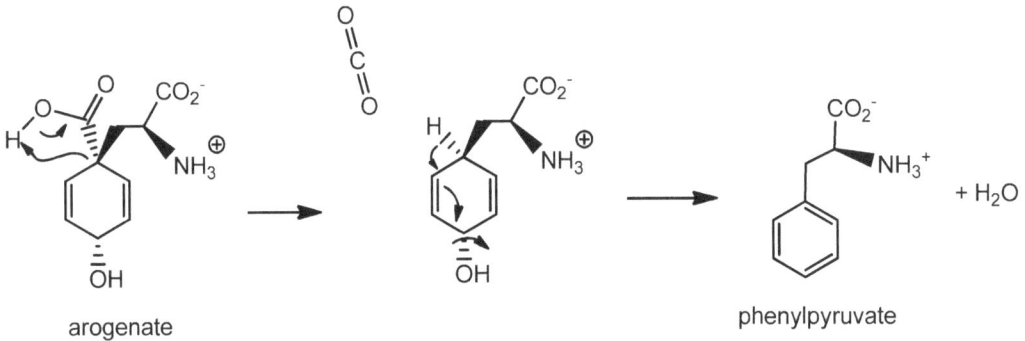

Fig. (212). Biosynthetic pathway for L-phenylalanine formation.

A number of six ADT subfamilies have been identified in plants three of them using more efficiently arogenate than prephenate (ADT1, 2 and 6) while ADT3, 4 and 5 essentially only arogenate [216].

7. TYROSINE BIOSYNTHESIS

Tyrosine is an aromatic amino acid, which can be prepared in the body from phenylalanine and therefore is no considered essential unlike phenylalanine, which can't be synthesized in the body and has to be incorporated through the diet. Tyrosine is very important neurotransmitter precursor of epinephrine, norepinephrine and dopamine and his deficiency can produce low pressure, depression and other disorders with serious consequences. In the shikimic pathway is produced by arogenate which is converted to tyrosine by the catalysis with arogenate dehydrogenase, producing NADH, although also NADPH has been have identified as cofactor (Fig. **213**).

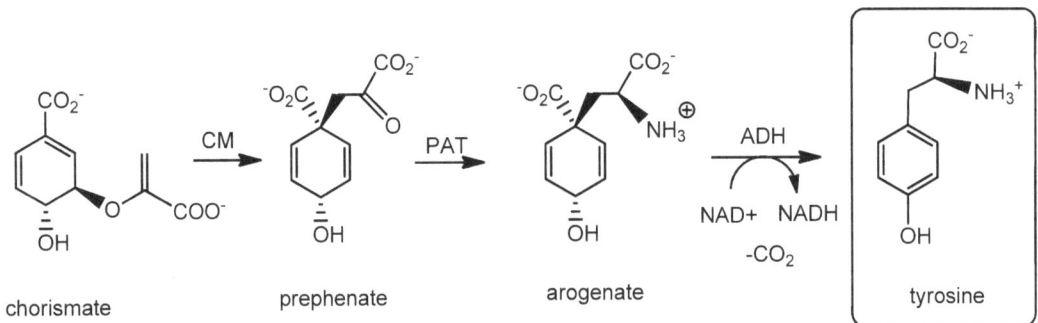

ADH = arogenate dehydrogenase

Fig. (213). The biosynthesis of tyrosine from chorismate.

7.1. Arogenate Dehydrogenase (ADH)

Arogenate dehydrogenase is an oxidoreductase, which converts arogenate to tyrosine, involving a decarboxylation, and dehydration process, for which requires either NAD or NADP to carry out the transformation (Fig. **214**).

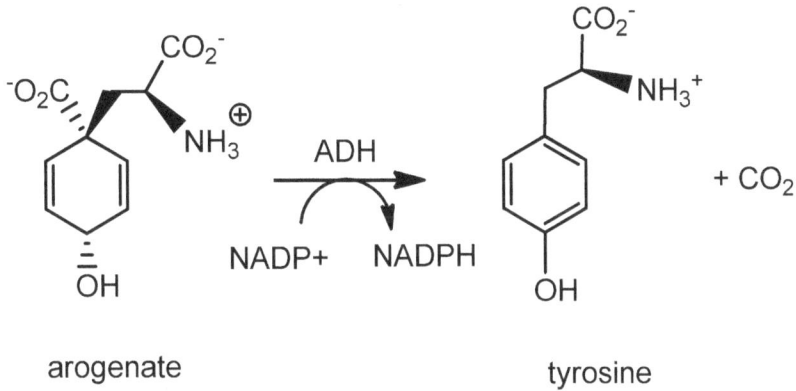

Fig. (214). The arogenate dehydrogenase reaction.

A consistent reaction mechanism proposes a decarboxylation step followed by hydride migration from arogenate to NAP^+ providing tyrosine and NADPH (Fig. **215**) [216, 217].

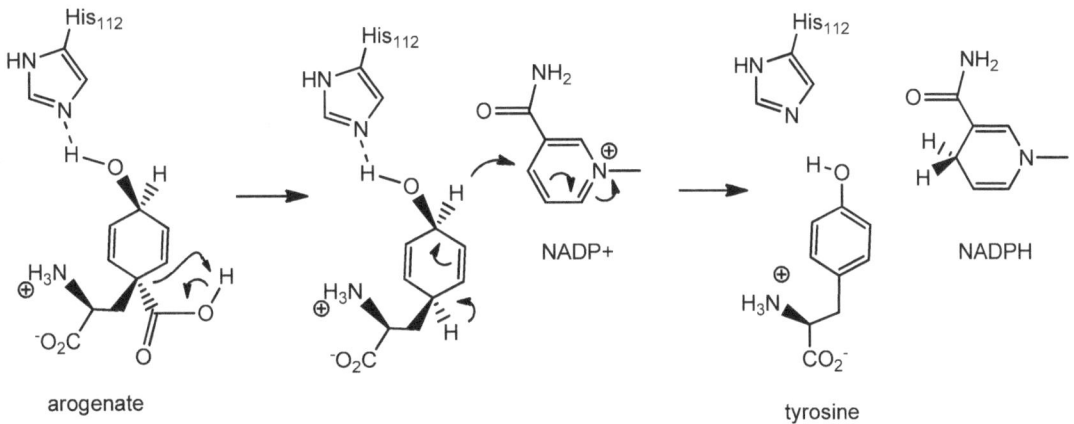

Fig. (215). Proposed mechanism for tyrosine formation from arogenate.

The crystal structure of arogenate dehydrogenase from *Synechocystis* shows a tetrameric structure and at the active site it can be seen the residues interacting with arogenate and NADP$^+$ molecules (Fig. **216**) [216].

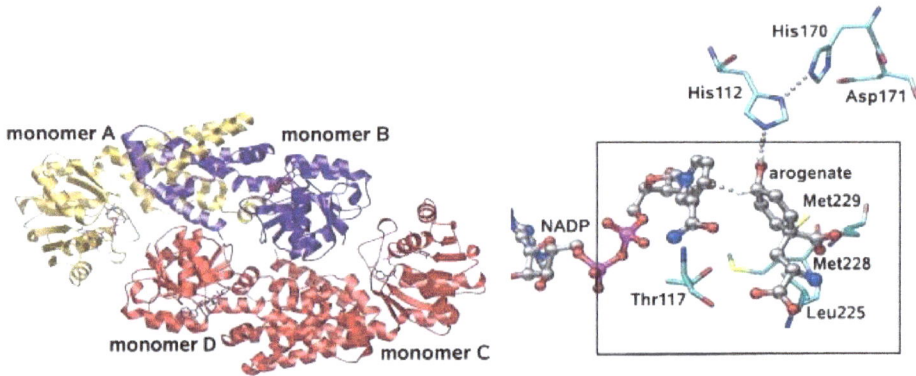

Fig. (216). The crystal structure of arogenate dehydrogenase from *Synechocystis* in complex with arogenate and NADP$^+$ (PDB: 2F1K).

8. SERINE BIOSYNTHESIS

Serine is a non-essential amino acid, which contains a hydroxymethyl group as R substituent, and can be either synthesized by the human or incorporated through the diet. It is the precursor of glycine and cysteine and participates in the biosynthesis of purines and pyrimidines.

Serine is synthesized from 3-phosphoglycerate (3-PG), which is oxidized by the enzyme phosphoglycerate dehydrogenase (PGDH) in the presence of NAD$^+$ to give 3-phosphohydroxypyruvate (PHP). The next step consisted in the formation of 3-phosphoserine (P-Ser) catalysed by the enzyme phosphoserine aminotransferase (PSAT), and finally converted to serine catalysed by the enzyme phosphoserine phosphatase (PSPH), as shown in Fig. (**217**) [218].

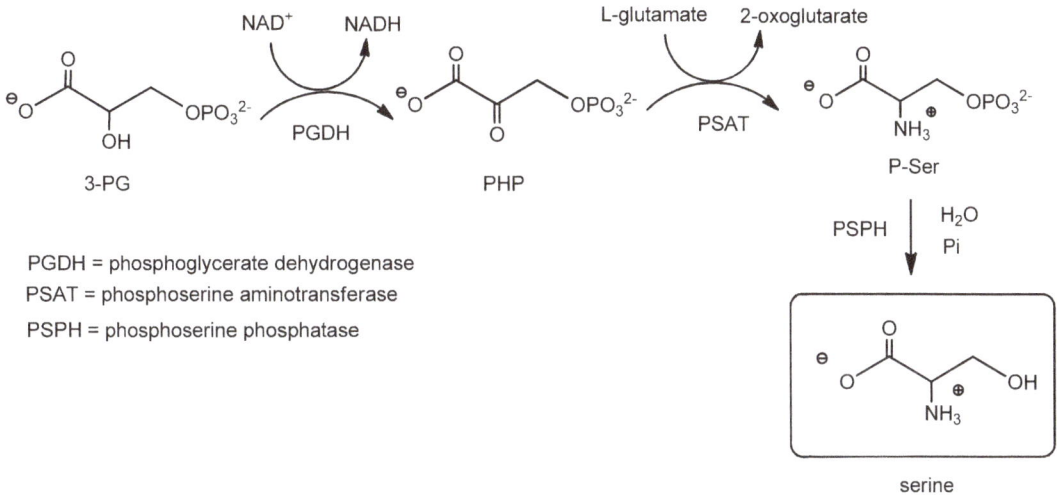

Fig. (217). The serine biosynthetic pathway.

8.1. Phosphoglycerate Dehydrogenase (PGDH)

This enzyme catalyses de initial step for the synthesis of serine consisting in the conversion of 3-phosphoglycerate (3-PG) to produce 3-phosphohydroxypyruvate (PHP) requiring NAD^+ as cofactor (Fig. **218**).

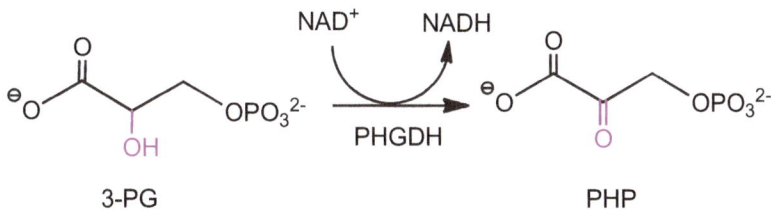

Fig. (218). The phosphoglycerate dehydrogenase reaction.

PGDH is a homotetramer oxidoreductase, having each subunit three domains identified as the serine regulatory binding (RBD), the cofactor or nucleotide binding (NBD) and the substrate binding domains (SBD) [219]. Human PGDH (Fig. **219**) has been identified as a target enzyme in certain types of cancer such as breast and melanoma since its knockdown resulted in reduced cancer cell growth [220].

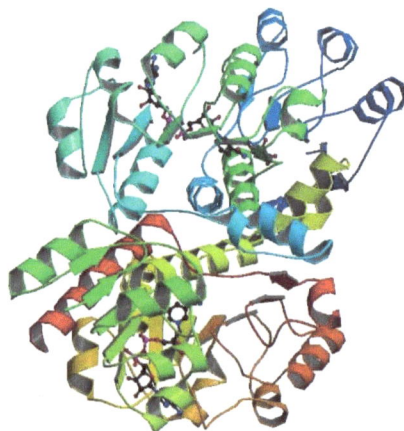

Fig. (219). Crystal structure of human PGDH (PDB: 5N6C).

8.2. Phosphoglycerate Dehydrogenase Inhibitors

High-throughput screening identified PGDH inhibitors, and their potency evaluated by using cancer cell lines with high serine biosynthetic activity. Some of the inhibitors are indole containing SPR, and piperazine-1-thiourea derivatives (NCT-502 and NCT-503) with the lowest IC_{50} values, also showing promising absorption, distribution, metabolism and excretion (ADME) properties. Other candidates less potent are CBR-5884, Raze inhibitor, α-ketothioamide, and PKUMDL-WQ 2201 and 2101 (Fig. **220**) [221].

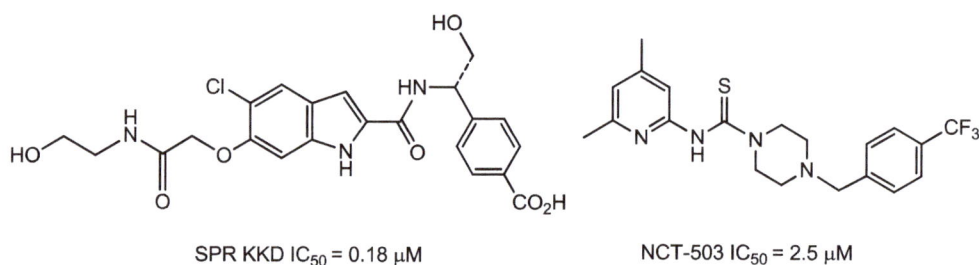

SPR KKD IC_{50} = 0.18 μM   NCT-503 IC_{50} = 2.5 μM

Fig. 220 cont.....

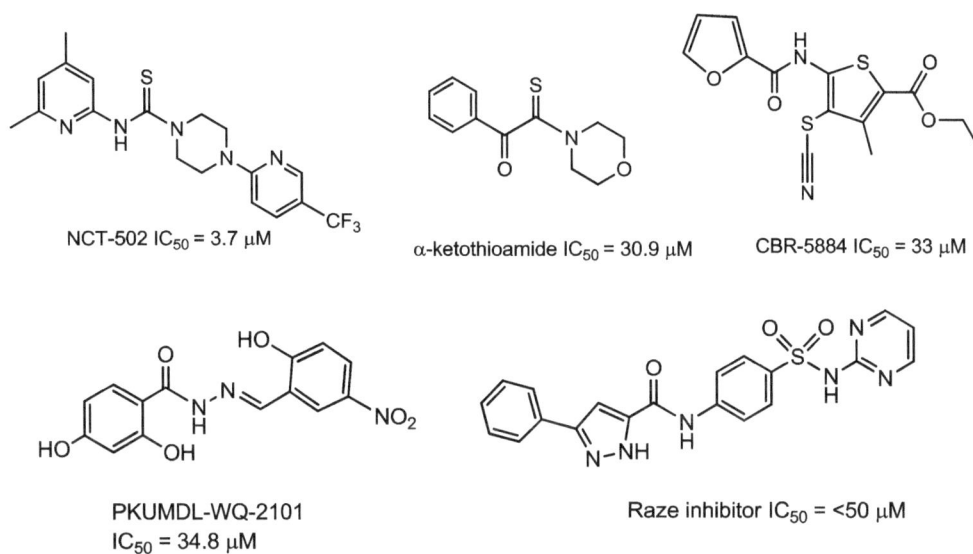

Fig. (220). Chemical structure of phosphoglycerate dehydrogenase inhibitors.

8.3. Phosphoserine Aminotransferase (PSAT)

PSAT is an aminotransferase involved in the coupling reaction between phosphohydroxypyruvate playing the role of amino acceptor and L-glutamate as amino donor, to produce the target molecule L-phosphoserine and 2-oxoglutarate as by-product (Fig. **221**).

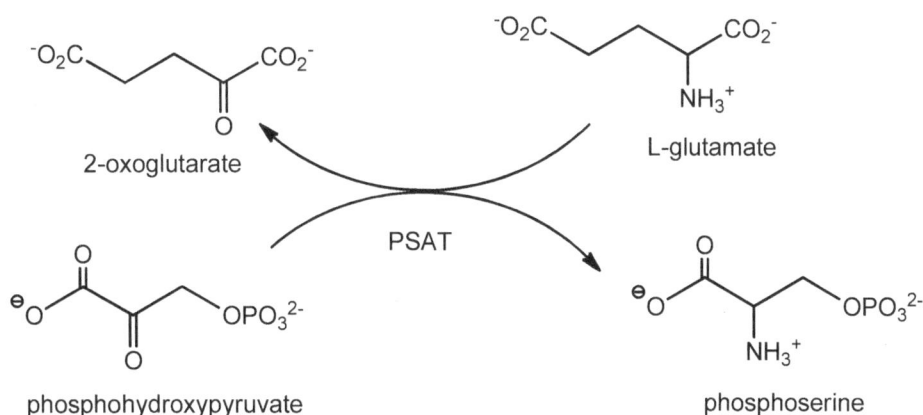

Fig. (221). The phosphoserine aminotransferase reaction.

The aminotransferase reaction can be visualized as a two-step process, the first involving pyridoxal-glutamine iminium salt formation as a result of glutamine and pyridoxal condensation, followed by tautomerization and hydrolysis to give 2-oxo amino acid and amino pyridoxal (step 1) [222]. The second step proceeds between pyridoxamine and phosphohydroxypyruvate as an amino group acceptor, forming the iminium salt, with participation of lysine as base taking a hydrogen. The resulting carbanion promotes resonance to generate the quinonoid form in equilibrium with the iminium intermediate, which is finally hydrolysed to yield phosphoserine and pyridoxal (Fig. **222**).

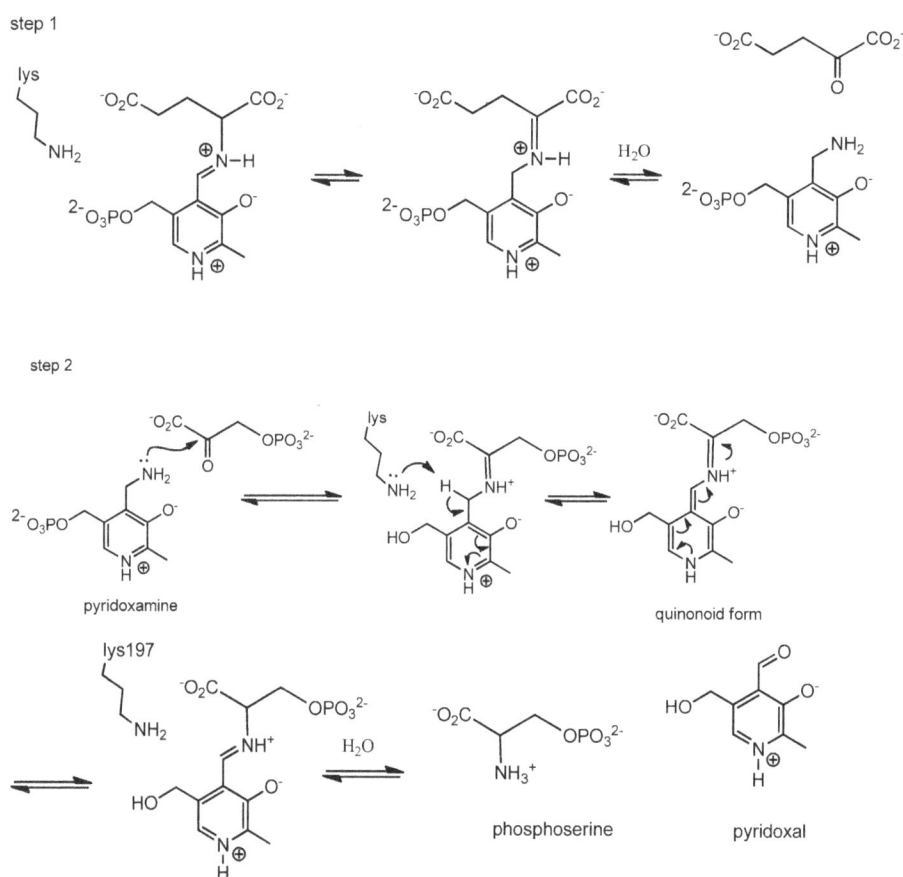

Fig. (222). Proposed mechanism for the phosphoserine formation.

Crystallographic studies of phosphoserine aminotransferase from *Mycobacterium tuberculosis* shows the architecture of the active site with the cofactor pyridoxal embedded y stabilized with different residues, being the interaction of the aldehyde group with lysine residue in agreement with the proposed mechanism (Fig. **223**) [223].

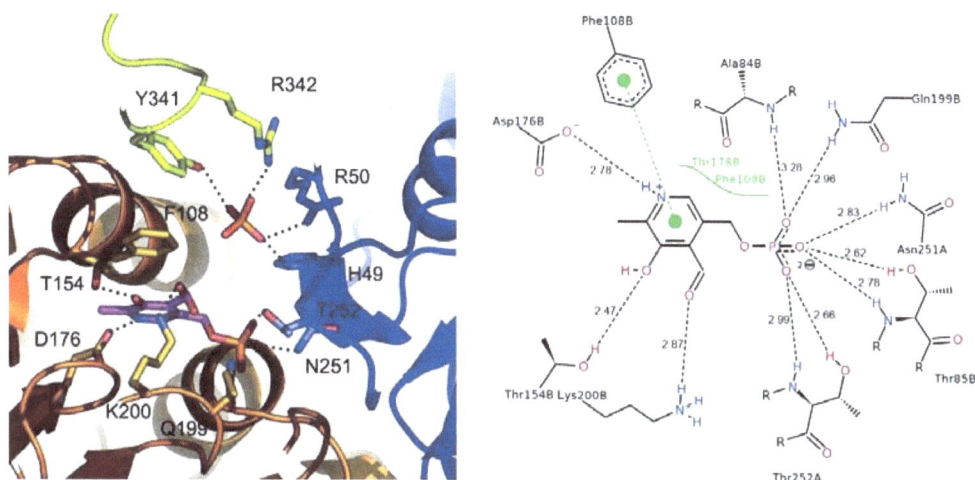

Fig. (223). Crystal structure and diagram at the active site of *Mt*PSAT (PDB: 2FYF).

8.4. Phosphoserine Aminotransferase Inhibitors

High throughput screening of PSP inhibitors have been described and tested against *Mycobacterium tuberculosis*, identifying clorobiocin and rosaniline as bactericidal in infected macrophages. Likewise candidate NSC76027 was the most potent compound *in vitro* enzymatic assays with IC_{50} values of 3.8 µM, while NSC693172 displayed IC_{50} value of 4 µM but high toxicity (Fig. **224**) [224].

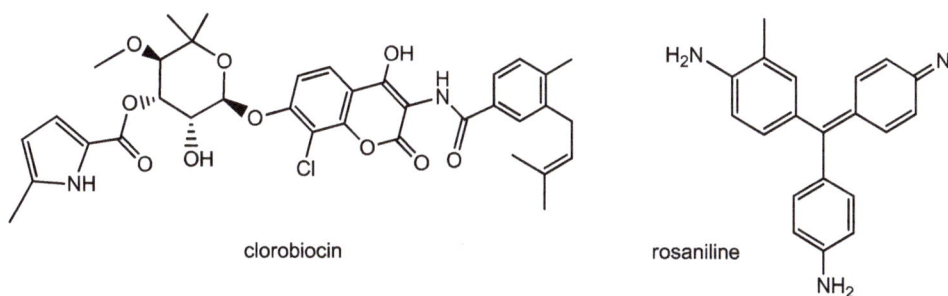

clorobiocin   rosaniline

Fig. 224 cont.....

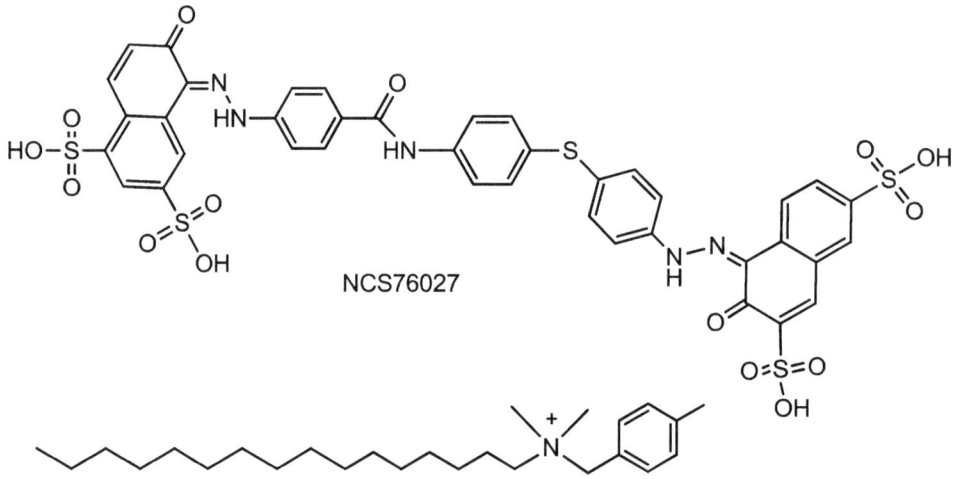

Fig. (224). Chemical structure of phosphoserine aminotransferase inhibitors.

8.5. Phosphoserine Phosphatase (PSP)

PSP is an enzyme involved in the hydrolysis of the phosphoester group being the last step in the synthesis of serine, converting 3-phosphoserine into L-serine (Fig. **225**).

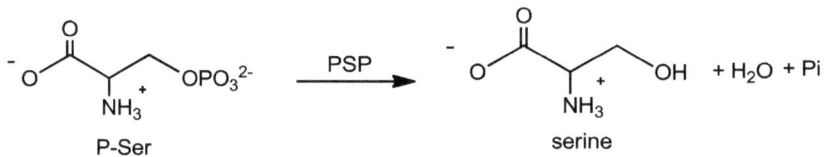

Fig. (225). The phosphoserine phosphatase reaction.

The removal of the phosphate group follows two possible mechanism, the unimolecular decomposition or associative and the dissociative mechanism (Fig. **226**) [225].

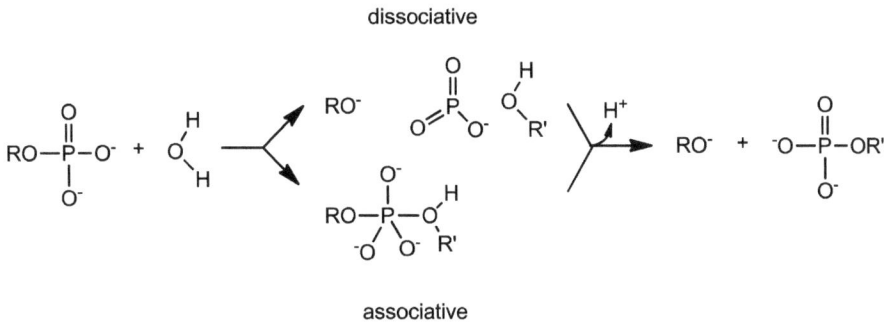

Fig. (226). The associative and dissociative mechanism of dephosphorylation reaction.

The dephosphorylation process involves the interaction between Asp residue with PSP with the participation of Mg^{2+}, to produce an interaction between the carboxylate and the phosphate group, leading to the release of serine. The phosphate group attached to the carboxylate is then hydrolysed producing phosphate and aspartate (Fig. **227**).

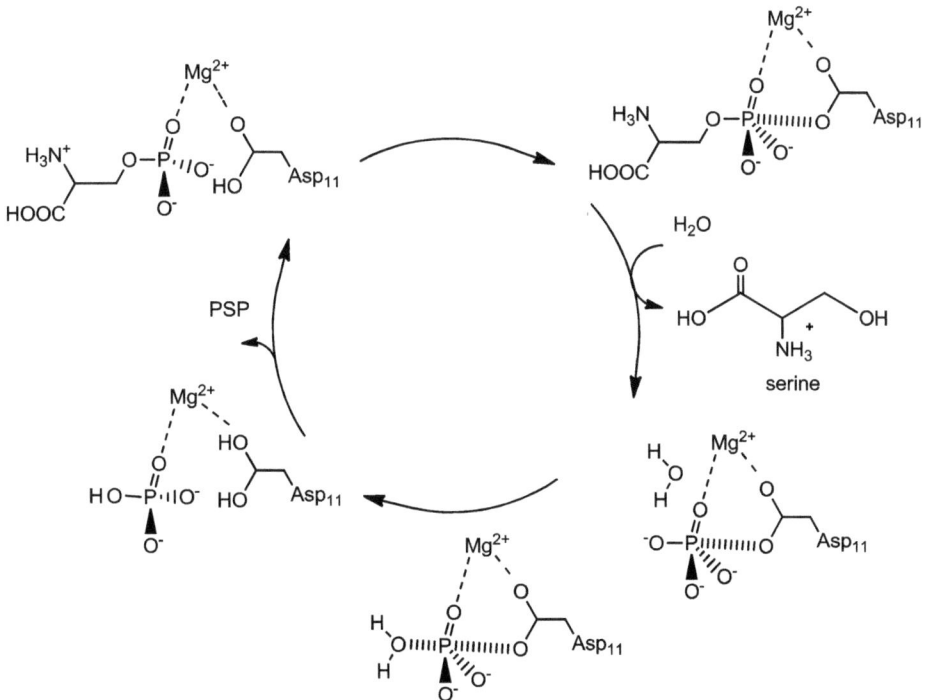

Fig. (227). Proposed dephosphorylation mechanism involving Mg^{2+} and Aspartate.

The PSP cycle within the active site was determined using mutants with transition state analogues modelled with AlF_3 –PSP and BeF_3-PSP complexes useful to confirm the mechanism proposed (Fig. **228**).

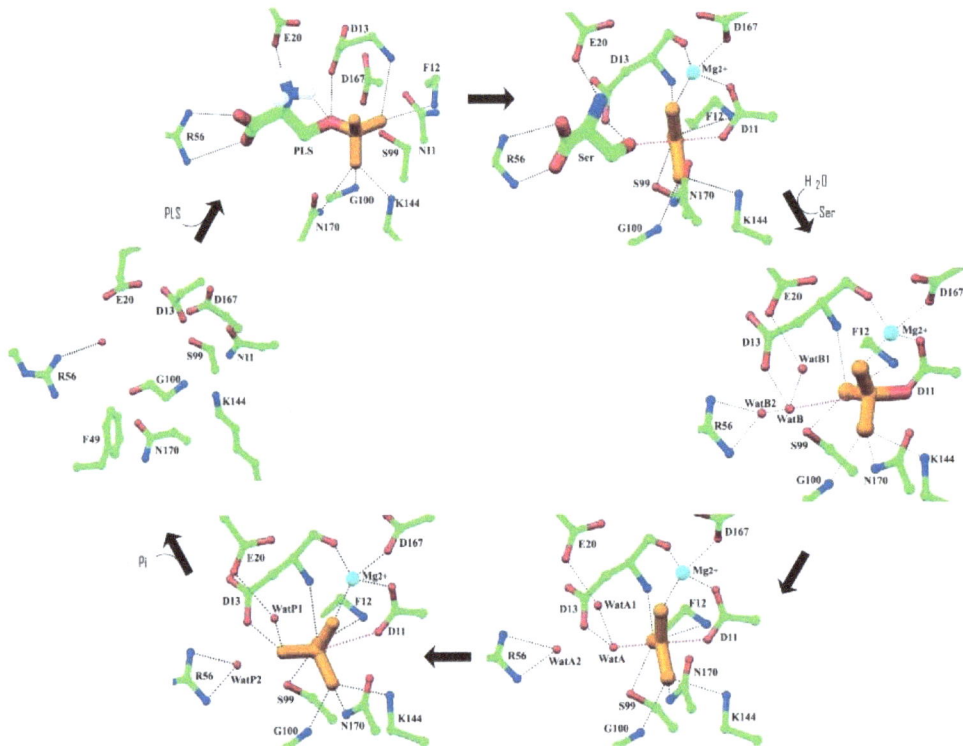

Fig. (228). Ball and stick diagram modelled with AlF_3–PSP and BeF_3-PSP complexes.

8.6. Phosphoserine Phosphatase Inhibitors

The pyrimidine analogue 5-fluorouracil (5-FU) used in the colonic cancer therapy has shown to induce formation of reactive oxygen species (ROS) leading to cell death, and to inhibit PSP activity according to a study which results in enhanced anticancer efficacy [226].

Other inhibitors presenting high potency were p-chloromercuriphenylsulfonic acid IC_{50} 9.3 μM), glycerophosphorylcholine, hexadecylphosphocholine, phosphoryl-choline, and N-ethylmaleimide (Fig. **229**) [227].

5-fluorouracil   **CMPSA**   **GPC**

AP3   **AP4**

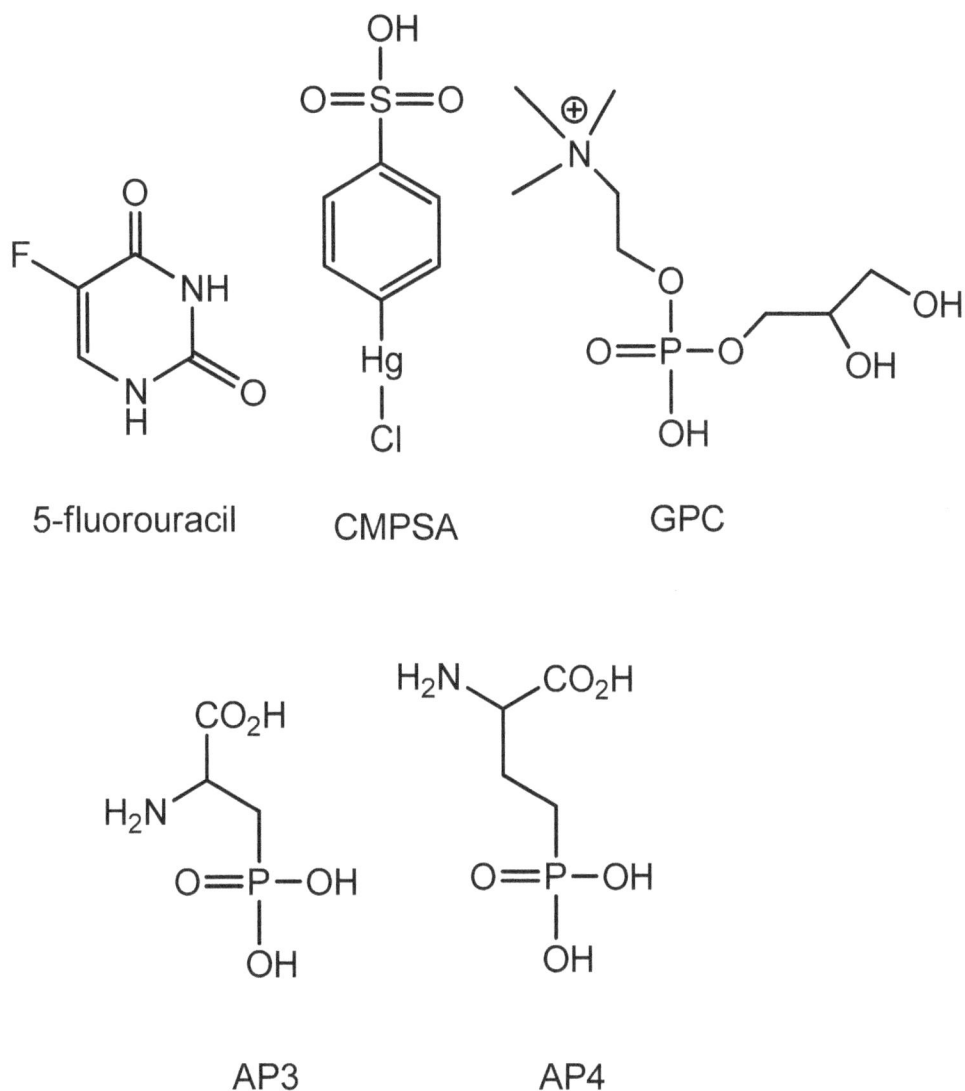

Fig. (229). Chemical structure of phosphoserine phosphatase inhibitors.

9. THREONINE BIOSYNTHESIS

Threonine is an amino acid with importance as serine and glycine precursor, and involved in a number of processes such as immune system, collagen production and muscle support. It is biosynthesized in bacteria, fungi and plants, but not in vertebrates, and therefore must be obtained from the diet. In plants and microorganisms, is synthesized from L-aspartate according to Fig. (**230**), consisting in phosphorylation of the carboxylic group under the catalysis of the enzyme aspartate kinase (ASK) resulting in the formation of L-aspartyl phosphate. The next step catalysed by the enzyme aspartyl semialdehyde dehydrogenase (ASD), involving a dephosphorylation and partial reduction first to form aspartyl semialdehyde and further to the reduced form to provide L-homocysteine, under the catalysis of the enzyme homocysteine dehydrogenase (HDH). Subsequent phosphorylation mediated by homoserine kinase (HK) provides L-homoserine phosphate, which undergoes the last step involving a replacement reaction and phosphate removal.

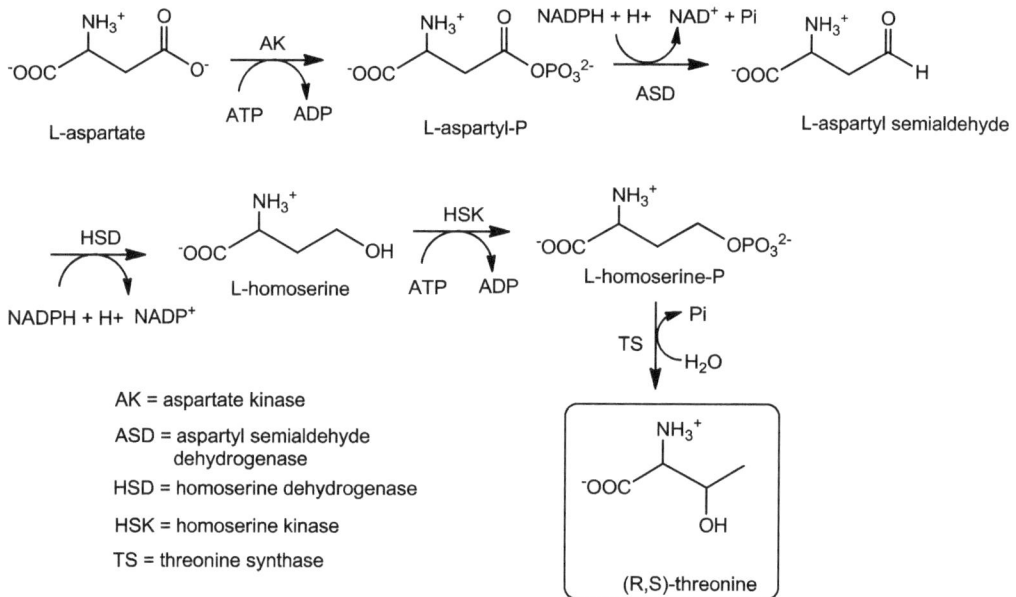

Fig. (230). Biosynthesis of threonine.

9.1. Aspartate Kinase (AK)

Aspartate kinase participates in the biosynthesis of lysine, methionine and threonine, catalysing the first step consisting in the phosphorylation of L-aspartate to produce L-aspartyl-3-phosphate (Fig. **231**).

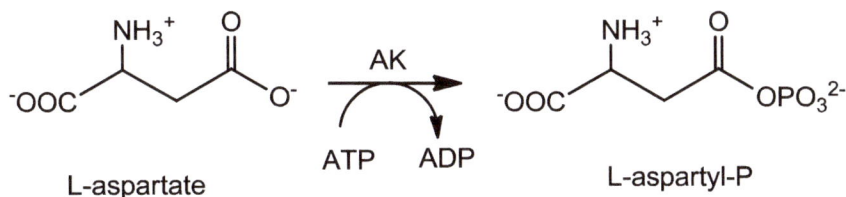

Fig. (231). The aspartate kinase reaction.

The structural architecture of aspartate kinase from *Corynebacterium glutamicum* (AK), *Arabidopsis thaliana* (AK-I), *Bacillus subtilis* (AK-II) and *Escherichia coli* (AK-III), have been elucidated, observing high homology up to 95% between AK and AK-II [152, 291]. The monomer containing the α and β subunits and the threonine region in spheres is illustrated in the ribbon diagram, and the binding site of *Mycobacterium tuberculosis Mt*AKb with the threonine interacting with the residues established (Fig. **232**) [228].

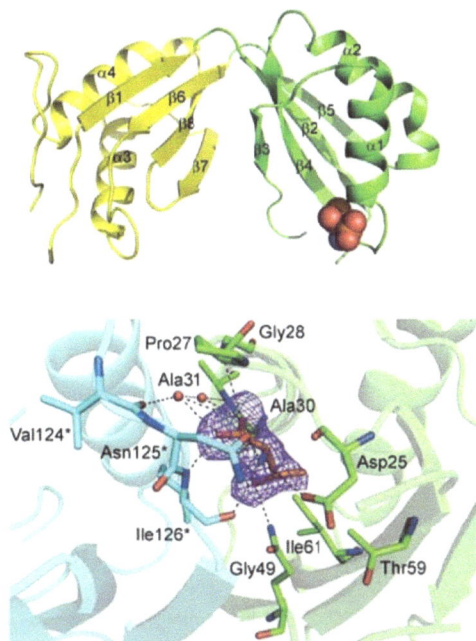

Fig. (232). Ribbon representation of aspartate kinase from Mycobacterium tuberculosis (PDB: 4GO5).

Likewise, the overall structure of the arginine kinase complex containing $\alpha_2\beta_2$ heterotetramer from *Corynebacterium glutamicum* CgAK showing the active site of threonine and lysisne at the β subunit shown in blue. In arrows the brown spheres correspond to the threonine site and the blue spheres the lysine site (Fig. **233**) [229].

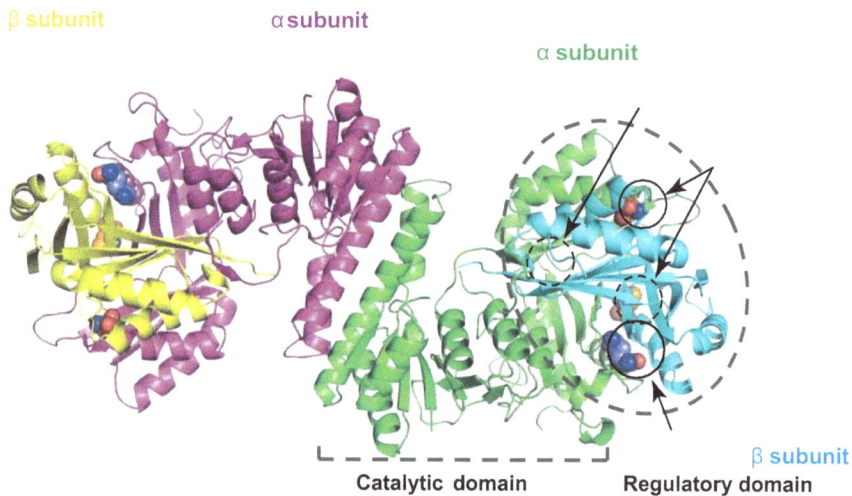

Fig. (233). Ribbon diagram of the arginine kinase complex from *Corynebacterium glutamicum* CgAK showing the active site of threonine and lysisne (PDB: 2DTJ).

9.1.1. Catalytic Domain Regulatory Domain

In general the kinase mechanism follows two transition states, called associative and dissociative. In the associative transition state the nucleophile establish a bond with the phosphorous prior the leaving group is expel, and in the dissociative the bond between the phosphorous atom and the leaving group is largely broken prior the bond formation between the nucleophile and the phosphorous involving a metaphosphate-like intermediate (Fig. **234**) [230].

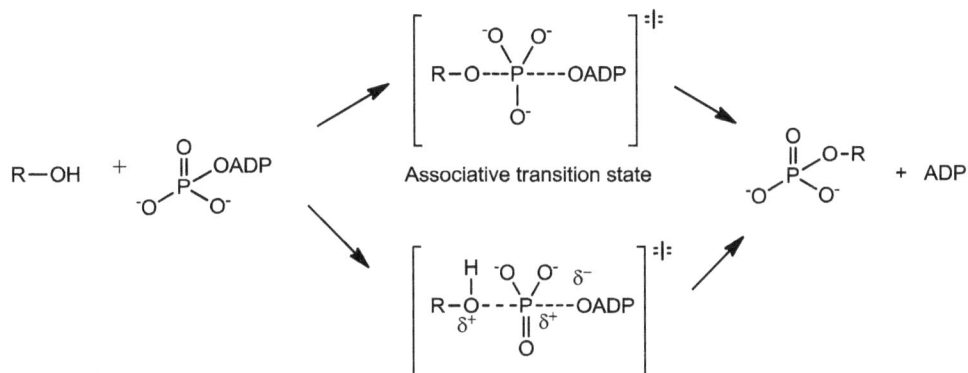

Fig. (234). General transition state mechanism of kinases.

9.2. Aspartate Kinase Inhibitors

The inhibition of aspartate kinase became of interest in drug design because of his implication in the synthesis of L-threonine, L-isoleucine and L-methionine the so-called aspartate family operating in bacteria, fungi and plants.

Antifungal activity has been found for 7-chloro-4([1,3,4]thiadiazol-2-yl sulfonyl)-quinoline derivatives 1-3, inhibiting AK (Hom3p) from *S. cerevisiae* with IC_{50} in µM range, although in liquid media did not affect growth neither in *E. coli*, or Candida strains (Fig. **235**) [231].

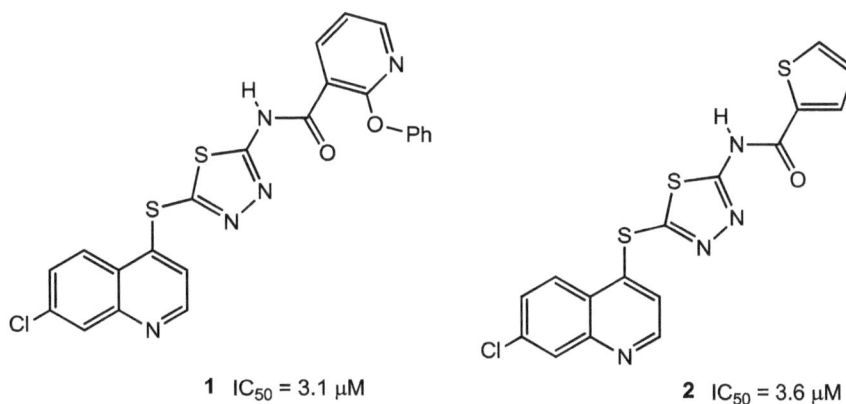

1 IC_{50} = 3.1 µM   **2** IC_{50} = 3.6 µM

Fig. 235 cont.....

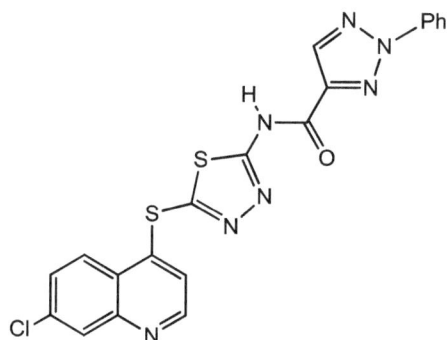

3 $IC_{50} = 1.6\ \mu M$

Fig. (235). Chemical structure and inhibition constants of thiadiazol-2-yl sulfonyl quinoline derivatives.

9.3. Aspartate-β-Semialdehyde Dehydrogenase (ASD)

Is a NADP-dependent enzyme with the role of catalysing the reductive phosphorylation of aspartyl-β-phosphate to produce aspartate β-semialdehyde (Fig. **236**).

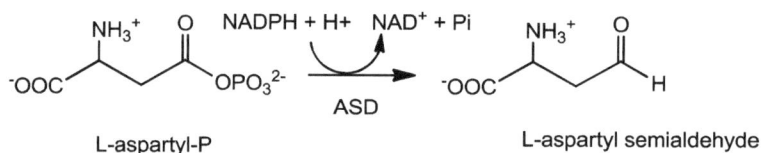

Fig. (236). The aspartate-β-semialdehyde dehydrogenase reaction.

The proposed mechanism starts with a nucleophilic attack from cysteine residue to the ketone of L-aspartyl-P to give a tetrahedral intermediate generating an alkoxide, which will promote the expelling of the phosphate group. Next, the NADPH mediates a hydride reduction of the thioester producing a tetrahedral intermediate and the alkoxide formed will effects the cleavage of the cysteine residue affording the aspartate β-semialdehyde (Fig. **237**) [232].

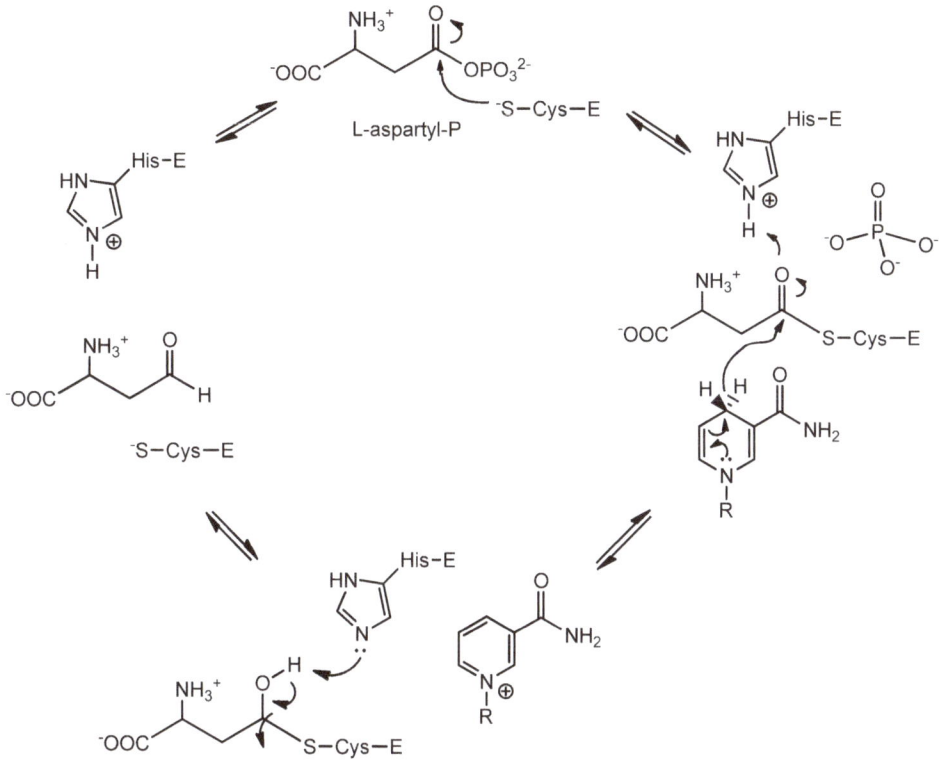

Fig. (237). Proposed mechanism for aspartate β-semialdehyde formation.

The overall structure of ASD from *S. pneumoniae* has been solved as homodimer and the active site complexed with 2',5'-ADP, and NADP showing the main interactions including electrostatic and hydrogen bonding (Fig. **238**) [233].

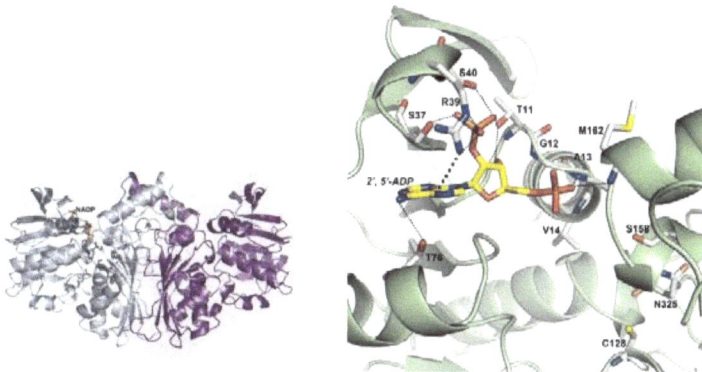

Fig. 238 cont.....

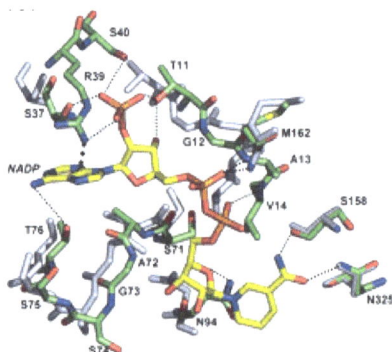

Fig. (238). Crystal structure of ASD from *S. pneumoniae* and binding site region (PDB: 2GZ3).

9.4. Aspartate-β-Semialdehyde Dehydrogenase Inhibitors

Aspartyl phoshonate derivatives were synthesized and evaluated as ASD inhibitors against *Escherichia coli* observing different patterns and moderate inhibition values (Fig. **239**) [234, 235]

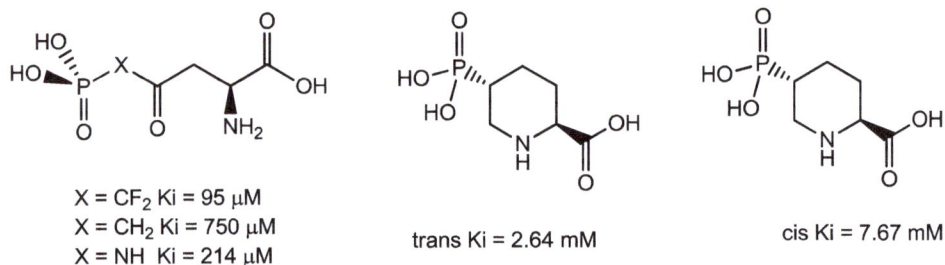

X = CF$_2$ Ki = 95 μM
X = CH$_2$ Ki = 750 μM
X = NH Ki = 214 μM

trans Ki = 2.64 mM

cis Ki = 7.67 mM

Fig. (239). Chemical structure and inhibition values of aspartyl phoshonate derivatives.

L-cystine was evaluated as ASD inhibitor, observing a 100% inhibition at 10 μM concentration and assays with L-[^{35}S]cystine were conducted to determine that the residue involved in the inhibition was ^{135}Cys (Fig. **240**) [236].

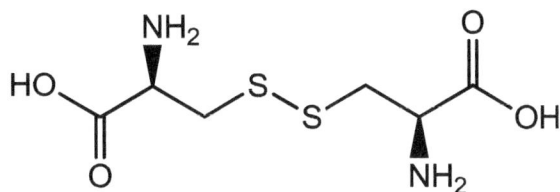

Fig. (240). Chemical structure of L-cystine.

9.5. Homoserine Dehydrogenase (HSD)

This regulatory enzyme participates in the third step of threonine biosynthesis converting L-aspartyl semialdehyde into L-homoserine in the presence of NAD(P)H as reducing cofactor (Fig. **241**).

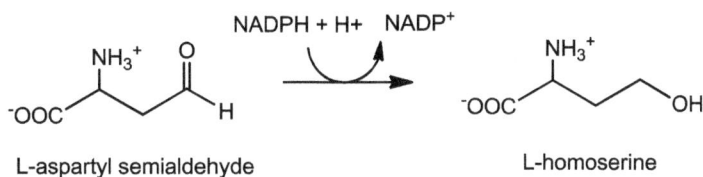

Fig. (241). The homoserine dehydrogenase reaction.

The reaction mechanism of HSD proposes the participation of Ly99, Lys195, and Asp191 in the reduction of L-aspartyl semialdehyde by NADPH to yield L-homoserine and $NADP^+$ (Fig. **242**) [237].

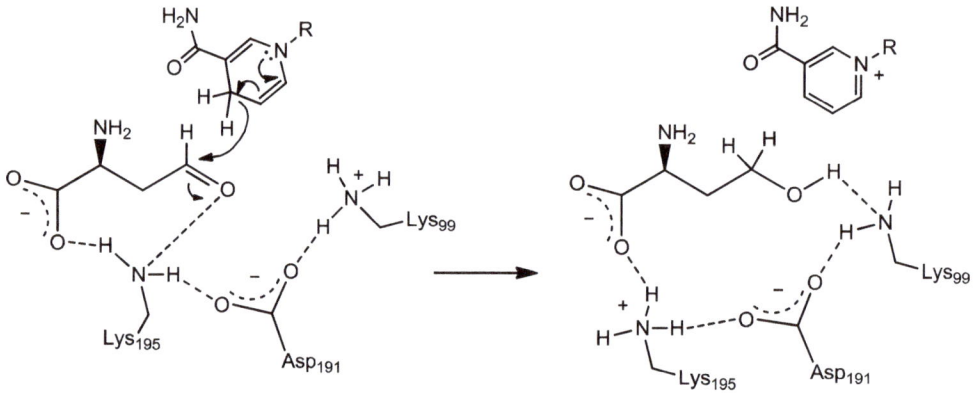

Fig. (242). Proposed mechanism for the conversion of L-aspartyl semialdehyde to L-homoserine.

The overall structure of homoserine dehydrogenase from *Thermus thermophiles* (TtSDH) has been determined, in which the substrate binding domains and the hydrogen bond interactions between NADPH and the residues are represented (Fig. **243**) [237].

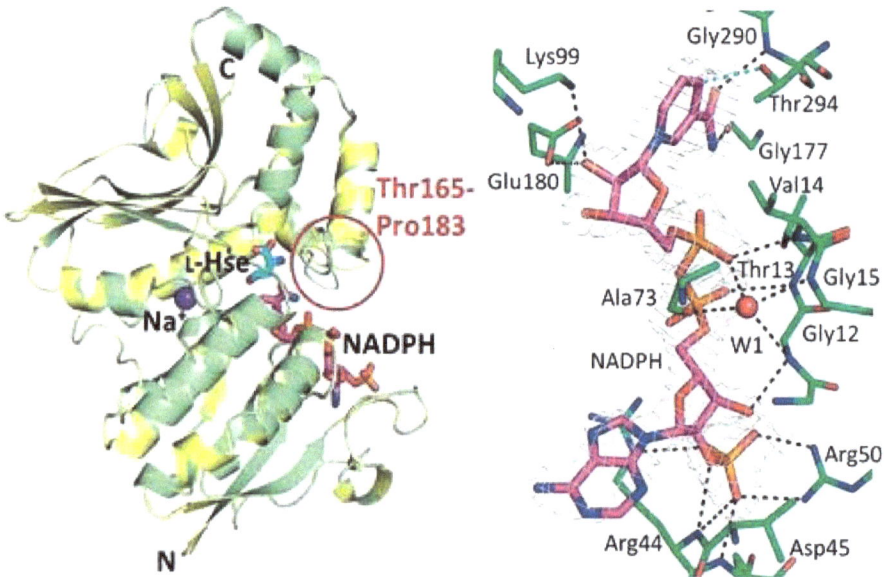

Fig. (243). Ribbon diagram and binding site region of homoserine dehydrogenase from *Thermus thermophiles* (PDB: 6A0R).

9.6. Homoserine Dehydrogenase Inhibitors

The natural amino acid derivative (S)-2-amino-4-oxo-5-hydroxypentanoic acid (RI-331) also known as 5-hydroxy-4-oxo-norvaline (HON) isolated from *Streptomyces sp* was described as potent antifungal agent for the treatment of systemic mycoses with K_i of 2 µM [238, 239]. Additionally, high-throughput screen of a library result in the identification of 3-tert-butyl-4-hydroxyphenyl sulphide displaying an IC_{50} value of 7.3 µM (Fig. **244**).

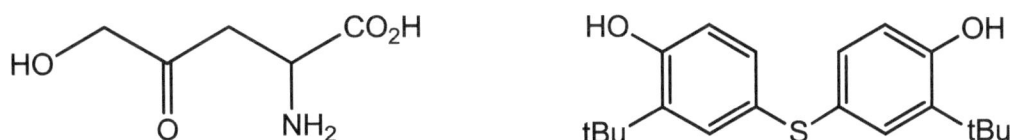

Fig. (244). Chemical structure and inhibition constants of (S)-2-amino-4-oxo-5-hydroxypentanoic acid 3-ter--butyl-4-hydroxyphenyl sulphide.

9.7. Homoserine Kinase (HSK)

This enzyme belongs to the superfamily of GHMP kinases, involved in the fourth step of the threonine biosynthesis consisting in the phosphorylation of L-homoserine to yield L-homoserine phosphate, requiring ATP as a phosphate source (Fig. **245**).

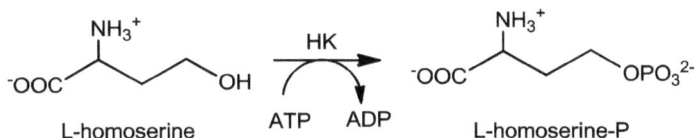

Fig. (245). The homoserine kinase reaction.

The main interactions between HSK and the substrate L-homoserine are established through Arg235, Asn17, Asp23, Asp140 residues, and with ATP through interactions with Ser101, Asn62, Lys87, Val63, Thr183, Glu130 and Mg^{2+} ion. Once the residues establish the polar contacts with the L-homoserine and ATP, the hydroxyl group of L-homoserine attack the terminal phosphate group giving as result the phosphorylation of the primary alcohol (Fig. **246**) [240].

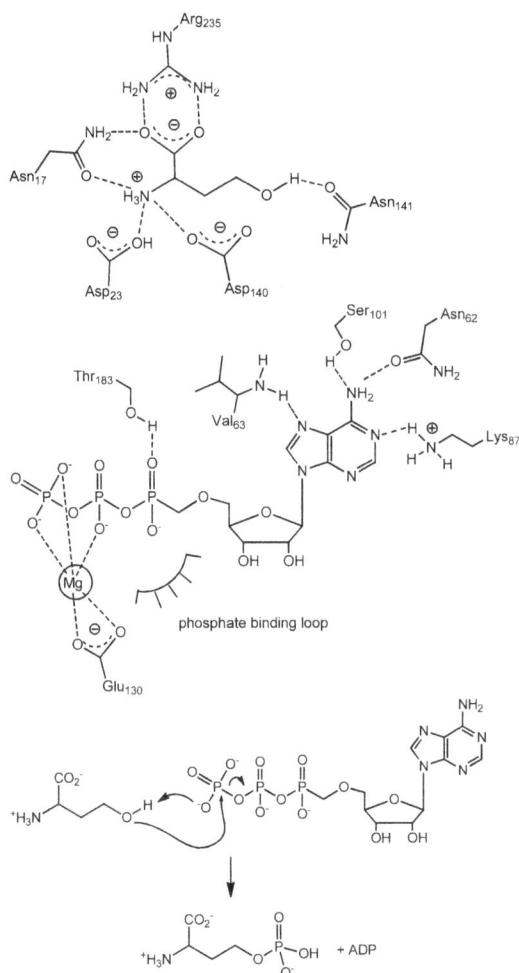

Fig. (246). The main interactions and reaction mechanism for L-homoserine phosphorylation reaction.

9.8. Homoserine Kinase Inhibitors

Inhibition of homoserine kinase from *S. cerevisiae*, *S. pombe* and *C. neoformans* were identified from libraries Prestwick, ChemDiv and BIOMOL, and the hit compounds were validated by the IC_{50} and confirmed by Ki determination. Thus, the compounds 1-4 having coumarin, pyrazole, benzamide and flavonoid rings were determined as competitive inhibitors (Fig. **247**) [241].

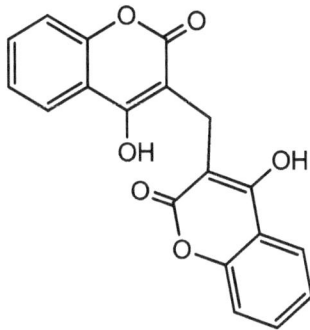

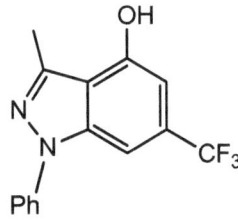

1 Ki = 0.46 μM

2 Ki = 8.6 μM

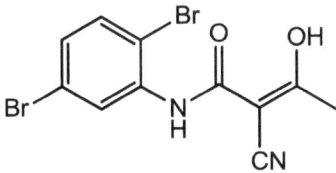

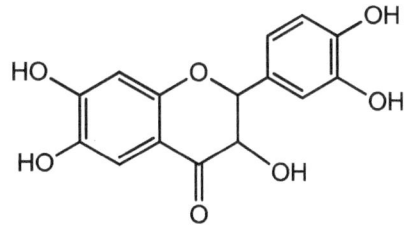

3 Ki = 12 μM

4 Ki = 27 μM

Fig. (247). Chemical structure and KS inhibition constants of compound 1-4.

9.9. Threonine Synthase (TS)

Threonine synthase is involved in the last step of the threonine biosynthesis, consisting in the dephosphorylation reaction and conjugated addition of water to the beta position producing L-threonine (Fig. **248**).

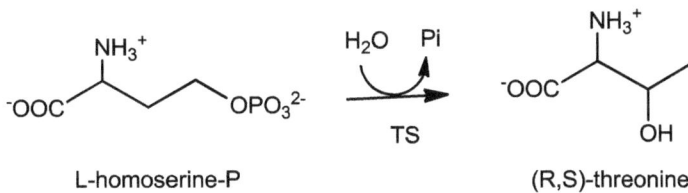

L-homoserine-P   (R,S)-threonine

Fig. (248). The threonine synthase reaction.

The main interactions of the external aldimine composed by L-homoserine-P-PLP and residues Asp, Arg and Ser oriented with the phosphate group, Lys toward the imminiun position and Thr, Ser with the carboxylate (Fig. **249**) [242].

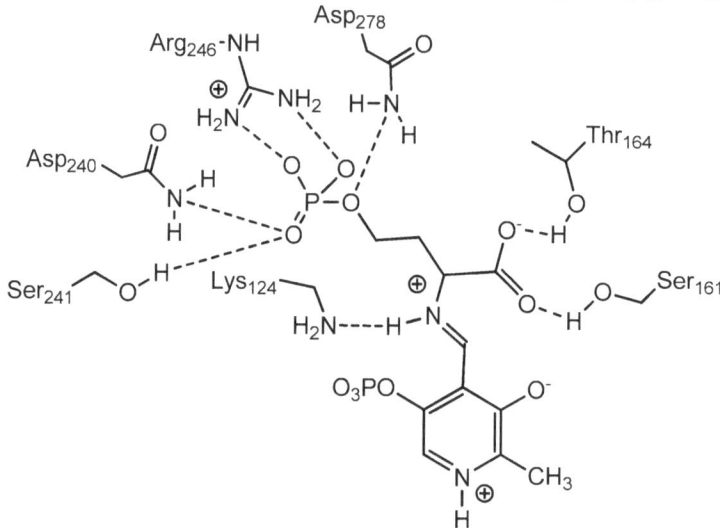

Fig. (249). Interactions established between the residues with L-homoserine-P-PLP external aldimine.

The key intermediates are the external aldimide I, vinylglycine ketamine intermediate III, and E-aminocrotonate IV. More extended and detailed mechanism has been also described (Fig. **250**) [243].

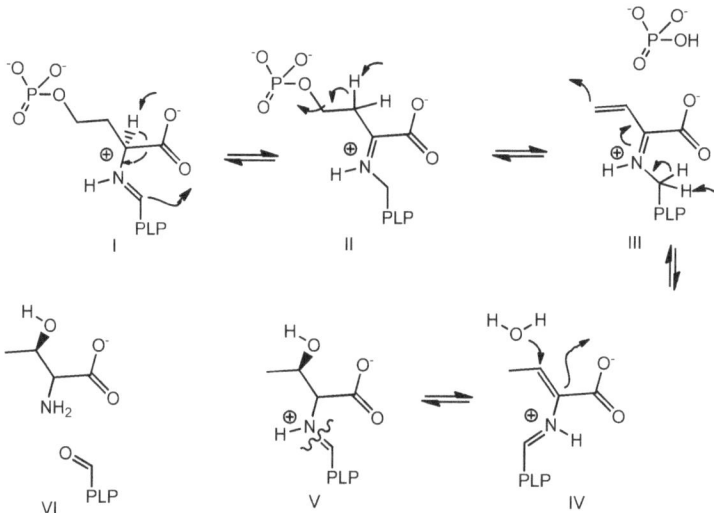

Fig. (250). Observed intermediates found in threonine formation.

9.10. Threonine Synthase Inhibitors

Naturally occurring phosphonate antibiotics rhizocticins (RZs) and plumbemycins (PBs) were isolated from culture media of *Bacillus subtilis* and *Streptomyces plumbeus* (Fig. **251**), and described as threonine synthase inhibitors. The antifungal activity of RZ-A against *Candida albicans, Saccharomyces cerevisiae*, was evaluated, observing antifungal inhibition at MIC = 0.35 µg/ml, although antimicrobial was not significant for PB and RZ [244].

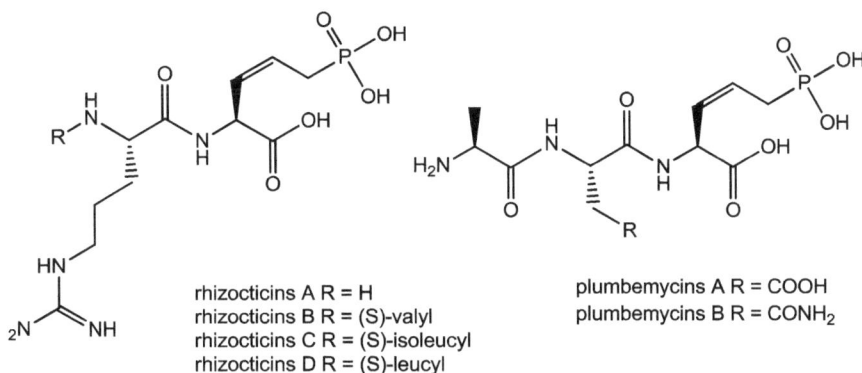

rhizocticins A R = H
rhizocticins B R = (S)-valyl
rhizocticins C R = (S)-isoleucyl
rhizocticins D R = (S)-leucyl

plumbemycins A R = COOH
plumbemycins B R = CONH$_2$

Fig. (251). Chemical structure of phosphonate antibiotics rhizocticins (RZs) and plumbemycins (PBs).

10. CYSTEINE BIOSYNTHESIS

Amino acid cysteine contains a thiomethyl group as substituent which is very important in the disulfide bond formation in proteins, responsible of folding, and stability. Also participates in the formation of glutathione which is an important antioxidant that help to prevent cellular damage due the reactive oxygen species (ROS) formation.

It is biosynthetically prepared from serine, and occurs in bacteria and plants through series the steps involving transesterification between serine and acetyl-CoA, catalysed by serine acetyltransferase (SAT), providing *O*-acetylserine. The next step is catalysed by *O*-acetylserine sulfhydrylase (O-ASS) and implies the generation of SH- coming from H$_2$S or thiosulfate as a source, effecting a nucleophile displacement of acetate ion giving as result the amino acid L-cysteine (Fig. **252**) [245, 246].

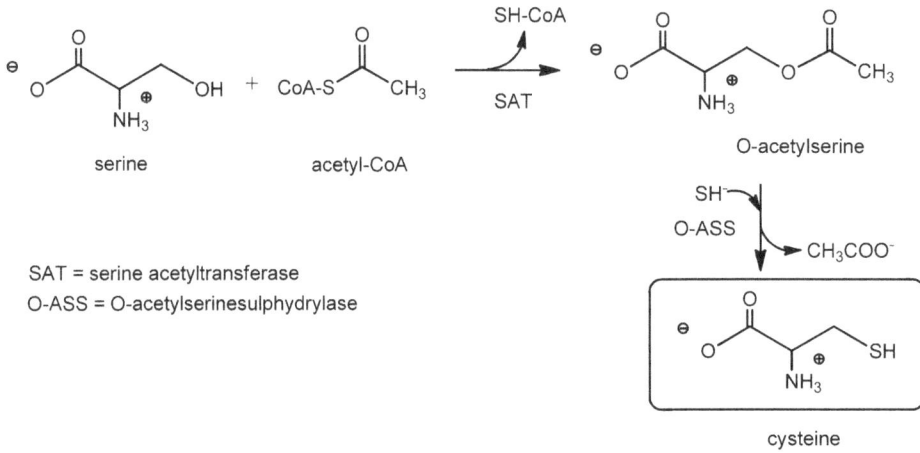

Fig. (252). The L-cysteine biosynthetic pathway from serine.

10.1. Serine Acetyltransferase (SAT)

This enzyme catalyses the initial step in the synthesis of cysteine consisting in the conjugation of serine with acetyl-CoA to generate *O*-acetylserine (Fig. **253**).

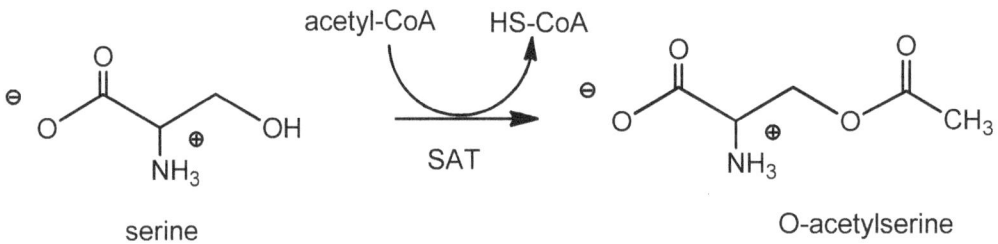

Fig. (253). The serine acetyltransferase reaction.

Enzyme SAT belongs to a O-acetyltransferase subfamily composed by a dimer of trimers (right panel), each monomer (left panel) consisting on an amino-terminal α-helical domain and carboxyl-terminal left-handed β-helix (Fig. **254**) [247].

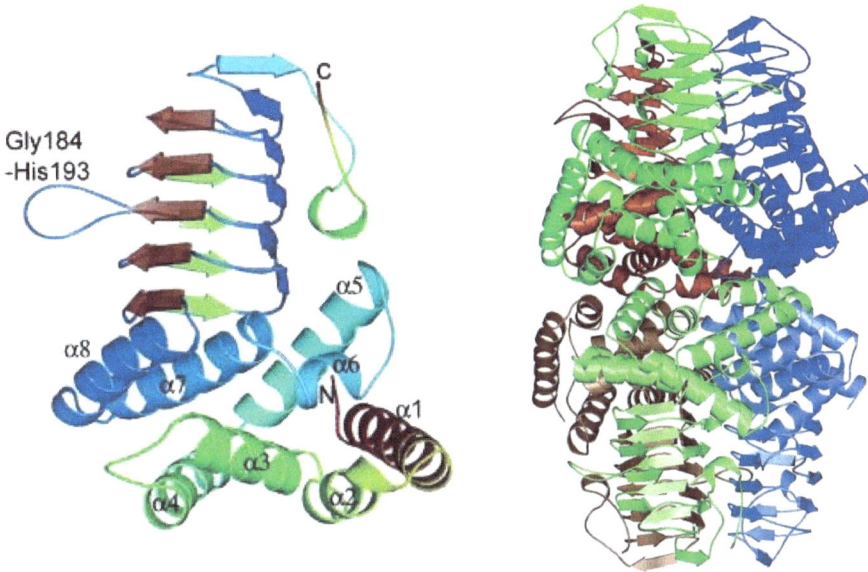

Fig. (254). Ribbon representation of *E. coli* serine acetyltransferase as monomer and dimer of trimers (PDB: 1T3D).

The proposed mechanism involves His158 which takes a hydrogen from the alcohol, giving an alkoxide that effects a nucleophilic substitution on acetyl-CoA, passing through a tetrahedral intermediate which ultimately release CoA-SH forming *O*-acetylserine. The ball and stick illustration shows the cysteine binding pocket with the residue surrounding the cysteine substrate (Fig. **255**) [247].

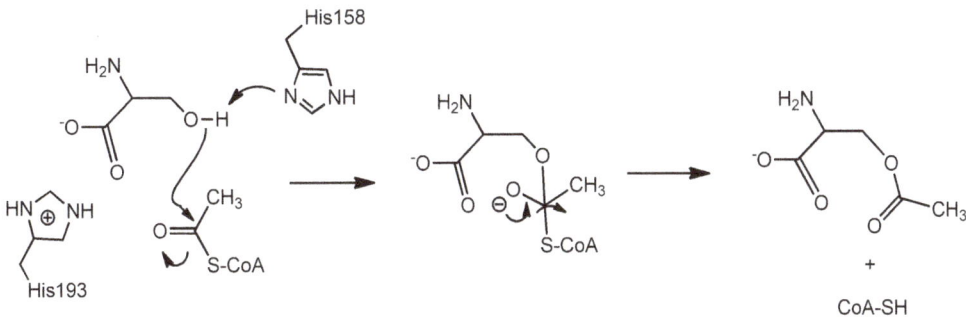

Fig. 255 cont.....

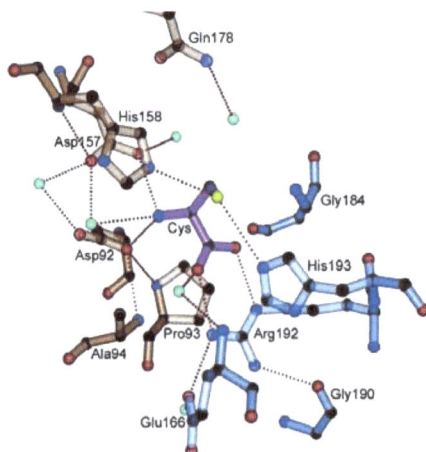

Fig. (255). Proposed mechanism and ball and stick diagram of serine acetyltransferase reaction.

10.2. Serine Acetyltransferase Inhibitors

Small molecule serine acetyltransferase inhibitors with the ability for blocking proliferation of *Entamoeba histolytica* trophozoites have been reported. The selected compounds for inhibition test were the fuse heterocycles NCI128884, NCI29607 and NCI653543 displaying 24%, 38% and 50% (IC_{50} = 72 µM), respectively (Fig. **256**) [248].

Fig. (256). Chemical structure of serine acetyltransferase inhibitors.

10.3. *O*-acetylserine sulfhydrylase (OASS)

This enzyme is not present in human and belongs to a pyridoxal phosphate (PLP) family, which catalyse the transformation of *O*-acetylserine to cysteine, and requires bisulfide ion as sulfur source (Fig. **257**).

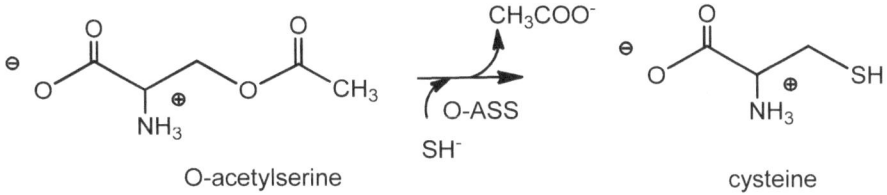

Fig. (257). The *O*-acetylserine sulfhydrylase reaction.

The proposed mechanism proceeds with the iminium salt formation between PLP and a lysine residue followed by nucleophilic addition of *O*-acetylserine producing a conjugate *O*-acetylserine-pyridoxal-enzyme forming a geminal diamine intermediate. A proton located at the quiral center is removed by a lysine residue, promoting a bond cleavage, which result in the release of the acetate group providing a double bond at the conjugate pyridoxal-enzyme germinal diamine intermediate. The addition of bisulfide anion to the double bond leads to the cysteine formation, and the pyridoxal-enzyme conjugate, which is finally cleavage affording cysteine and pyridoxal-enzyme enamine intermediate (Fig. **258**) [249].

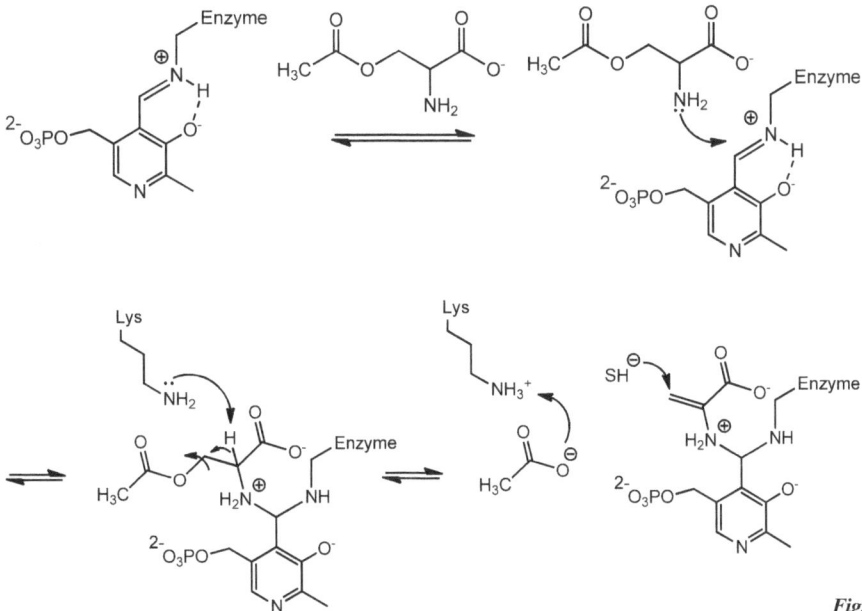

Fig. 258 cont.....

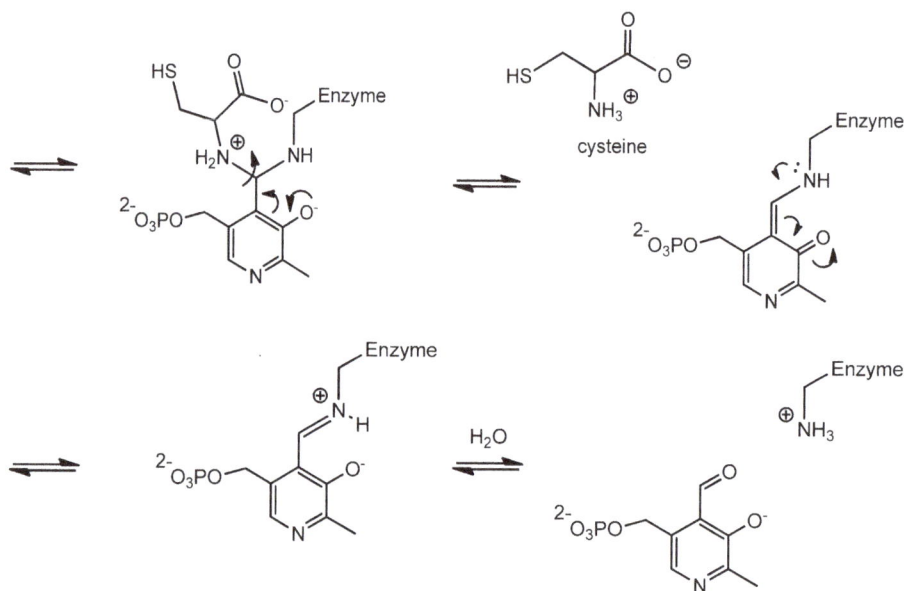

Fig. (258). The proposed mechanism for the OASS reaction.

In bacteria have been identified two isoenzymes *O*-acetylserine sulfhydrylase-A and *O*-acetylserine sulfhydrylase-B with similar binding sites. Structural comparison of OASS-A and OASS-B from *Salmonella typhimurium* (PDB 1OAS and 2JC3) gave an overall identity of 40.32% and a similarity 56.51% (Fig. **259**) [250].

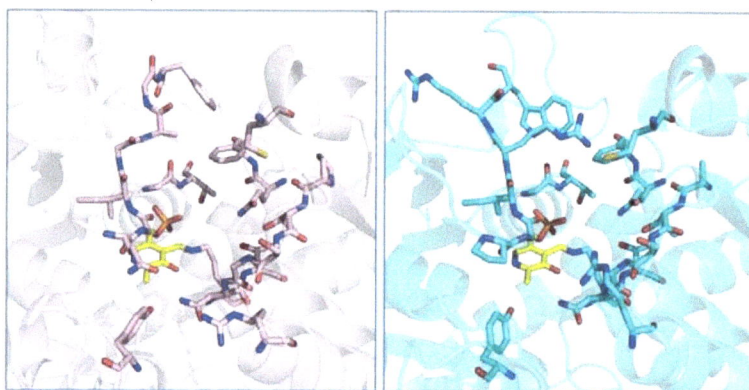

Fig. (259). Structural comparison of OASS-A and OASS-B from *Salmonella typhimurium* (PDB 1OAS and 2JC3).

10.4. *O*-acetylserine Sulfhydrylase Inhibitors

The fact that OASS is only present in prokaryotes and protozoa make this enzyme a suitable target for the development of antibiotics. A thorough study on pentapeptides has been launched including *in silico* and docking analysis, with the aim to determine the predicted affinity and to correlate with experimental dissociation constants. The C-terminal peptides MNLNI, MNWNI, MNYDI, MNENI (Fig. **260**) were subjected to combined docking-scoring analysis to predict the free binding energy interacting with HiOAAA-A active site and the free energy predictions verified with experimental determinations [251].

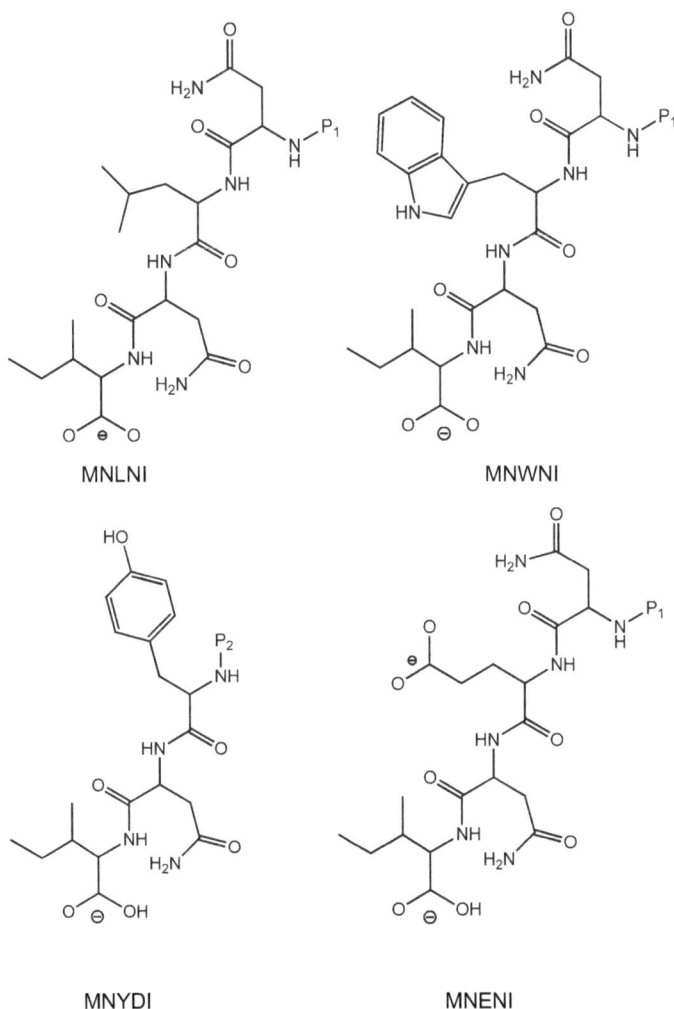

Fig. (260). Chemical structure of C-terminal peptides MNLNI, MNWNI, MNYDI, MNENI.

The inhibitory effect of fluoro alanine (F-Ala) and trifluoro alanine (triF-Ala) on OASS-A and OASS-B was evaluated, observing for F-Ala a competitive inhibition with the substrate along with slow degradation, while for triF-Ala an irreversible inactivation at 50 mM during more than 70 h [252]. Another class of small molecule inhibitors described were the trans-2-substituted-cyclopropa-e-1-carboxylic acids (CPCA) displaying the highest IC_{50} of 700 μM for the compound bearing the alkene substituent (Fig. **261**) [253].

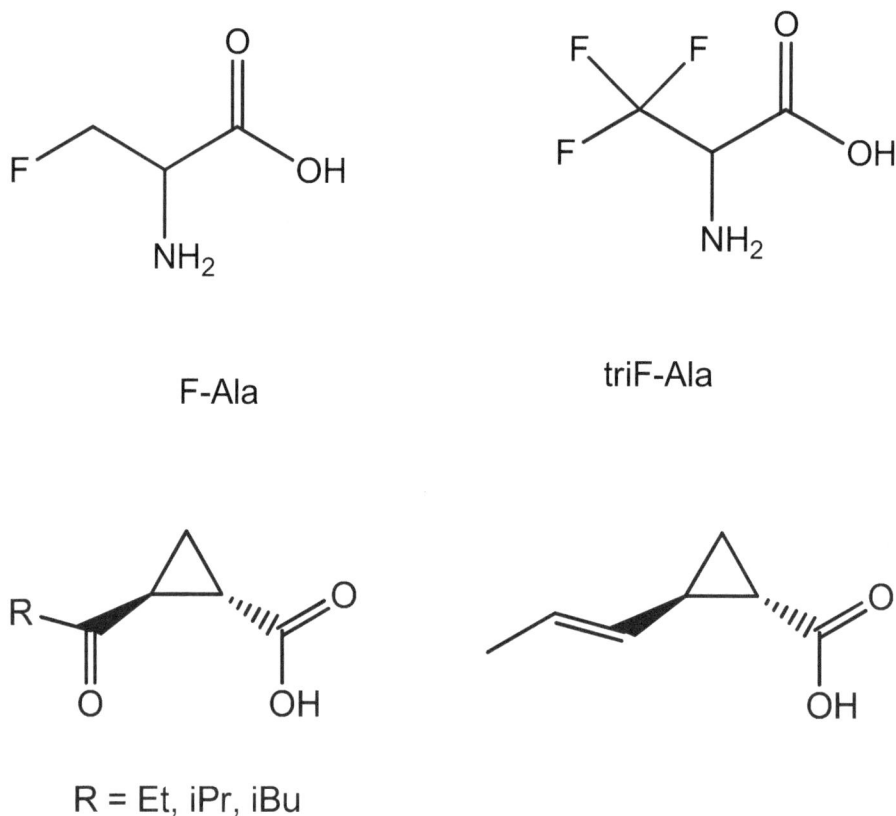

F-Ala

triF-Ala

R = Et, iPr, iBu

Fig. (261). Chemical structure of *O*-acetylserine sulfhydrylase inhibitors.

Another route for cysteine biosynthesis is known as reverse-transsulfuration (RTS) which occurs in fungi and mammals [254] consisting in the attachment of methionine with ATP, catalysed by methionine adenosy transferase (MAT) to form S-adenosylmethionine. Further de-methylation reaction takes place under the the catalysis of MAT-methylase giving as result S-adenosycysteine. The next step involves the cleavage and release of homocysteine from adenosine catalysed by the enzyme S-adenosyl homocysteine hydrolase. The resulting homocysteine may follow two pathways, being on one side the completion of the full cycle producing methionine under the catalysis of methionine synthase (MS) or alternatively to condense with serine to produce cystathionine which finally will be converted to cysteine by the catalysis of cystathionine gamma-lyase (CGL) (Fig. **262**).

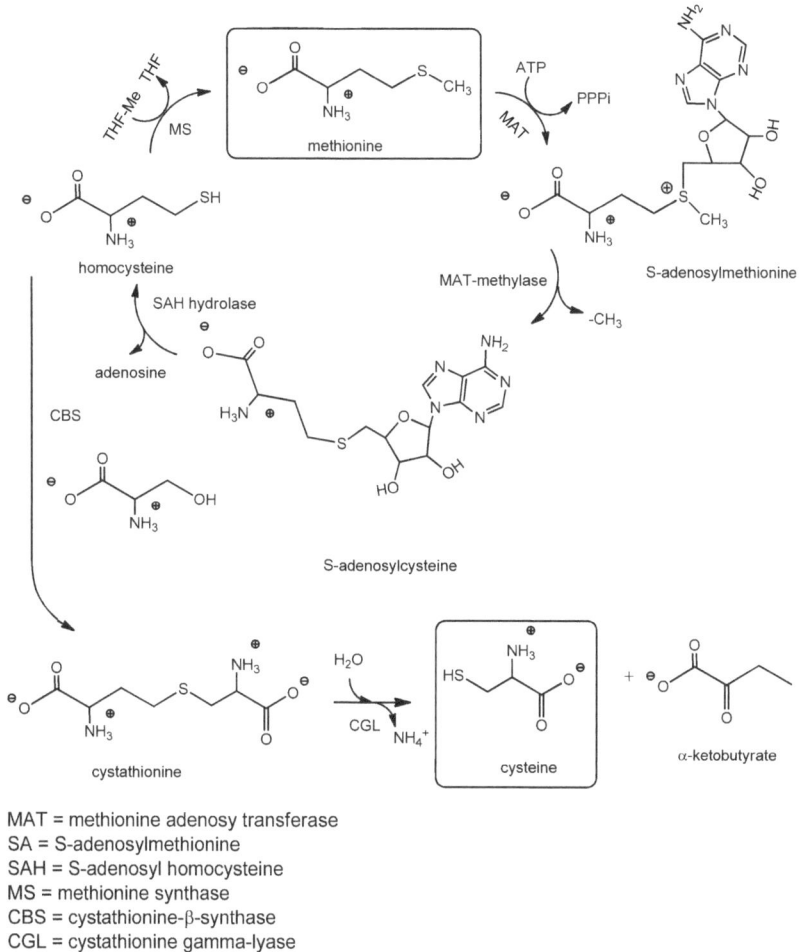

MAT = methionine adenosy transferase
SA = S-adenosylmethionine
SAH = S-adenosyl homocysteine
MS = methionine synthase
CBS = cystathionine-β-synthase
CGL = cystathionine gamma-lyase

Fig. (262). The reverse-transsulfuration (RTS) cycle for cysteine biosynthesis.

L-cysteine is also correlated with the formation of important molecules such as glutathione, which play an important rule in the control of reactive oxygen species (ROS), and amino sulfonic acid taurine playing a wide variety of functions in the central nervous system (Fig. **263**) [255].

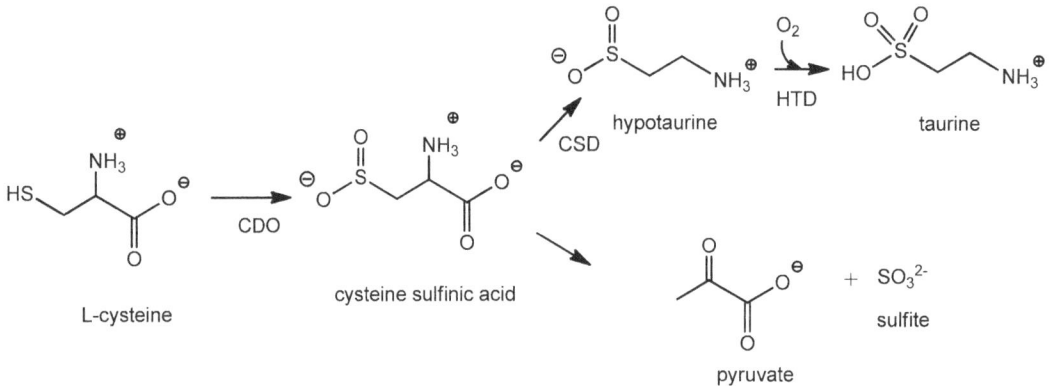

CDO = cysteine deoxygenase
CSD = cysteinesulfinate decarboxylase
HTD = hypotaurine dehydrogenase

Fig. (263). Taurine biosynthesis from L-cysteine.

10.5. Methionine Adenosy Transferase (MAT)

It is an enzyme involved in the formation of S-adenosylmethionine considered the most important donor present in living organism, with high impact in processes such as synthesis of neurotransmitters, phospholipids and polyamide biosynthesis [256, 257].

The reaction in which MAT participates involves the condensation of methionine with ATP affording S-adenosylmethionine, pyrophosphate (PPi) and orthophosphate (Pi) (Fig. **264**).

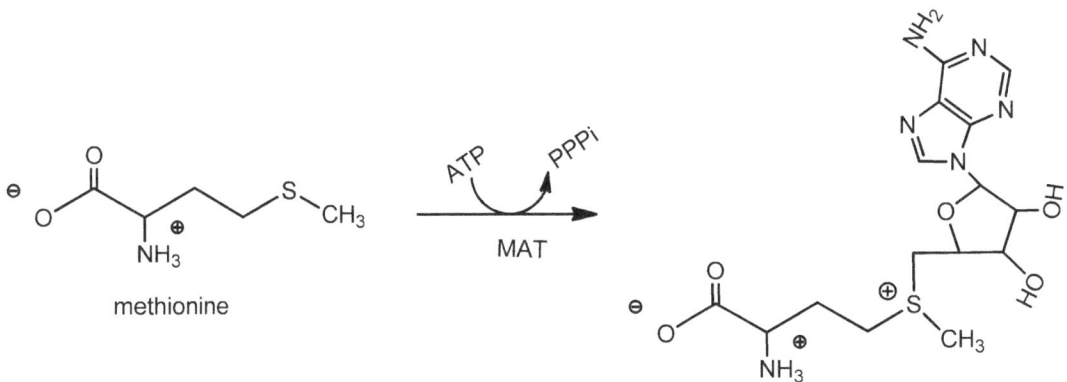

Fig. (264). The methionine adenosy transferase reaction.

Molecular evidence support a mechanism based on SN_2 nucleophilic attack from methionine sulphur at the 5th position of ribose, producing the exit of the triphosphate group as leaving group.

The active site of MAT I complexed with ATP and methionine shows the interactions with the residues and the sulphur coordinated to Mg^{2+} cation (Fig. **265**) [258].

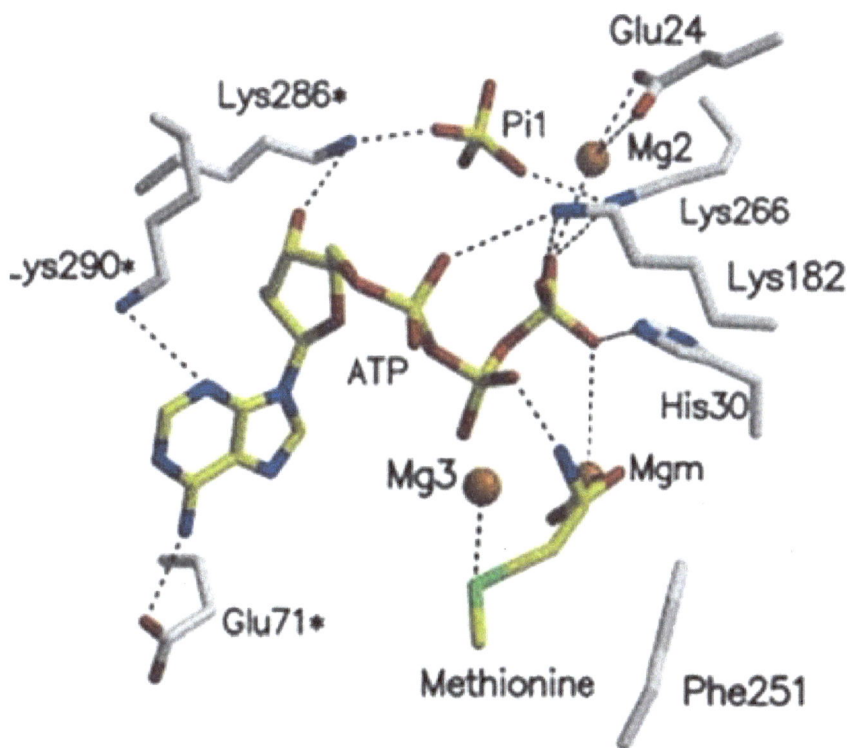

Fig. (265). Ball and stick diagram at the binding site showing the MAT residues interacting with the methionine-ATP complex (PDB: 1O90).

10.6. Methionine Adenosy Transferase Inhibitors

Combinatory analysis using molecular docking and virtual screening of libraries led to the identification of two potent MAT inhibitors denoted as antiMAT-1 and antiMAT-2 producing complete inhibition at 1 mM concentration (Fig. **266**) [259].

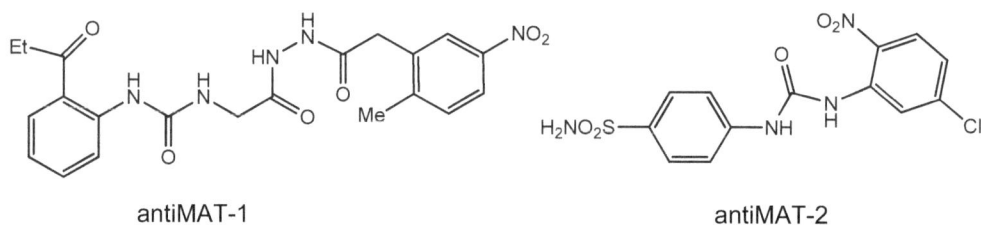

Fig. (266). Chemical structure of MAT inhibitors.

10.7. S-adenosyl homocysteine hydrolase (SAH)

This enzyme participates in the break off of the thiol bond on S-adenosylhomocysteine, giving as result the release homocysteine and adenosine (Fig. **267**).

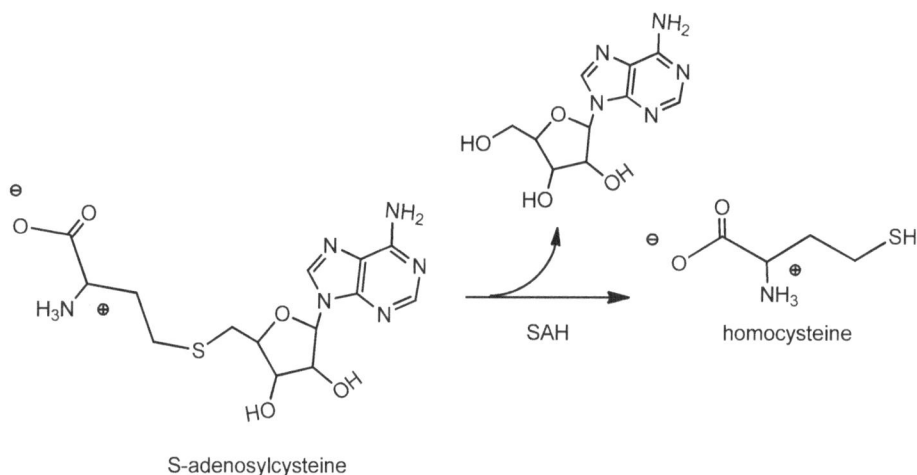

Fig. (267). The S-adenosyl homocysteine hydrolase reaction.

The mechanism proposes an initial NAD^+ promoted oxidation of ribose's at the 3th position to form 3'-keto intermediate. The next step involves an enolate and exocyclic double bond formation, which promotes the homocysteine release. Next, a Michael type addition of water will produce the alcohol and a final reduction of the 3'-keto will provide homocysteine and adenosine (Fig. **268**) [260].

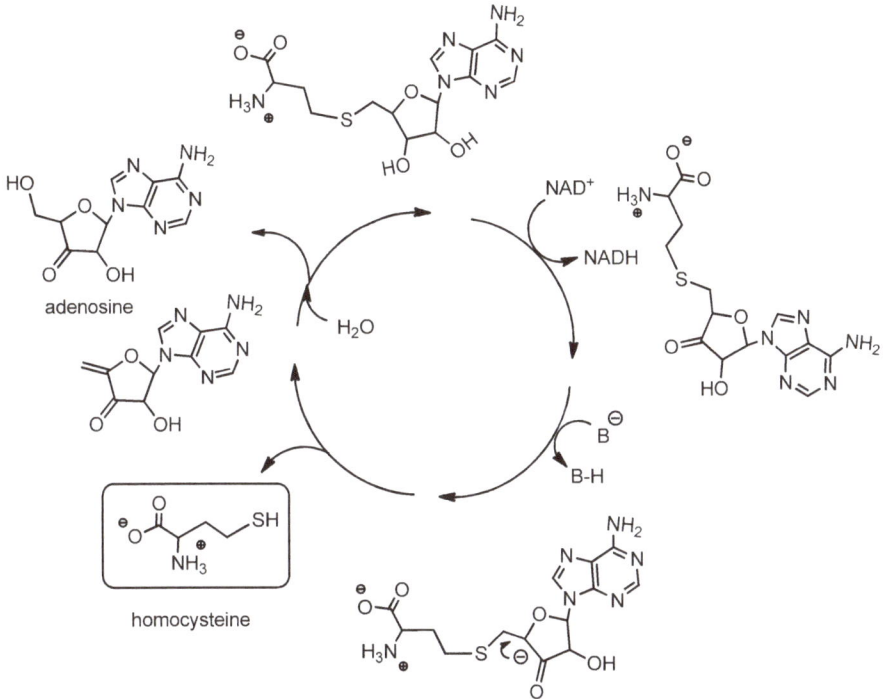

Fig. (268). Proposed mechanism for homocysteine formation.

The crystal structure of human S-adenosylhomocysteine hydrolase (SAH) complexed with carbocyclic nucleoside haloneplanocin A and inhibitor Neplanocin A were described, showing the tetrameric structure, indicating the catalytic, and the cofactor domains. Also the map of hydrophobic interaction and hydrogen bond contacts with the residues are indicated in Fig. (**269**) [261, 262].

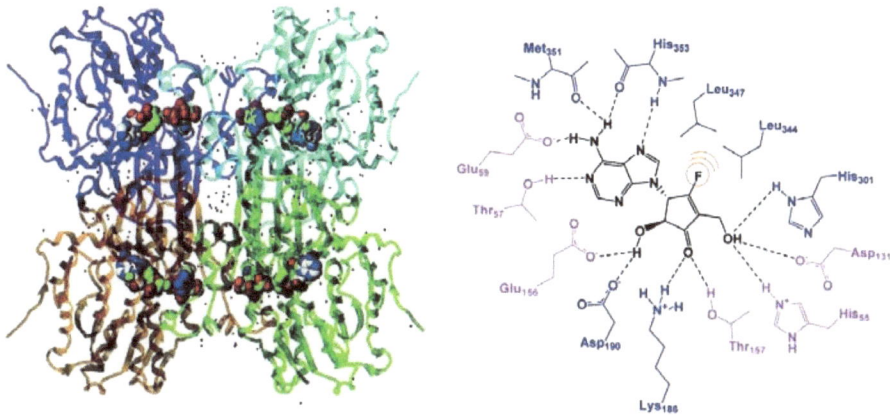

Fig. 269 cont.....

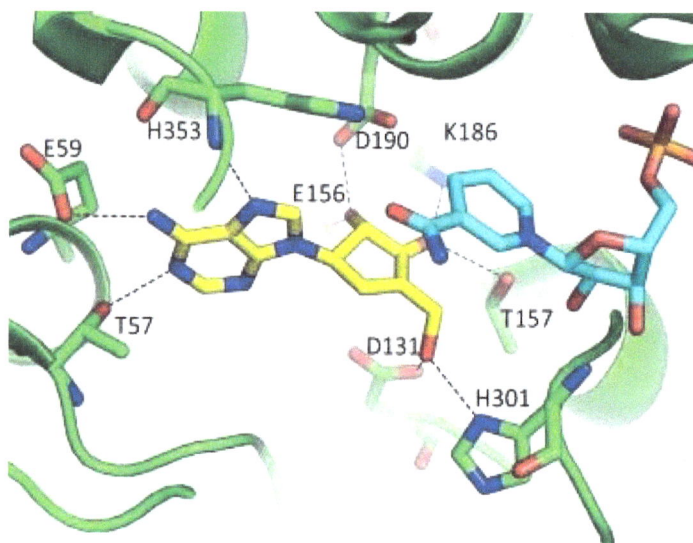

Fig. (269). The crystal structure of human S-adenosylhomocysteine hydrolase complexed with carbocyclic nucleoside haloneplanocin A and inhibitor Neplanocin A (PDB: 3NJ4 and PDB: 1LI4).

10.8. S-Adenosyl Homocysteine Hydrolase Inhibitors

SAH has been a target enzyme for developing small molecule inhibitors because of his implication with Alzheimer disease, due increased serum homocysteine levels detected in patients [263].

Modified nucleosides have been reported to inhibit SAH, such as D-Eritadenine (DEA) and natural product Neplanocin A [261] with IC_{50} values of 7nM and 1.5 nM, although cytotoxic synthoms appear after long treatments. Other modified adenosine analogues evaluated had less significant SAH inhibition activity with IC_{50} values 55 and 250 nM, but improved cell IC_{50} values with 1.5 nM each (Fig. **270**).

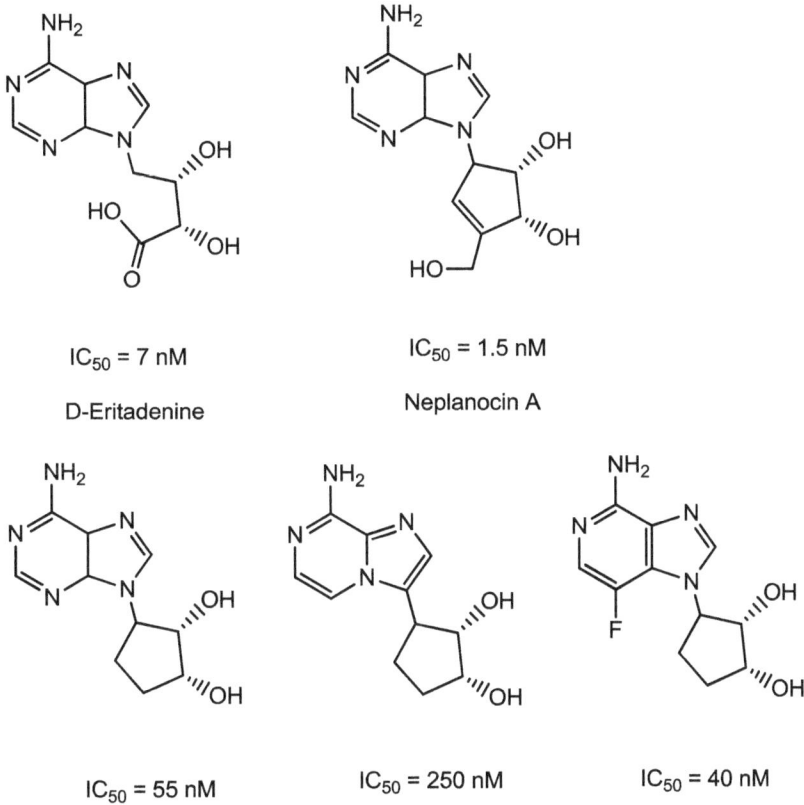

IC$_{50}$ = 7 nM

D-Eritadenine

IC$_{50}$ = 1.5 nM

Neplanocin A

IC$_{50}$ = 55 nM

IC$_{50}$ = 250 nM

IC$_{50}$ = 40 nM

Fig. (270). Chemical structure of S-adenosyl homocysteine hydrolase inhibitors.

10.9. Cystathionine-β-Synthase (CBS)

This enzyme is a pyridoxal 5′-phosphate (PLP) dependent cofactor closely related to cysteine synthase [264] having the role of catalysing the coupling reaction between serine and homocysteine to produce cystathionine (Fig. **271**).

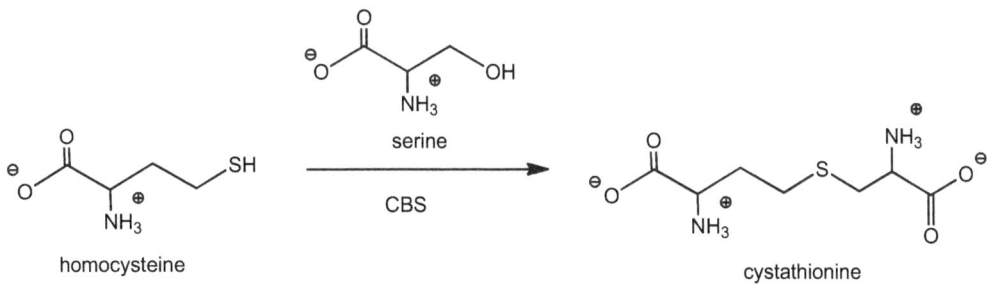

serine

CBS

homocysteine

cystathionine

Fig. (271). The cystathionine-β-synthase reaction.

The reaction mechanism involves the nucleophilic addition of lysine residue to pyridoxal to form the imine which reacts with condense with serine giving E-GD-I. Further release of the lysine residue promoted by the nucleophilic attack resulted in the serine-pyridoxal imine E-ser which establish resonance with the quinonoid form, and after dehydration produce unsaturated aminoacrylate intermediate E-AA. The homocysteine aminoacid effects an addition to give the conjugate E-cyst which is transformed to the aldimide E-GD-II, and then a final cleavage assisted by the lysine residue to yield cystathionine and pyridoxal-lysine imine intermediate which will be able to repeat another cycle (Fig. **272**) [265].

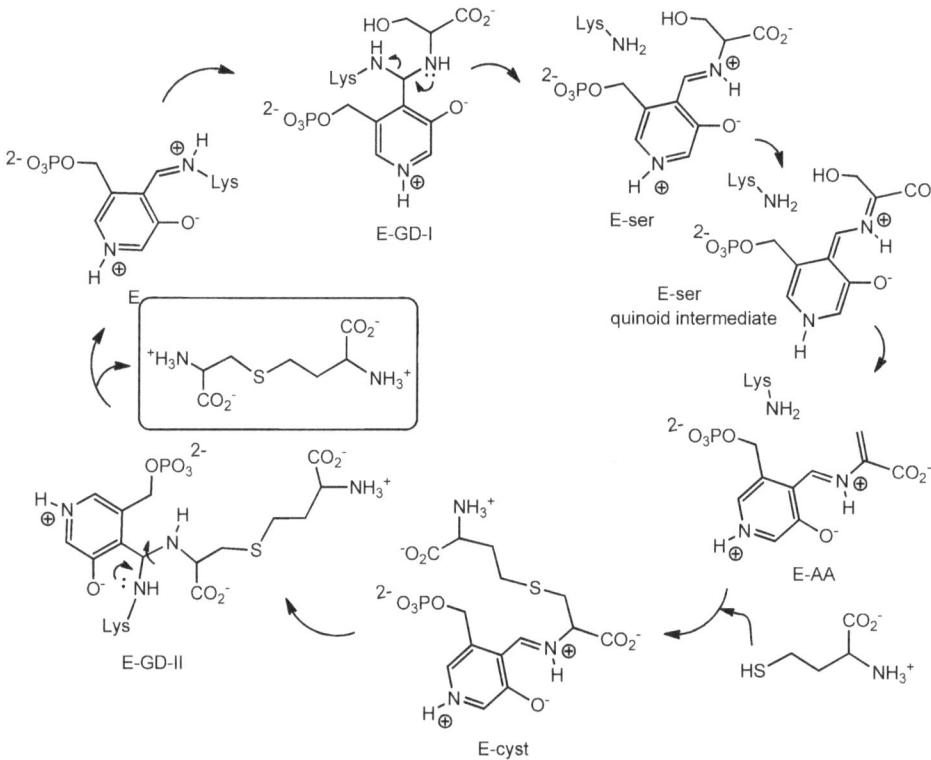

Fig. (272). Proposed reaction mechanism for cystathionine formation.

The structure of cystathionine β-synthase is composed by a heme-binding module, and a pyridoxal phosphate domain hosting the active site. In human CBS the active site has been localized at a cleft between the N-and C-terminal domains in a narrow channel (Fig. **273**) [266].

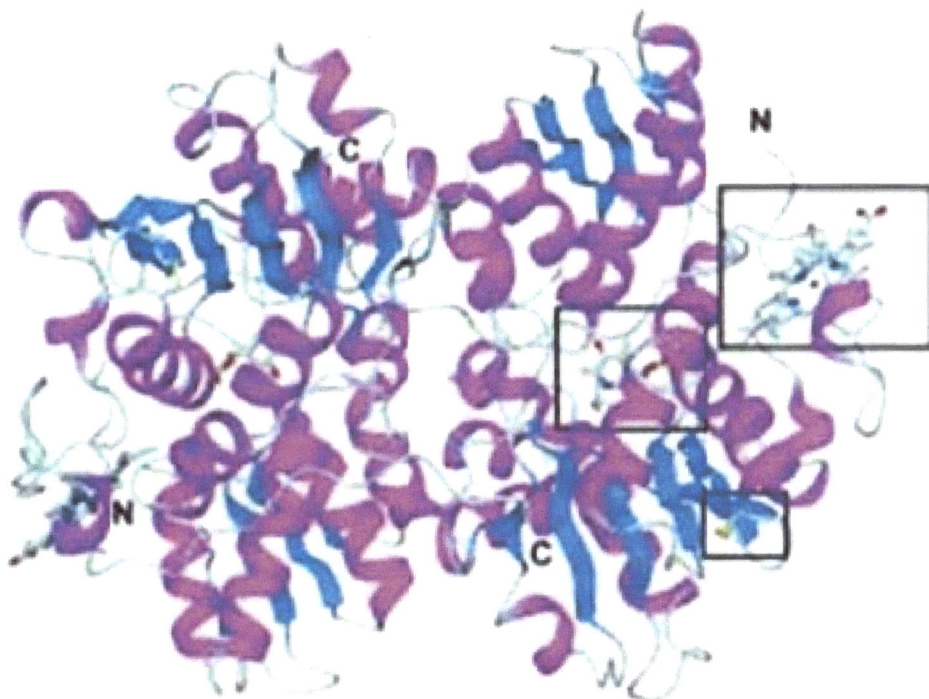

Fig. (273). Crystal structure of human cystathionine β-synthase (PDB: 1JBQ).

10.10. Cystathionine-β-Synthase Inhibitors

CBS and cystathionine γ-lyase enzymes are largely the main producers of hydrogen sulphide (H_2S) considered a gasotransmitter recognized as a signalling molecule in central nervous system (CNS), associated as a risk factor for cardiovascular alterations, artherosclerosis, stroke, and cancer [267, 268]. The reactions leading to H_2S production are cysteine with homocysteine to generate cystathionine, two cysteine molecules self-condensing to give lanthionine, and cysteine hydration to give serine (Fig. **274**) [268].

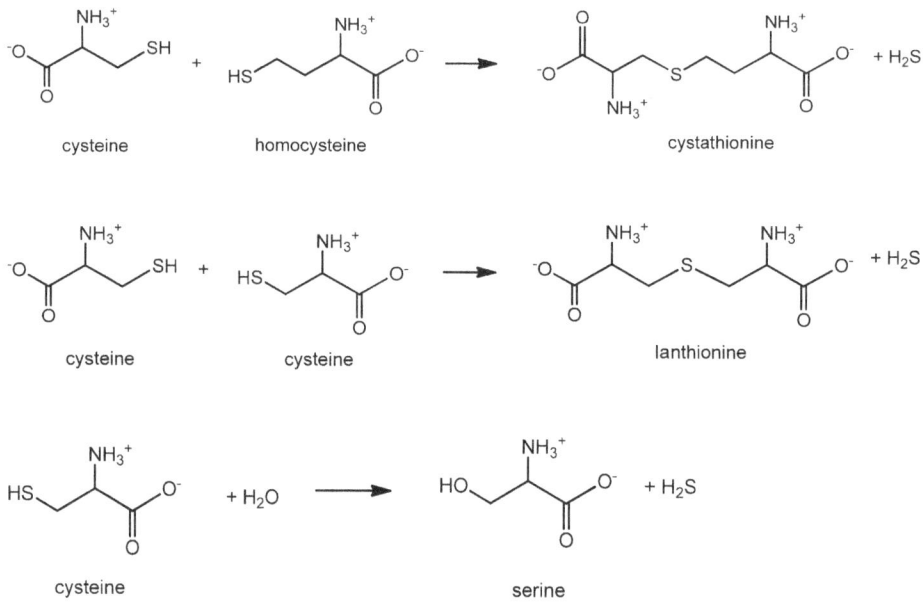

Fig. (274). Reaction leading to gasotransmitter H_2S formation.

Due its involvement in the production of H_2S, CBS has been considered an strategy in the treatment of ischemic injury, stroke, and cancer [269] To achieve this purpose H_2S inhibitors such as aminooxyacetic acid and trifluoroalanine has been evaluated as CBS inhibitors (Fig. **275**) [270].

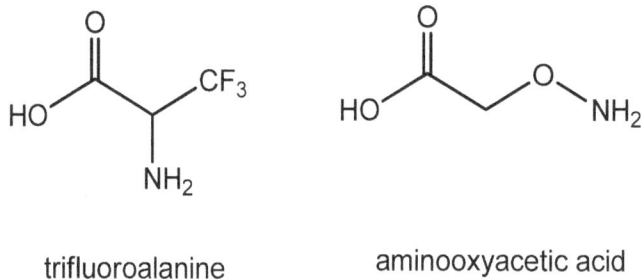

trifluoroalanine   aminooxyacetic acid

Fig. (275). Chemical synthesis of cystathionine-β-synthase inhibitors.

A recent report describes a correlation in overexpression of CBS with proliferation and migration of colorectal cancer cells, and propose the natural product sikokianin C (Fig. **276**), isolated from the roots of *Wikstroemia sikokiana*, as a potent competitive inhibitor [271].

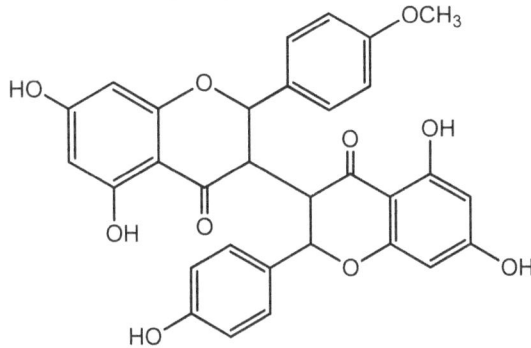

Fig. (276). Chemical structure of sikokianin C.

10.11. Cystathionine Gamma-Lyase (CGL)

This enzyme is present in the methionine pathway, and it is involved in the last step of the transsulfuration pathway consisting in the transformation of cystathionine to produce cysteine. It is together with CBS a hydrogen sulphide produced and it is also involved in the coupling reaction of cysteine to produce lanthionine (Fig. **277**).

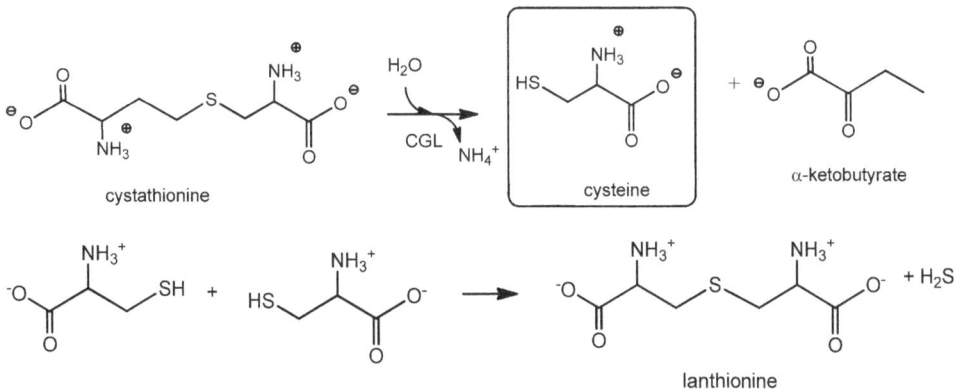

Fig. (277). The cystathionine gamma-lyase reaction.

A mechanism explaining the production of H_2S involves cystathionine gamma-lyase, pyridoxal 5-phosphate (PLP) and cysteine, starting with the addition of L-cysteine to the complex enzyme-PLP, to form the cysteine-PLP complex. Next, a lysine residue acting as a base produce an anion establishing pyridinium-quinonoid forms, driving the reaction to the acrylamine intermediate and H_2S. Final cysteine addition to the double bond and hydrolysis produce lanthionine and pyridoxal 5-phosphate (Fig. **278**) [272].

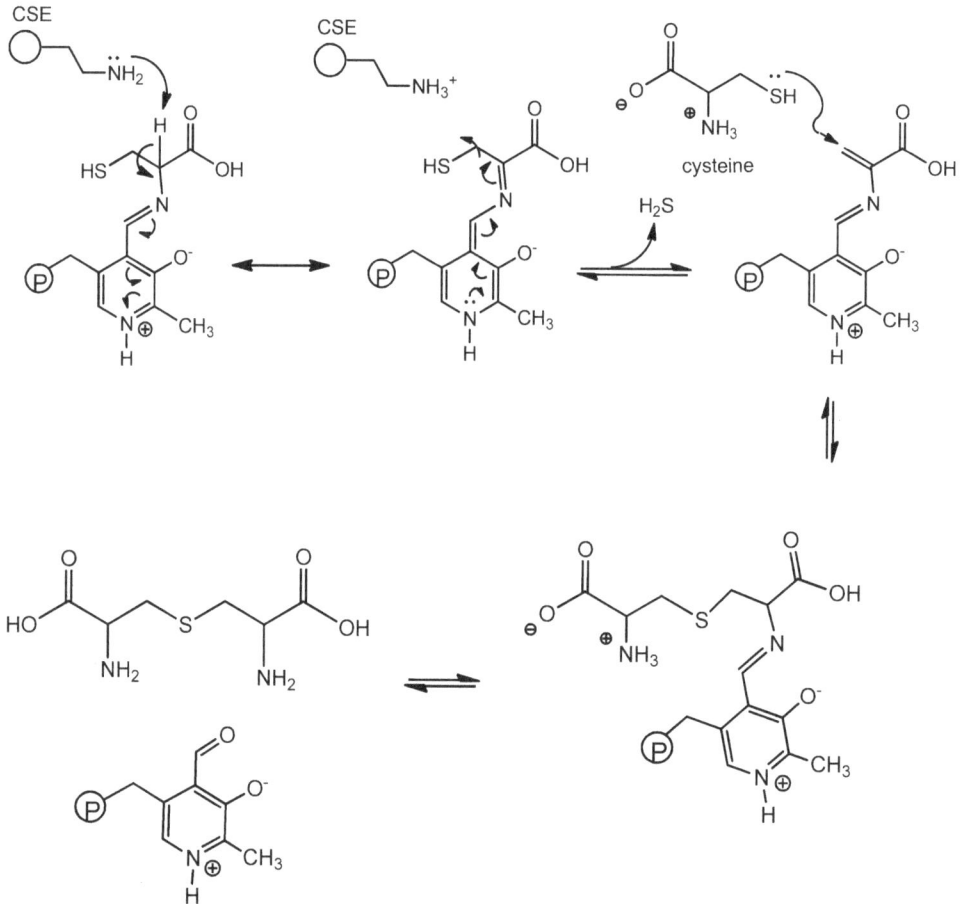

Fig. (278). A proposed mechanism for lanthionine formation.

The crystal structure of human cystathionine γ-lyase in the apo form, complexed with inhibitor PGA was reported, revealing the position at the active site and the residues Arg119, Arg62, Tyr114, and Glu339 establishing interactions (Fig. **279**) [272].

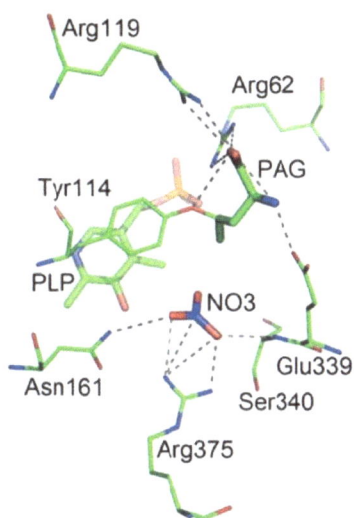

Fig. (279). The crystal structure of human cystathionine γ-lyase in the apo form, complexed with inhibitor PGA (PDB: 3COG).

10.12. Cystathionine Gamma-Lyase Inhibitors

Propargylglycine (PAG) and β-cyanoalanine are modified amonoacids described as specific γ-cystathionase inhibitors [270], and more recently aminoethoxyvinylglycine described as analogue displaying higher potency (Fig. **280**) [273].

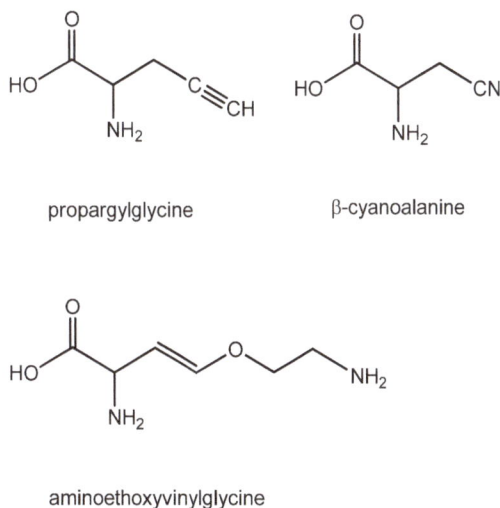

propargylglycine   β-cyanoalanine

aminoethoxyvinylglycine

Fig. (280). Chemical structure of cystathionine gamma-lyase inhibitors.

10.13. Cysteine Deoxygenase (CDO)

This enzyme is responsible for the cysteine catabolism and its activity is dependent on protein diet, therefore high protein content in diet will increase CDO [274]. The CDO is involved in the oxidation of L-cysteine into L-cysteine sulfinic acid (CSA) as shown in Fig. (**281**), a key intermediate in the synthesis of taurine, and pyruvate from L-cysteine. It is identified as a non-heme Fe(II) oxygenase using the iron center and it is present in mammals, and some yeast and bacteria.

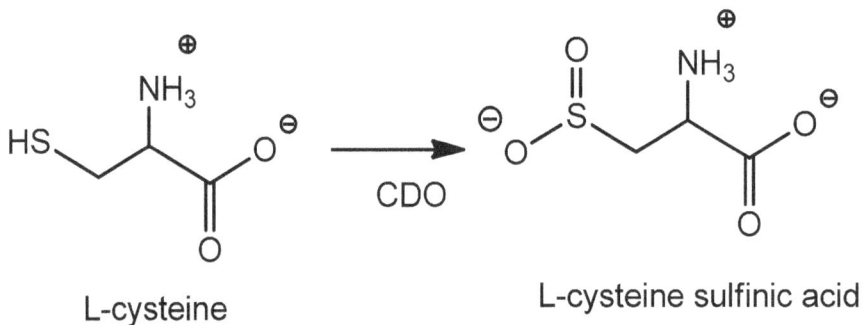

Fig. (281). The cysteine deoxygenase reaction.

The use of cysteine as redox active substrate to form Fe(III)-peroxy intermediate result in four possible mechanism, with each intermediate containing different oxygenated species such as radical peroxide which are responsible of the sulphur oxidation as illustrated in Fig. (**282**) [275].

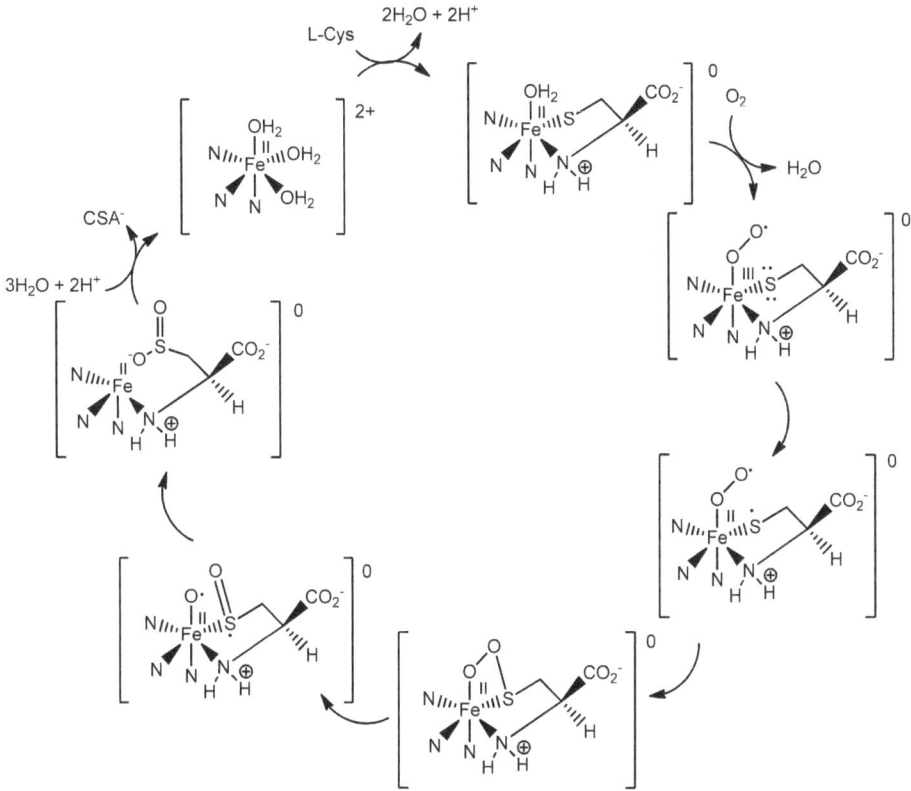

Fig. (282). The iron-peroxy-cysteine intermediates leading to cysteine oxidation.

10.14. Cysteine Sulfinic Acid Decarboxylase (CSAD)

Is a rate-limiting enzyme involved in taurine biosynthesis converting cysteine sulfinic acid into hypotaurine (Fig. **283**) [276].

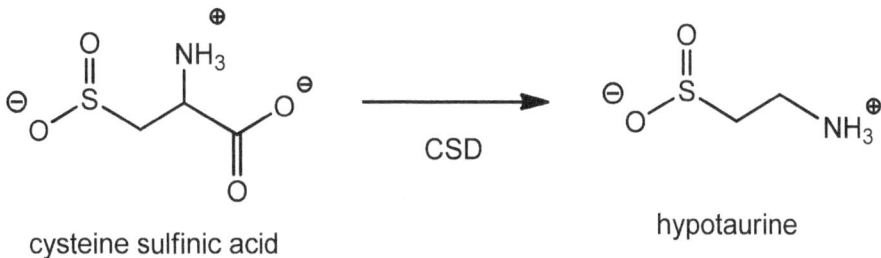

cysteine sulfinic acid   hypotaurine

Fig. (283). The cysteine sulfinic acid decarboxylase reaction.

The crystal structure of human cysteine sulfinic acid decarboxylase (CSAD) in complex with PLP are described, however detailed information about its architecture and binding site are not published (Fig. **284**).

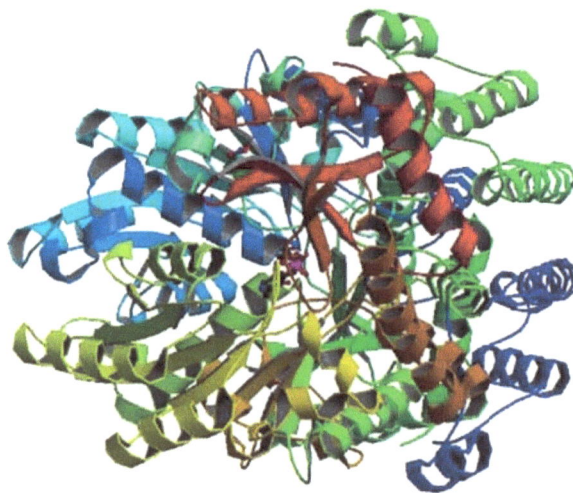

Fig. (284). The crystal structure of human cysteine sulfinic acid decarboxylase (PDB: 2JIS).

10.15. Cysteine Sulfinic Acid Decarboxylase Inhibitors

Small molecule inhibitors have been synthesized, submitted to docking studies, and tested *in vitro*. The compounds evaluated as CSAD inhibitors (Fig. **285**) are bis-carboxymethyl-trithiocarbonate having the highest affinity (K_i = 70 uM, IC_{50} = 0.12 mM) [277].

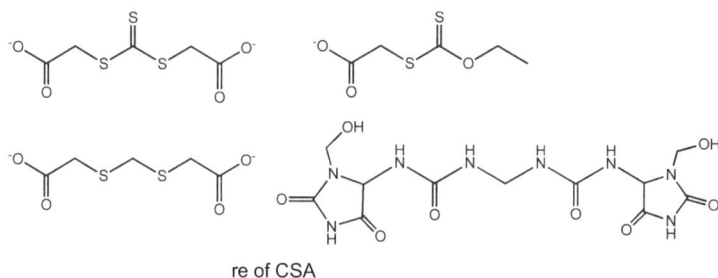

re of CSA

Fig. (285). Chemical structure of CSAD inhibitors.

11. Tryptophan Biosynthesis

Tryptophan is an indole containing essential amino acid needed by the animals to produce key molecules such as hormone melatonin, and neurotransmitter serotonin [278]. It is biosynthetically prepared from shikimic acid precursor chorismate, being transformed to antranilate by the enzyme antranilate synthase, and then to 5-phosphoribosyl-antranilate catalysed by phosphoribosylantranilate synthase, and further transformation to CdRP by the phosphoribosylantranilate isomerase. The indole ring is formed by the enzyme indole-3-glycerolphosphate synthase, giving place to indole-3-glycerol phosphate, and ultimately to tryptophan by the catalysis of tryptophan synthase (Fig. **286**).

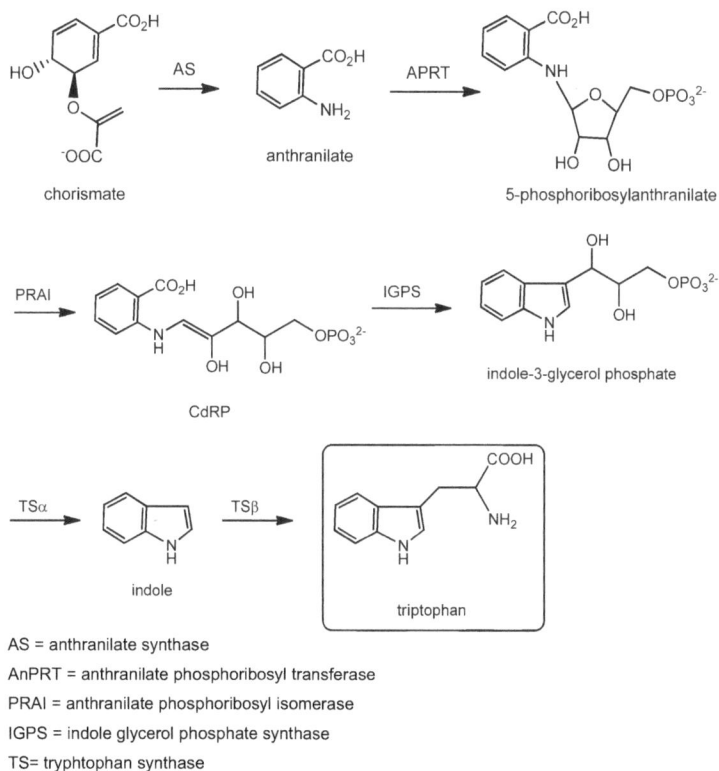

AS = anthranilate synthase
AnPRT = anthranilate phosphoribosyl transferase
PRAI = anthranilate phosphoribosyl isomerase
IGPS = indole glycerol phosphate synthase
TS= tryphtophan synthase

Fig. (286). Biosynthesis of tryptophan from chorismate.

11.1. Anthranilate Synthase (AS)

The first step in the tryptophan pathway starts with the transformation of chorismate to anthranilate through the intermediate 2-amino-2-deoxyisochorismate (ADIC) by the catalysis of anthranilate synthase (Fig. **287**), which is composed by two subunits, a glutamine aminotransferase (TrpG) where ammonia is produced, and TrpE responsible for the ADIC and anthranilate formation.

Fig. (287). The anthranilate synthase reaction.

A consistent mechanism propose initially the hydrolysis of glutamine to glutamic acid producing ammonia which effects a nucleophilic attack to chorismate C-2 position. The addition of ammonia promotes the delocalization of the double bond forcing the hydroxyl group to be ejected, producing as result 2-amino-2-deoxyisochorismate (ADIC). The final step consists in a beta elimination reaction of the carboxyenol group to produce anthranilate and pyruvate (Fig. **288**). For the second step concerning the elimination of pyruvate two possible mechanism have been considered, the acid/base elimination, and [1,5] sigmatropic rearrangement, however isotopic experiments favours the acid/elimination route [279 - 281].

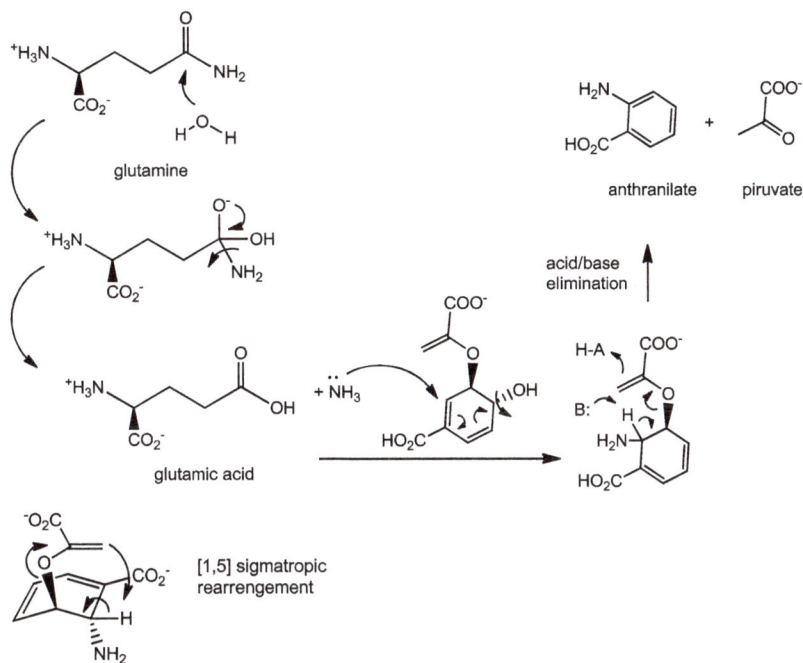

Fig. (288). Proposed mechanism for anthanilate formation from chorismate.

The heterodimeric structure of AS of *Serratia marcescens* showing TrpE and TrpG (large and small subunits) with the glutamine binding site shown in blue and the chorismate binding site in magenta (Fig. **289**) [282].

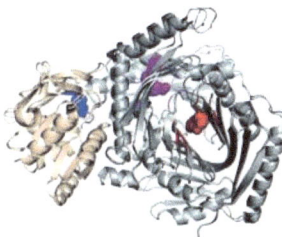

Fig. (289). The heterodimeric structure of AS of *Serratia marcescens* (PDB: 1I7Q).

11.2. Anthranilate Synthase Inhibitors

Small molecule analogues have been designed and evaluated as anthranilate synthase inhibitors. The cyclohexene ring was modified and the analogues were evaluated against *S. marcescens* obtaining inhibition constants ranging from micromolar to millimolar (Fig. **290**) [283].

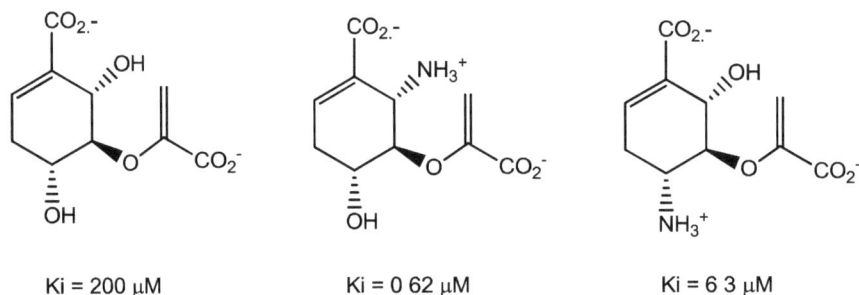

Ki = 200 µM   Ki = 0 62 µM   Ki = 6 3 µM

Fig. (290). Chemical structure and inhibition constant of anthanilate synthase inhibitors.

11.3. Anthranilate Phosphoribosyl Transferase (ANPRT)

This enzyme is involved in the second step of the tryptophan biosynthesis consisting in the attachment of a ribosyl group from 5-phospho-α-D-ribo-yl1-diphosphate (PRPP) to anthranilate to give 5-phosphoribosylantranilate (PRA) (Fig. **291**).

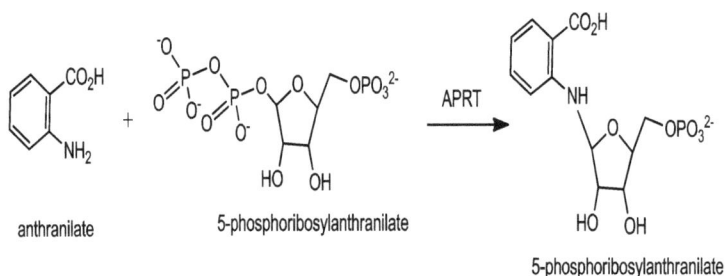

anthranilate   5-phosphoribosylanthranilate   5-phosphoribosylanthranilate

Fig. (291). The anthranilate phosphoribosyl transferase reaction.

The crystal structure of AnPRT from *Mycobacterium tuberculosis* was solved showing a homodimeric structure with N-terminal formed by six α-helices and larger C-terminal domain with eight α-helices around central seven stranded β-sheet. At the anthranilate binding sites it can be observed the region with PRPP represented in cyan, inhibitor 4-fluoroanthranilate (4-FA) in purple, the magnesium ions and the residues involved in the interactions (Fig. **292**) [284].

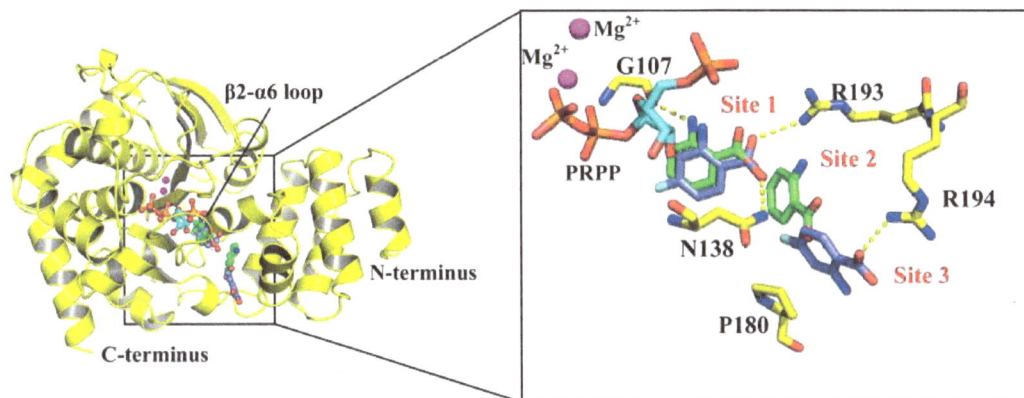

Fig. (292). Crystal structure of *Mtb*-AnPRT variant and binding site region showing PRPP substrate and 4-fluoro anthranilate inhibitor (PDB: 4N5V).

11.4. Anthranilate Phosphoribosyl Transferase Inhibitors

Anthranilate-like derivative 4-fluoroanthranilate (Fig. **293**) binds to Mtb-AnPRT and induce inhibition of M. tuberculosis growth [285].

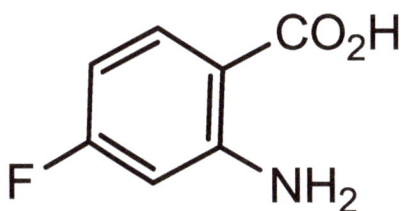

4-fluoro anthranilate

Fig. (293). Chemical structure of 4-fluoro anthranilate.

11.5. Phosphoribosyl Antranilate Isomerase (PRAI)

This enzyme is responsible for the conversion of N-(5-phospho-b-D-ribosyl) anthranilate (PRA) to 1-(2-carboxyphenylamino)-1-deoxy-D-ribulose 5-phosphate (CdRP) following an Amadori rearrangement (Fig. **294**).

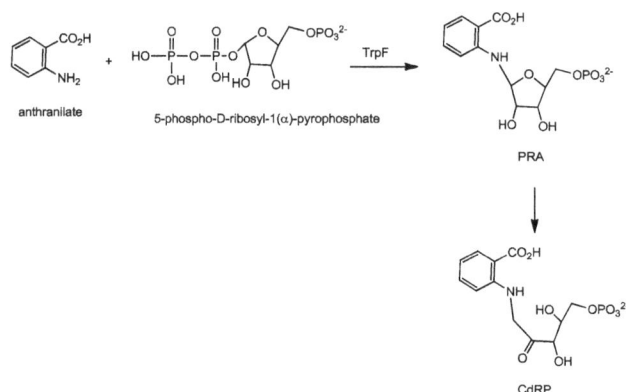

Fig. (294). The phosphoribosyl anthranilate isomerase reaction.

The mechanism by which the N-glycoside PRA is converted to CdRP is understood through the Amadori rearrangements consisting in the imine formation accompanied by ring opening, followed by tautomerization to give CdRP (Fig. **295**) [286].

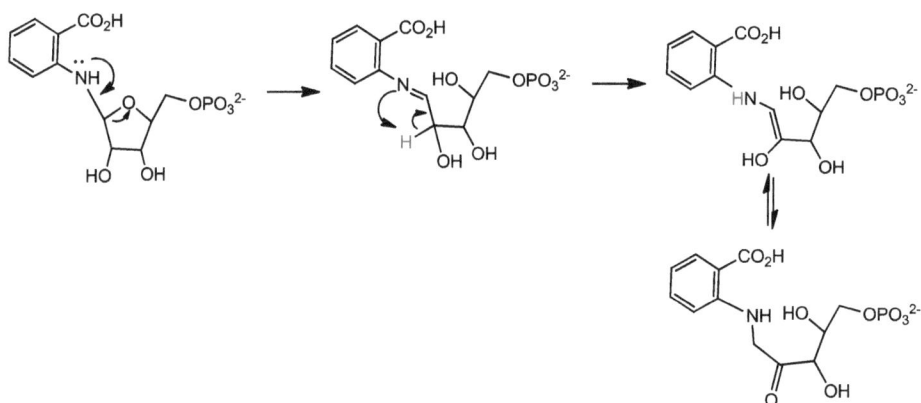

Fig. (295). Proposed mechanism for the conversion of N-(5-phospho-b-D-ribosyl) anthranilate (PRA) to 1-(2-carboxyphenylamino)-1-deoxy-D-ribulose 5-phosphate (CdRP).

The crystal structure of phosphoribosyl anthranilate isomerase from different microorganism has been elucidated (Fig. **296**), establishing as common features a monofunctional enzyme with a single polypeptide chain or together with TrpC in the same polypeptide chain. The structural comparison has been established between hyperthermophilic archaeon *Thermococcus kodakaraensis* (TkTrpF; PDB 5lhe), *Thermotoga maritime* (TmTrpF; PDB 1lbm), and *Pyrococcus furiosus* (PfTrpF; PDB 4aaj) showing sequence identity (100, 35 and 58%) [287].

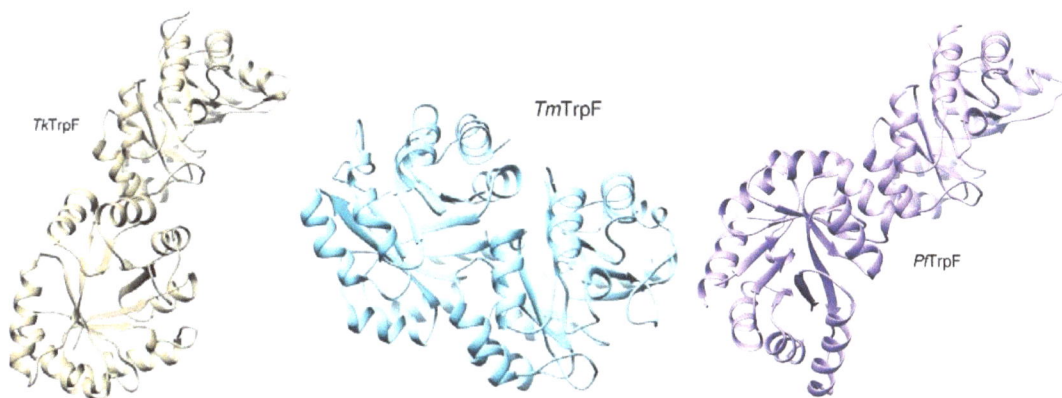

Fig. (296). The structural comparison of TrpF from *Thermococcus kodakaraensis* (PDB: 5lhe), *Thermotoga maritime* (PDB: 1lbm), and *Pyrococcus furiosus* (PDB: 4aaj).

11.6. Phosphoribosyl Anthranilate Isomerase Inhibitors

The reduced form of 1-(2-carboxyphenylamino)-1-deoxy-D-ribulose 5-phosphate (Fig. **297**) was tested against phosphoribosyl anthranilate isomerase (TrpF) from *Escherichia coli*, observing two different binding sites displaying K_i values of 0.2 µM and 12.5 µM indicating a high-affinity on one of the binding sites [288].

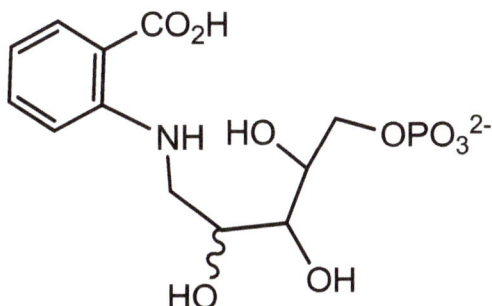

Fig. (297). Chemical structure of rCdRP.

11.7. Indole Glycerol Phosphate Synthase (IGPS)

The fifth reaction in the tryptophan biosynthesis is catalysed by the enzyme IGPS, involving the indole ring formation from1-(o-carboxyphenylamino)-1-deoxyribulose 5-phosphate (CdRP) to produce indole-3-glycerol phosphate (Fig. **298**).

Fig. (298). The indole glycerol phosphate synthase reaction.

This reaction can be considered to the Bischler-type indole synthesis where the amino group activates the ring and favors an electrophilic substitution with the ketone group to produce the indole formation, followed by decarboxylation and dehydration (Fig. **299**) [289 - 291].

Fig. (299). The interaction diagram and simplified mechanism for the indole-3-glycerol phosphate formation.

More than 20 crystal structures of bacterial IGPS have been elucidated what reflects the significance to provide novel strategies for the attenuation of bacterial infection. For instance the structure of IGPS from *Mycobacterium tuberculosis* is described, showing a typical $(\beta/\alpha)_8$-barrel structure and the binding region of CdRP (Fig. **300**) [292].

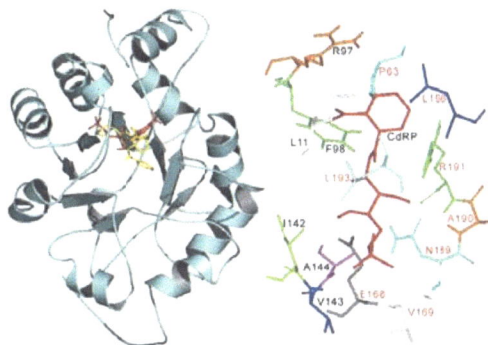

Fig. (300). Ribbon structure of IGPS from *Mycobacterium tuberculosis* and the binding region of CdRP (PDB: 3T40).

11.8. Indole Glycerol Phosphate Synthase Inhibitors

A heterocyclic compound having a central 1,3,5-triazine ring and azabicyclo hydrazone, pyrrolidine and aniline substituents (ATB107) has been identified as a potent inhibitor against *M. tuberculosis* H37Rv with an MIC of 0.1 µg.mL^{-1}. Another inhibitor bearing indole and pyrazole heterocycles described as ATB26 showed MIC values at 0.05 µg.mL^{-1} against the same Mycobacterium strains (Fig. **301**) [293].

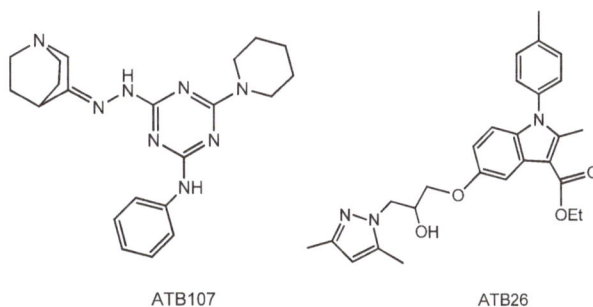

ATB107   ATB26

Fig. (301). Chemical structure of IGPS inhibitors 1,3,5-triazine ATB107 and indole ATB26 analogues.

11.9. Tryptophan Synthase (TS)

The tryptophan synthase complex is constituted by two domains: The α and β chains being separated in bacteria and plants, while in fungi they are fused. This enzyme is responsible of the conversion of indole-3-glycerol phosphate (IGP) to L-tryptophane (Fig. **302**). The first step occurs at the α chain consisting in the transformation of IGP to indole and glyceraldehyde phosphate (GAP) through a retroaldol type reaction, the second step proceeds at the β chain and requires tryptophan and L-serine.

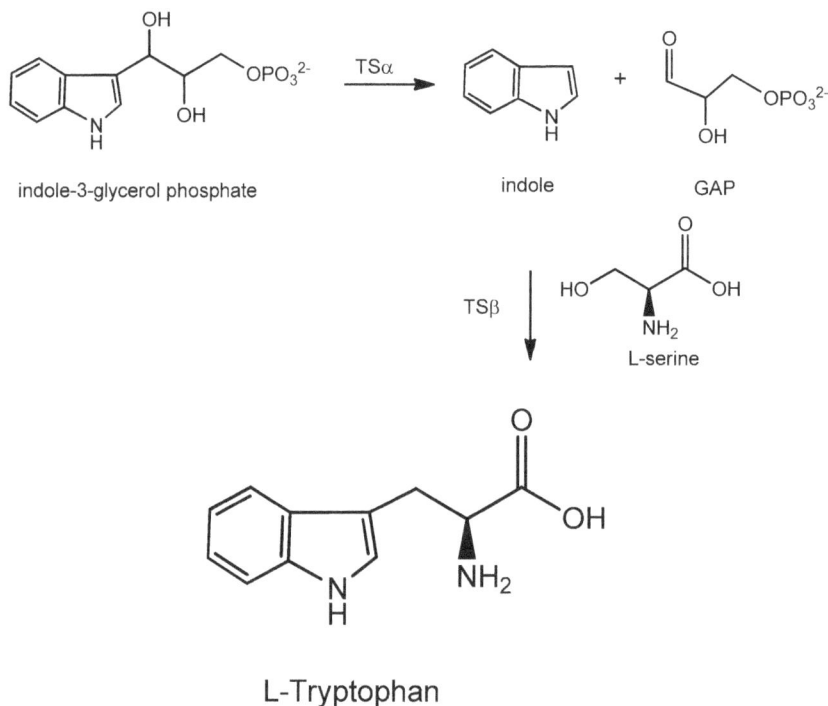

Fig. (302). The tryptophan synthase reaction.

The reaction steps leading to the formation of L-tryptophan involves the formation of intermediates gem-diamine (E-GD), external aldimine of L-Ser (E-Ser), external aldimine of aminoacrylate (E-AA), indole-quinoid intermediate (E-Q), external aldimine (E-Trp) and final product L-Trp as shown in Fig. (**303**) [294].

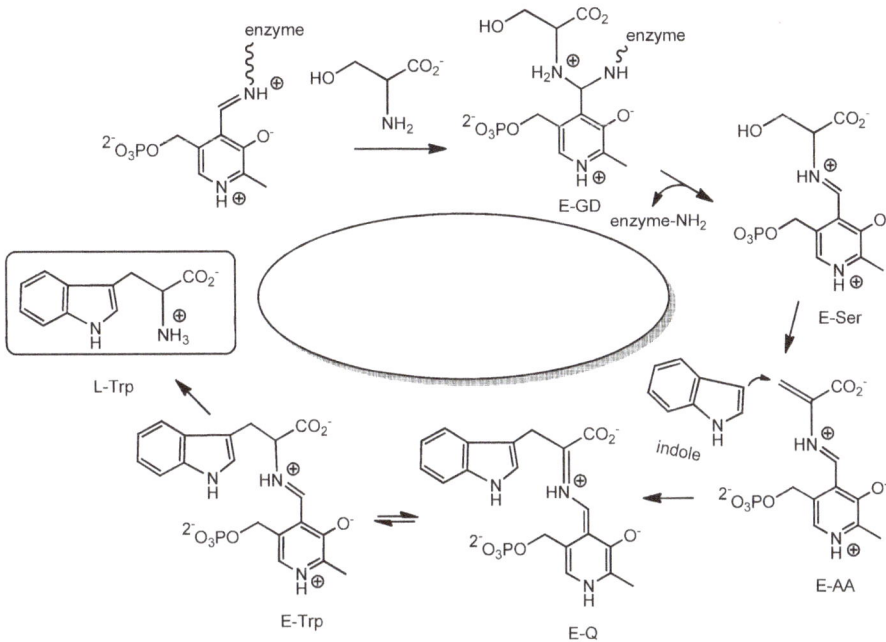

Fig. (303). Proposed diagram representing the intermediates for L-tryptophan biosynthesis.

The crystal structure of ancestral tryptophan synthase was determined, showing their α subunits in green and the β in blue. It is also represented as spheres glycerol-3-phosphate bound at α, the cofactor PLP bound at β. It is also observed as brown mesh a channel interconnecting both active sites (Fig. **304**) [295].

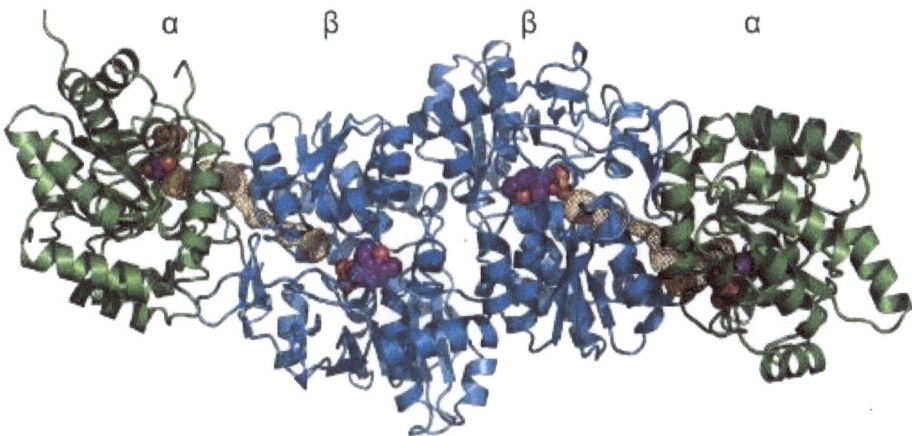

Fig. (304). The crystal structure of ancestral tryptophan synthase showing subunits, and the connecting channel (PDB: 5EY5).

11.10. Tryptophan Synthase Inhibitors

(1-Fluorovinyl)glycine was described as tryptophan synthase inhibitor against *Salmonella typhimurium*, suggesting that the inactivation is due the elimination HF and the resulting allene specie interacts with the enzyme-pyridoxal complex (E-PLP) leading to inactivation (Fig. **305**) [296].

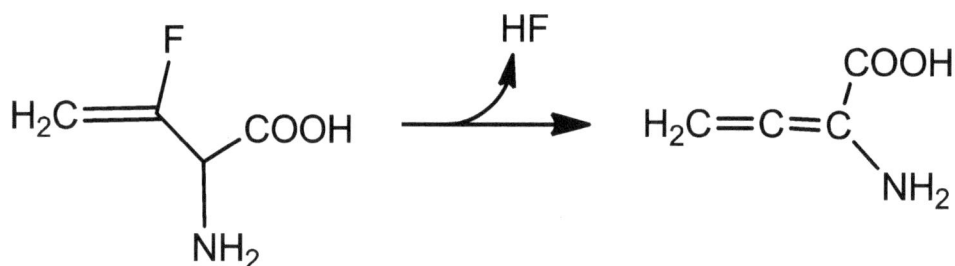

Fig. (305). Chemical structure of (1-fluorovinyl) glycine and reactive allene specie involved in the TS inhibition.

Sulfolane and indoline sulphonamides have been analysed as *Mycobacterium tuberculosis* inhibitors, evaluating tryptophan synthase as drug target against *Mtb* H37Rv. The results shows significant potency on *in vitro* assays, with low toxicity against HepG2 cells, and lipophilicity falling within the range with ClogP between 1.63 and 2.05 (Fig. **306**) [297].

12. HISTIDINE BIOSYNTHESIS

Histidine amino acid is an essential amino acid not synthesized by human and therefore needs to be supplied in the diet. It contains an imidazole heterocycle ring positively charged as substituent that serves as an acid and as base in he neutral form. Histidine is required in a number of vital processes such as blood production, protection of neurons and to protect the body against toxic heavy metals in particular lead or mercury.

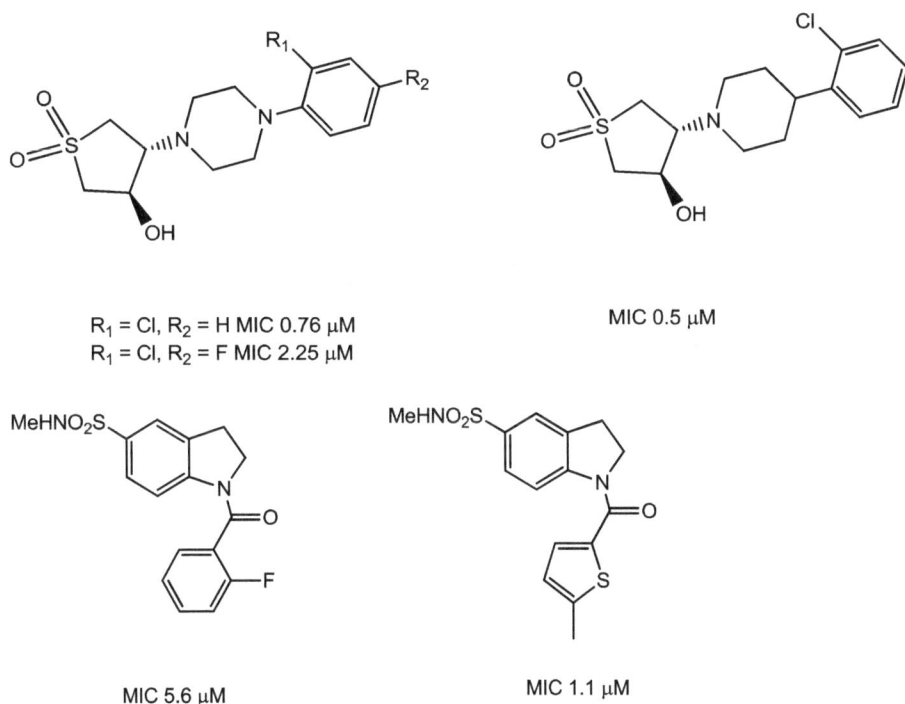

Fig. (306). Chemical structure and inhibition constants of sulfolane and indoline sulphonamides.

Its biosynthesis is a multi-step process starting with reaction between ATP and ribosyl diphosphate catalysed by HisG to give phosphoribosyl-ATP, which subsequently expel a diphosphate group to produce phosphoribosyl-AM by the catalysis of HisE enzyme. The next step involving the ring opening mediated by the enzyme HisI results in the formation of phosphoribosyl-formimino-AICAR-P, which is further transformer to phosphoribulosyl-formimino-AICAR-P under the HisA catalysis. The imidazole ring formation is achieved by the catalysis of HisF/HisH complex, to give imidazole-glycerol-P, followed by subsequent reaction sequence involving dehydroxylation, alcohol oxidation to ketone, transformation to amine, phosphate cleavage and oxidation of primary alcohol to carboxylic acid, leading ultimately to target amino acid histidine (Fig. **307**).

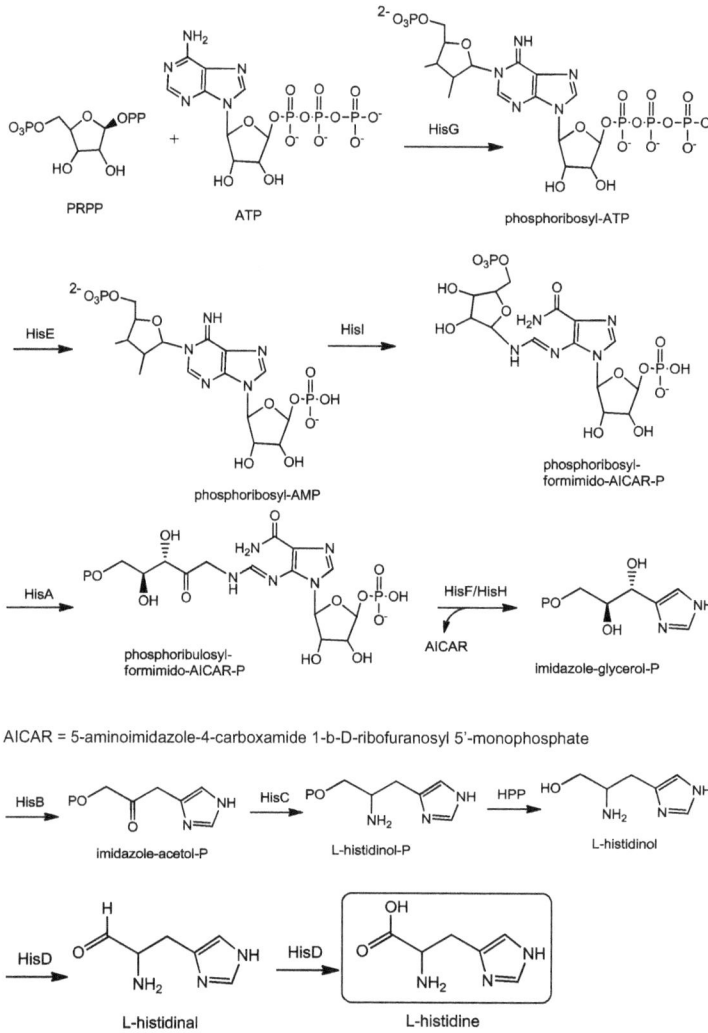

AICAR = 5-aminoimidazole-4-carboxamide 1-b-D-ribofuranosyl 5'-monophosphate

HisB = imidazole glycerol phosphate dehydratase
HisC = histidinol phosphate aminotransferase
HisD = histidinol dehydrogenase
HisE = phosphoribosyl-ATP pyrophosphohydrolase
HisF = cyclase
HisG = ATP phospho-ribosyltransferase
HisH = glutamine amidotransferase
HisI = phosphoribosyl-AMP cyclohydrolase
HPP = Histidinol phosphate phosphatase

Fig. (307). The biosynthetic pathway for L-histidine formation.

12.1. ATP Phosphoribosyl Transferase (HISG)

Considered a high energy process it is present in bacteria, fungi and plants but absent in mammals, and represent the first step in the biosynthesis of histidine starting with the condensation of adenosine triphosphate (ATP) with 5'-phosphoribosyl 1'-pyrophosphate (PRPP) in the presence of Mg^{2+}, yielding phosphoribosy-ATP (PR-ATP) and inorganic pyrophosphate (PPi) (Fig. **308**) [298].

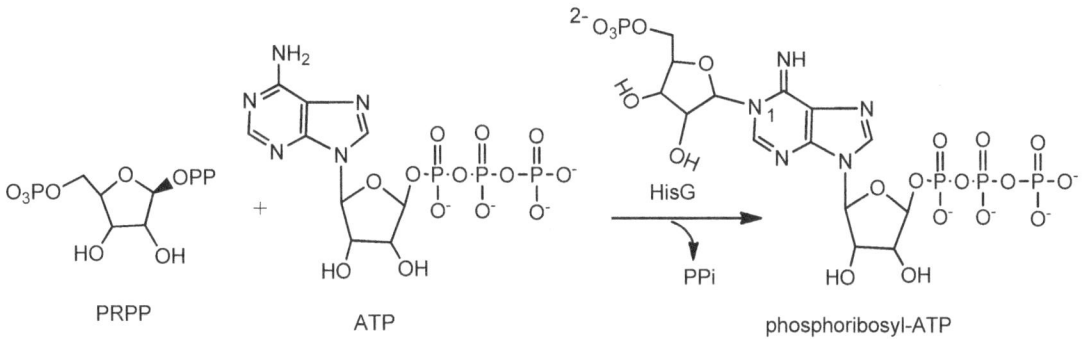

Fig. (308). The phosphoribosyl transferase reaction.

Crystallographic studies of HisG from *Lactococcus lactis* reveals that this enzyme exist in two forms termed the long-form as homohexamer with a C-terminal regulatory domain containing the allosteric binding site for histidine, and the short-form, a hetero-octamer containing two catalytic dimers which associate HisZ with the active site present in the cleft of domain I and II in both short- and long- forms. The (Fig. **Fig. 309**) shows the hetero-octameric His G from *L. lactis*, composed by four units and four regulatory subunits HisZ [299].

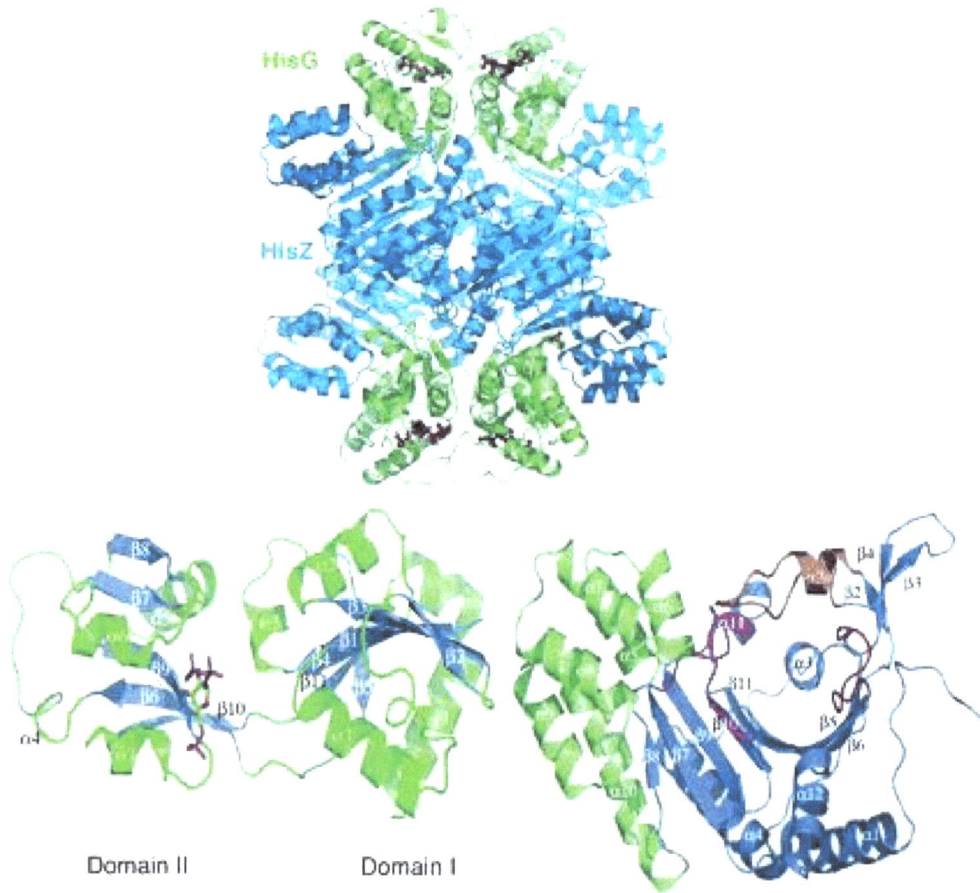

Fig. (309). Ribbon representation of ATP Phosphoribosyl transferase (PDB 1Z7M).

Further analysis on *B. subtilis* shows that ribose 5-phosphate coordinates with MgATP, free Mg^{2+}, and ligands, reaching the maximal activity when this ion is at optimal concentration. The stereo view shows the allosteric site occupied by methyleneADP of PRPP synthase, interacting through polar contacts with residues belonging to subunits A, B and D as shown (Fig. **310**) [300].

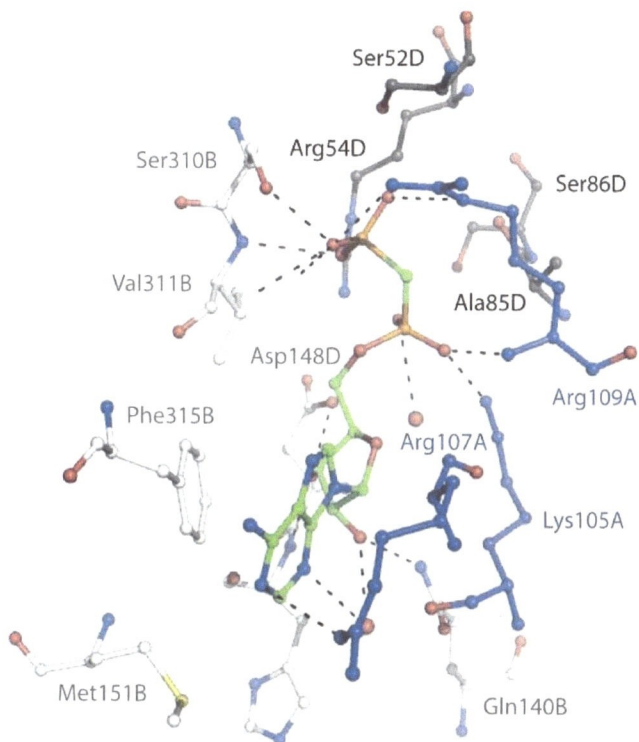

Fig. (310). The stereo view of HisG binding site showing the residues in a complex with methylene ADP (PDB: 1DKU).

12.2. ATP Phosphoribosyl Transferase Inhibitors

Low-molecular weight (<450) candidates have been analysed as HisG inhibitors, showing good potency particularly the dithiol naphtalen amine showing an IC_{50} value of 50 nM (Fig. **311**) [301].

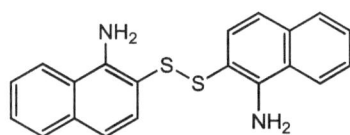

100 % inhibition at 1 and 100 µM

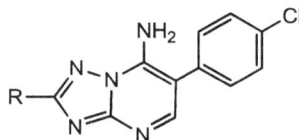

R = H 92 % inhibition at 100 µM
R = SMe 86 % inhibition at 100 µM

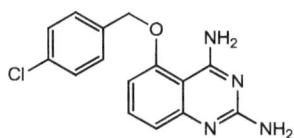

82 % inhibition at 100 µM

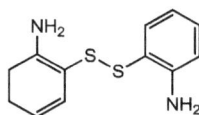

84 % inhibition at 100 µM

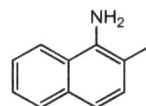

21 % inhibition at 100 µM

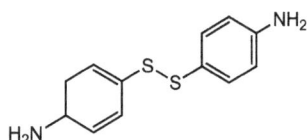

35 % inhibition at 1 µM

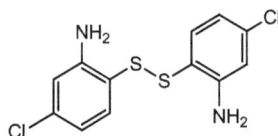

78 % inhibition at 1 µM

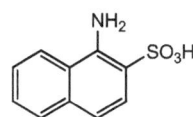

3 % inhibition at 100 µM

Fig. (311). Chemical structure and inhibition constants of ATP Phosphoribosyl transferase inhibitors.

12.3. Phosphoribosyl-ATP Pyrophosphohydrolase (HISE)

The second step of histidine biosynthesis is catalysed by the enzyme HisE, performing the hydrolysis of phosphoribosyl-ATP to give phosphoribosyl-AMP and pyrophosphate (Fig. **312**).

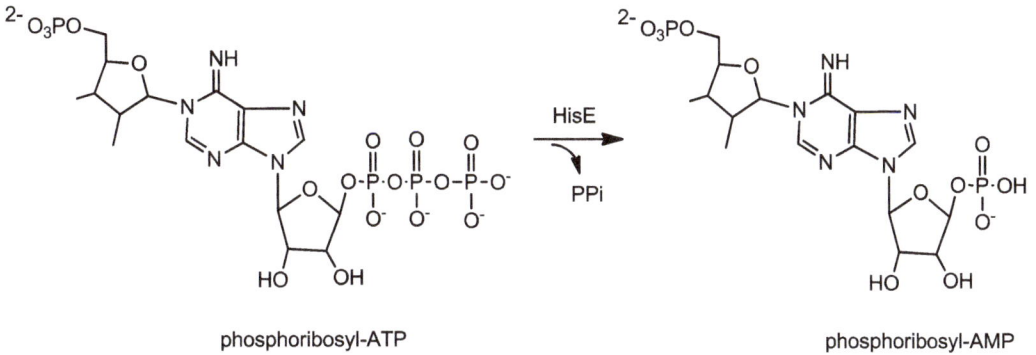

Fig. (312). The phosphoribosyl-ATP pyrophosphohydrolase reaction.

The protein HisE form *M. tuberculosis* is composed by five α-helices having as basic unit an homodimer, and although the active site was not determined directly by co-crystallization with ATP, AMP or other analogues, the superimposition of the backbones of *Mtb*HisE (PDB 1w2y; white) with other close related phosphohydrolases such as *S. solfataricus* MazG (PDB 1vmg; red), mouse RS21-C6 complexed with m5dCTP (PDB 2oig; green) and *T. cruzi* dUTPase (PDB1w2y; yellow) allowed to locate the putative binding region (Fig. **313**) [302].

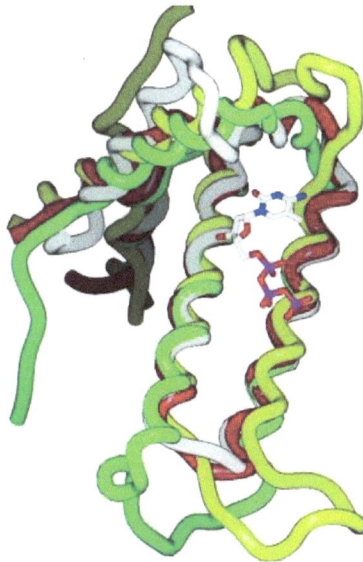

Fig. (313). Superimposition of backbones of *Mtb*HisE (PDB 1W2Y; white) with other close related phosphohydrolases such as *S. solfataricus* MazG (PDB: 1VMG); red), mouse RS21-C6 complexed with m5dCTP (PDB 2OIG; green) and *T. cruzi* dUTPase (PDB1W2Y; yellow) to determine the binding region.

12.4. Phosphoribosyl-ATP Pyrophosphohydrolase Inhibitors

Application of virtual screening using libraries from ZINC database did not show any small molecule inhibitor for HisE [303].

12.5. Phosphoribosyl-Amp Cyclohydrolase (Hisi)

The third step in the histidine biosynthesis is catalysed by the metalloenzyme enzyme HisI, which performs the ring opening of phosphoribosyl-AMP providing phsphoribosyl-formimido-AICAR-P (Fig. **314**).

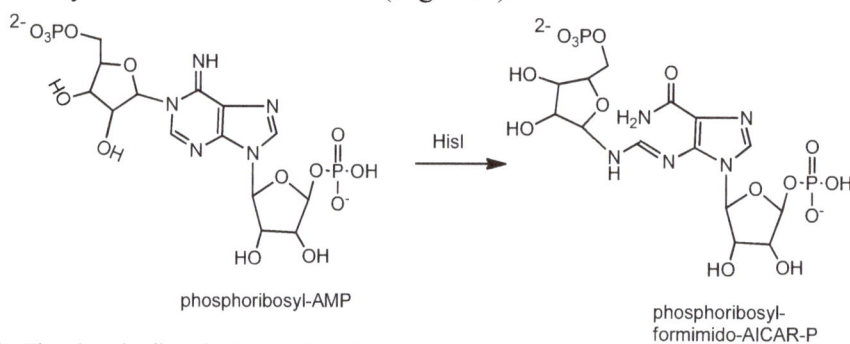

Fig. (**314**). The phosphoribosyl-AMP cyclohydrolase reaction.

The ribbon diagram of HisI from *Methanobacterium thermoautotrophicum*, and stereo view showing the PR-AMP substrate at the lowest energy are described, presenting hydrogen bond interactions with residues and metal coordination bonds with Zn^{2+} and Mg^{2+} ions (Fig. **315**) [304].

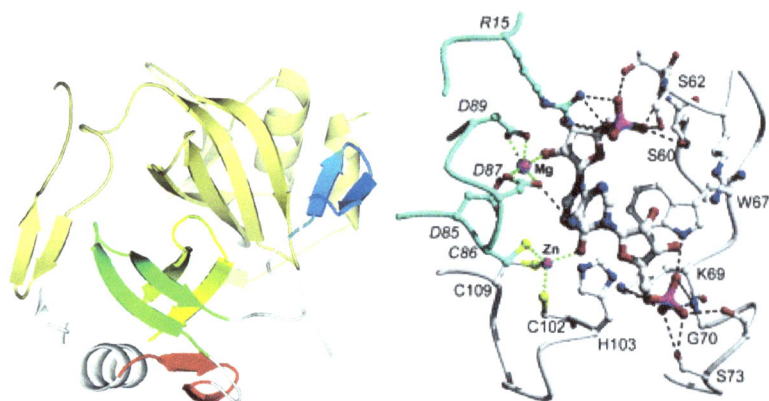

Fig. (**315**). Ribbon diagram and binding site region of HisI from *Methanobacterium thermoautotrophicum* (PDB: 1ZPS).

A mechanism based on experimental data, support the evidence of an initial state between three cysteine residues with Zn ion coordinated with water molecule, effecting a nucleophilic attack to the iminium, giving place to a tetrahedral intermediate. Next, the hydroxyl group attacks the tetrahedral carbon promoting the ring opening, which after proton exchanging with histidine residue affords phsphoribosyl-formimido-AICAR-P (Fig. **316**) [305].

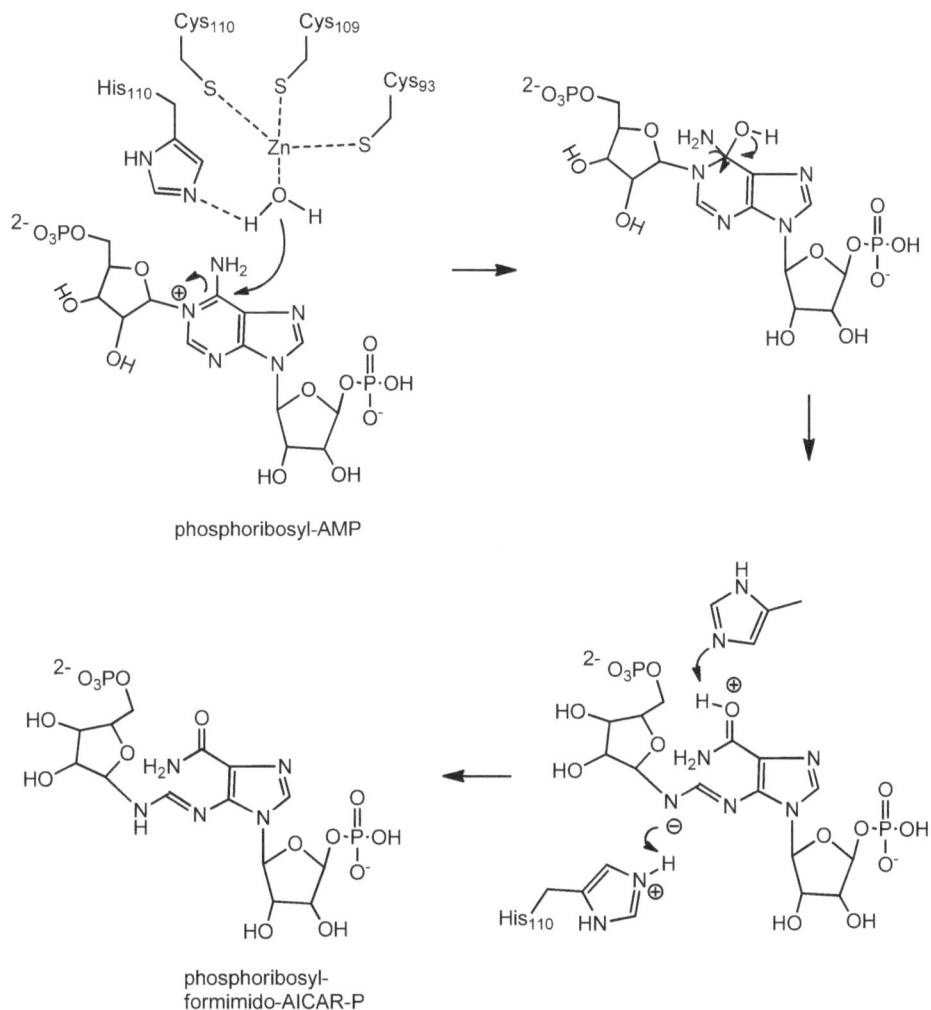

Fig. (316). Proposed mechanism for the phosphoribosyl-formimido-AICAR-P formation.

12.6. Phosphoribosyl-AMP Cyclohydrolase Inhibitors

Based on virtual screening and docking experiments a total of six small molecules displaying binding energies in the range of -8.1 to -9.1 kcal/mol were selected as suitable candidates for HisI inhibition against *B. melitensis*, proposing as result the compound with value (-9.1 kcal/mol), as the best candidate for inhibition test (Fig. **317**) [306].

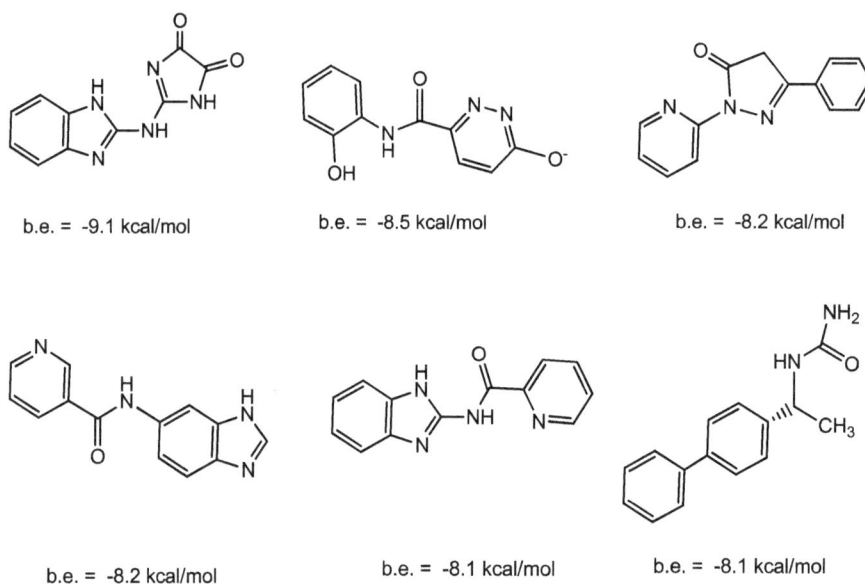

Fig. (**317**). Chemical structure and inhibition constants of phosphoribosyl-AMP cyclohydrolase inhibitors.

Other candidates for inhibition identified by virtual screening using ZINC library were analysed by docking and MD simulations, and their G-scores from Glide XP and binding energies from MM-PBSA and MM-GBSA were calculated (Fig. **318**) [307].

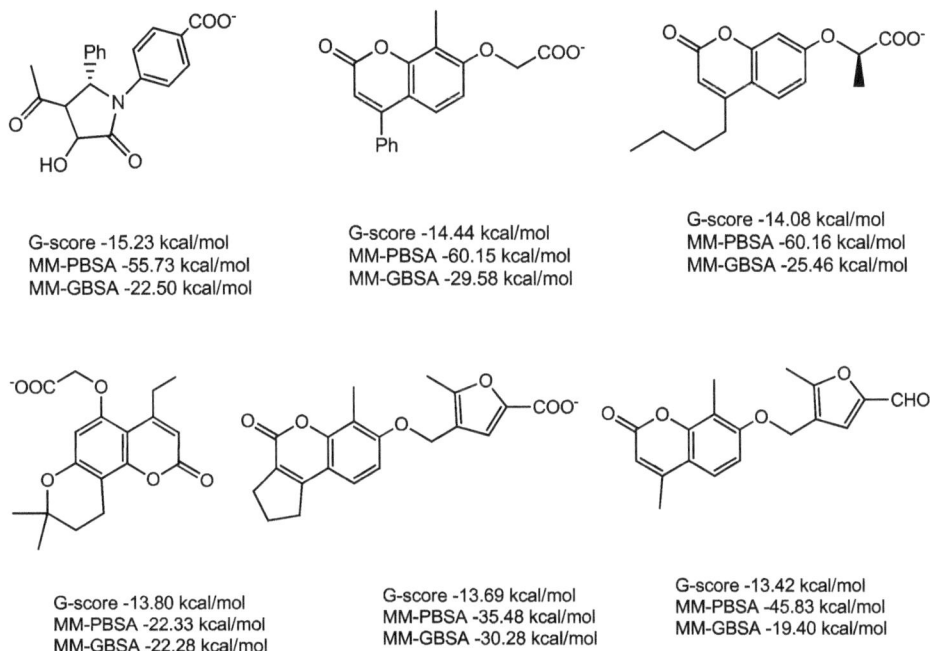

G-score -15.23 kcal/mol
MM-PBSA -55.73 kcal/mol
MM-GBSA -22.50 kcal/mol

G-score -14.44 kcal/mol
MM-PBSA -60.15 kcal/mol
MM-GBSA -29.58 kcal/mol

G-score -14.08 kcal/mol
MM-PBSA -60.16 kcal/mol
MM-GBSA -25.46 kcal/mol

G-score -13.80 kcal/mol
MM-PBSA -22.33 kcal/mol
MM-GBSA -22.28 kcal/mol

G-score -13.69 kcal/mol
MM-PBSA -35.48 kcal/mol
MM-GBSA -30.28 kcal/mol

G-score -13.42 kcal/mol
MM-PBSA -45.83 kcal/mol
MM-GBSA -19.40 kcal/mol

Fig. (318). Chemical structure, G-scores and binding energies of potential candidates obtained by virtual screening.

12.7. 5'Profar Isomerase (HISA)

The fourth step in the histidine biosynthesis is catalysed by HisA consisting in the conversion of phosphoribosyl-formimido-AICAR-P (5'ProFAR) to phosphoribulosyl-formimido-AICAR-P (PRFAR), involving the ribose ring opening followed by Amadori rearrangement (Fig. **319**).

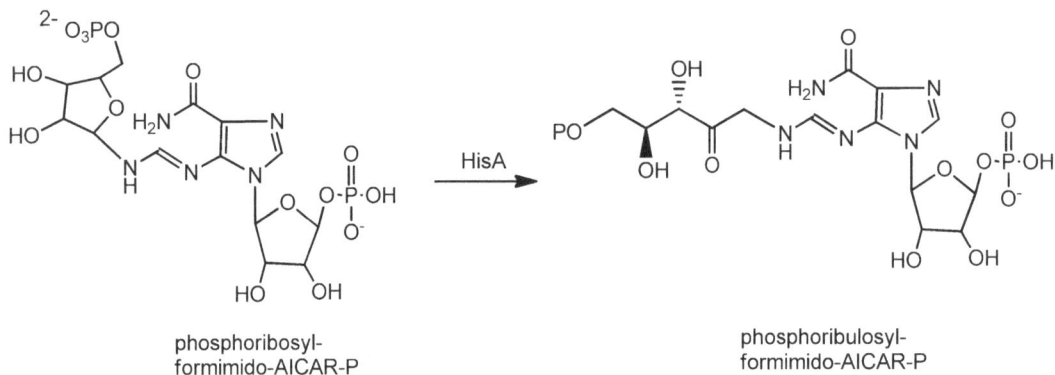

phosphoribosyl-
formimido-AICAR-P

phosphoribulosyl-
formimido-AICAR-P

Fig. (319). The 5'ProFAR isomerase reaction.

Correlation studies between HisA and phosphoribosyl anthranilate isomerase (APRI), another isomerase that fold into $(\beta/\alpha)_8$-barrel were conducted since they achieve similar transformations, however the encoding genes *his*A and *trp*F express two different single-substrate enzymes, and the active sites are substantially different in size [308]. Despite the structural differences, they carried out an isomerization reaction from PRA to CdRP, and from ProFAR to PRFAR according to the Amadori rearrangement in virtually the same way as it can be seen in the next step sequence starting from PRA and ProFAR precursors, leading to CdRP and PRFAR, respectively (Fig. **320**) [307].

The HisA from *Thermotoga maritime* (*Tm*HisA) complexed with a product analogue has been useful to clarify the mechanism however *Tm*HisA has little resemblance with most of bacteria and therefore another HisA from *Salmonella enterica* (*Se*HisA) has been used as model for enzyme evolution. Thus, the *Se*HisA complexed with ProFAR substrate show the interacting residues, the water molecules and the hydrogen bond from two stereo view perspectives (Fig. **321**) [308].

12.8. Imidazole Glycerol Phosphate Synthase (HISF/HISH)

The fifth step in the histidine biosynthesis consist in the conversion of phosphoribulosyl-formimido-AICAR-posphate into imidazole glycerol phosphate and AICAR as by product, following a two step mode, being the first a coupled bi-enzyme system of a glutaminase HisH to convert glutamine to glutamate, producing ammonia, and on the second a cyclase HisF which carry out the imidazole ring formation (Fig. **322**).

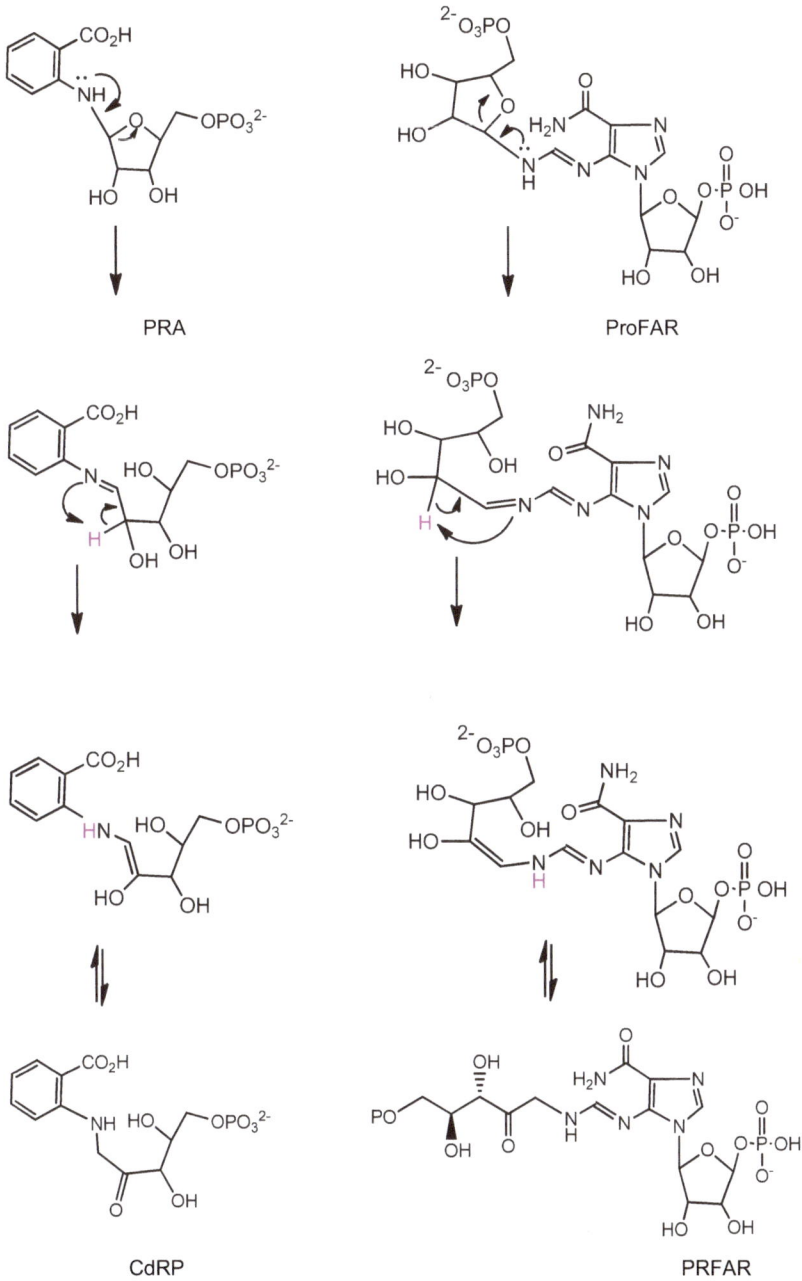

Fig. (320). Comparative sequence for CdRP, and PRFAR formation following the Amadori rearrangement.

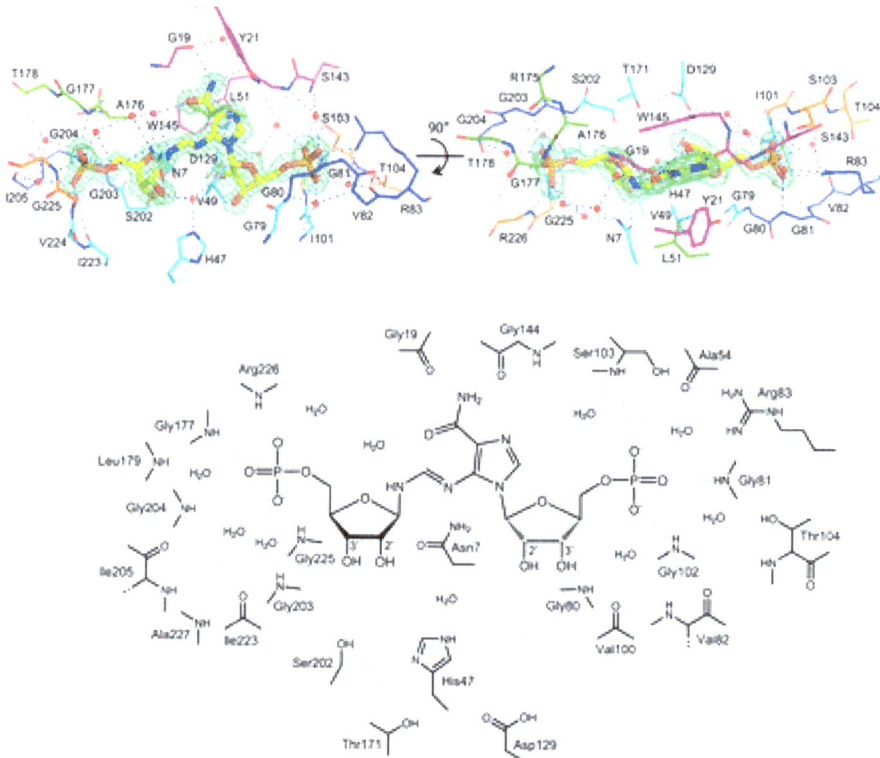

Fig. (321). The active site diagram of HisA from *Thermotoga maritime* and *Salmonella enterica* and interactions of ProFAR with the residues (PDB: 5AHE).

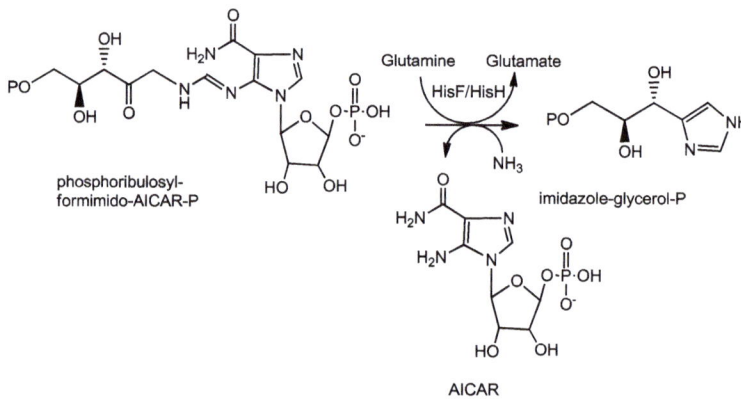

Fig. (322). The imidazole glycerol phosphate synthase reaction.

The crystal structure of ImPG synthase from *T. maritime* reveals two active sites separated by a tunnel transporting the ammonia generated by the glutaminase activity, and delivered to the cyclase active site where the imidazole formation occurs (Fig. **323**) [309].

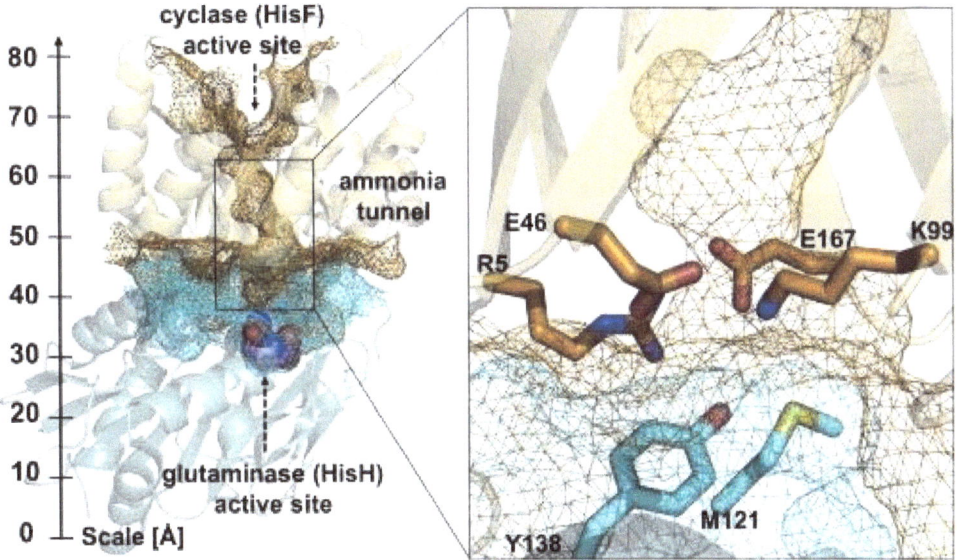

Fig. (323). Crystal structure of ImPG synthase from *T. maritime* showing the HisH and the HisF active sites (PDB: 2WJZ).

A consistent mechanism proposes two possible conformations assumed: (a) trans-eclipsed and (b) cis-extended forms of PRFAR and the residues implicated in the catalytic site (Fig. **324**) [310].

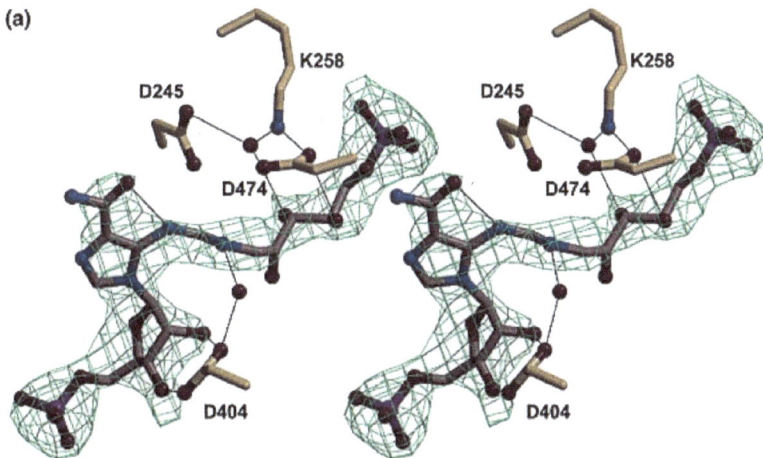

Fig. 324 cont.....

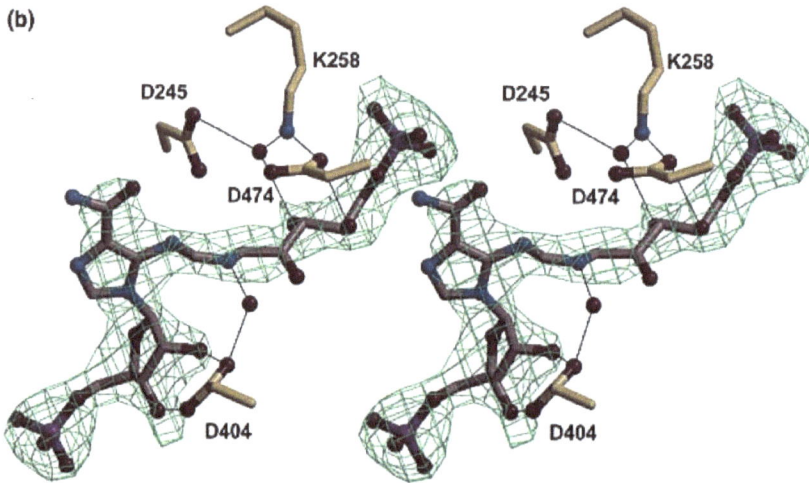

Fig. (324). The active site region showing the interaction of the residues with phosphoribulosyl-formimid--AICAR-posphate.

12.9. Imidazole Glycerol Phosphate Synthase Inhibitors

This enzyme has been a target for antibacterial, antifungal and herbicide agents, due their two active site domains connected by a tunnel where ammonia is transported to the cyclase catalytic site. The glutaminase domain (HisH) can be inactivated by isoxazole aminoacid analogue acivicin, which reacts with a cysteine residue of HisH domain of IGPS forming an imino-thioether linkage (Fig. **325**) [311].

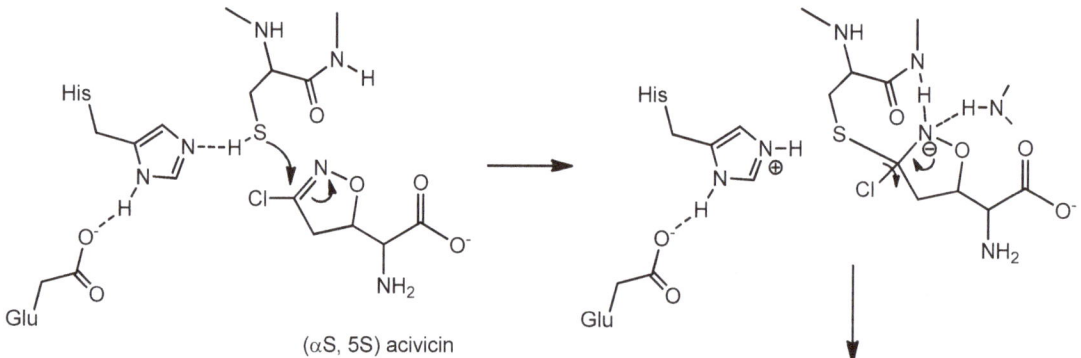

(αS, 5S) acivicin

Fig. 325 cont.....

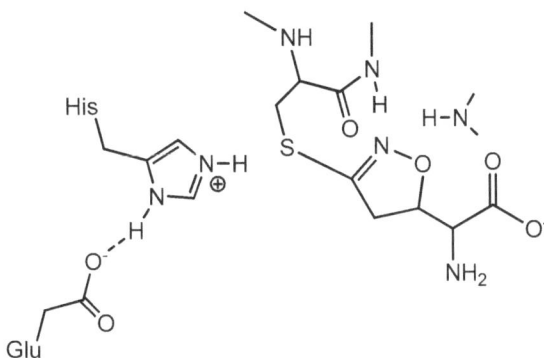

Fig. (325). Inactivation mechanism of acivicin to ImPG synthase.

12.10. Imidazole Glycerol Phosphate Dehydratase (Hisb)

The sixth step in the histidine biosynthesis corresponds to the conversion of imidazole glycerol phosphate (IGP) to imidazole acetol phosphate (IAP) by the enzyme HisB (Fig. **326**).

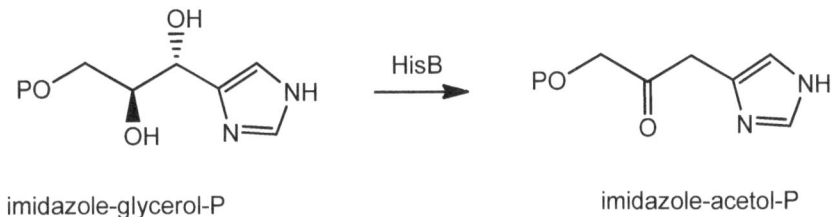

imidazole-glycerol-P   imidazole-acetol-P

Fig. (326). The imidazole glycerol phosphate dehydratase reaction.

The mechanism for the transformation of IGP to IAP has not been completely elucidated, however the formation of a diazafulvene and Δ^2-enol intermediates in the presence of Mn^{2+} coordinated with IGP and Glu residues has been demonstrate (Fig. **327**).

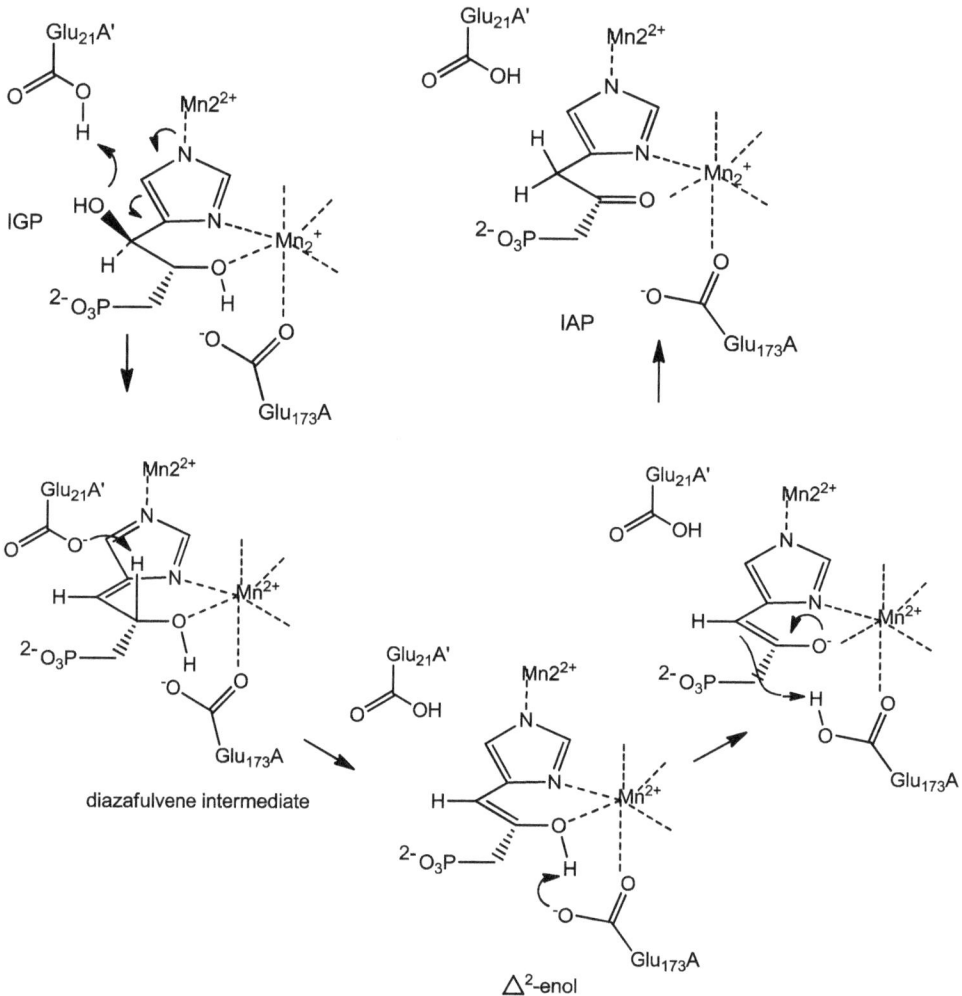

Fig. (327). Proposes mechanism fro the conversion of imidazole glycerol phosphate (IGP) to imidazole acetol phosphate (IAP).

The crystal structure of IGPD from Arabidopsis thaliana is composed by 24 identical subunits through a manganese cluster forming a shell with a large internal cavity. At the catalytic site the substrate is coordinated with the residues and two manganese ions (Fig. **328**) [312].

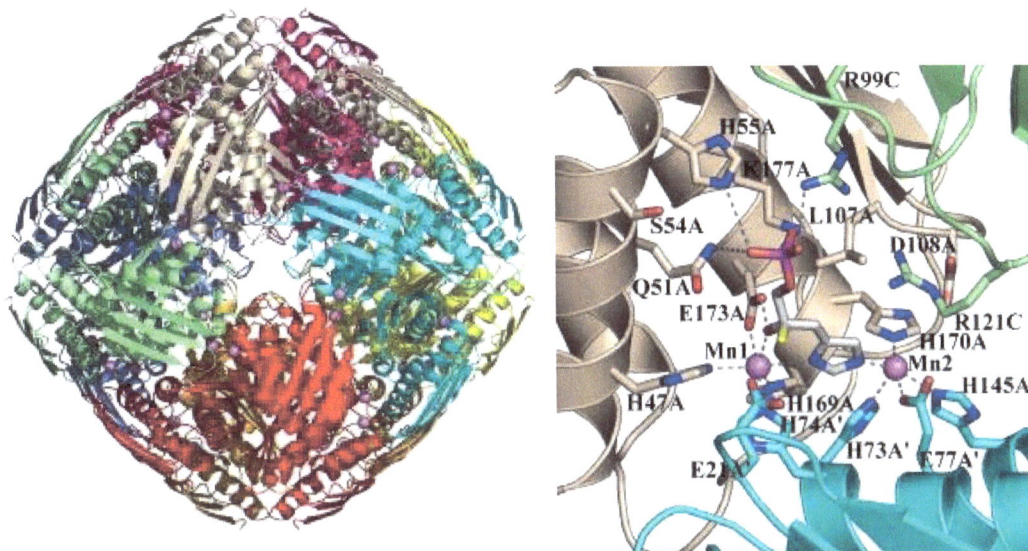

Fig. (328). The crystal structure and binding site region of IGPD from *Arabidopsis thaliana* (PDB: 2F1D).

12.11. Imidazole Glycerol Phosphate Dehydratase Inhibitors

IGPD is used as a target enzyme in the development of broad-spectrum herbicides with the aim of disrupting the histidine biosynthesis. Thus, triazolylphosphonates have been identified as a class of molecules displaying IGPD inhibition, among them 2-hydroxy-3-(1,2,4-triazol-1-yl) propylphosphonate displaying high potency (Fig. **329**) [313].

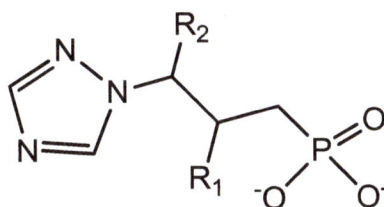

R_1 = OH, R_2 = H Ki <0.001 µM
R_1 = H, R_2 = OH Ki 0.011 µM

Fig. (329). Chemical structure of triazolylphosphonates displaying IGPD inhibition.

12.12. Histidinol Phosphate Aminotransferase (Hisc)

The seventh step in the histidine biosynthesis is performed by the enzyme HisC responsible for the conversion of imidazole acetol phosphate to L-histidinol phosphate, involving the transfer of the amino group from glutamate to L-histidinol-P, along with α-ketoglutarate as by-product (Fig. **330**).

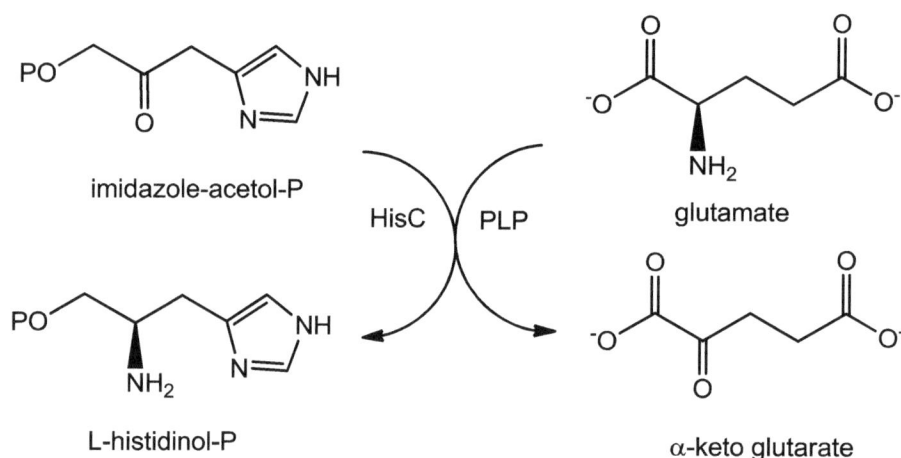

Fig. (**330**). The histidinol phosphate aminotransferase reaction.

HisC belong to the aminotransferase family together with aspartate, and tyrosine AT, which are dependent of pyridoxal-5'-phosphate (PLP) required to form the Schiff base and the internal aldimine between HisC, PLP and L-histidinol phosphate as it can be seen in (Fig. **331**) [314].

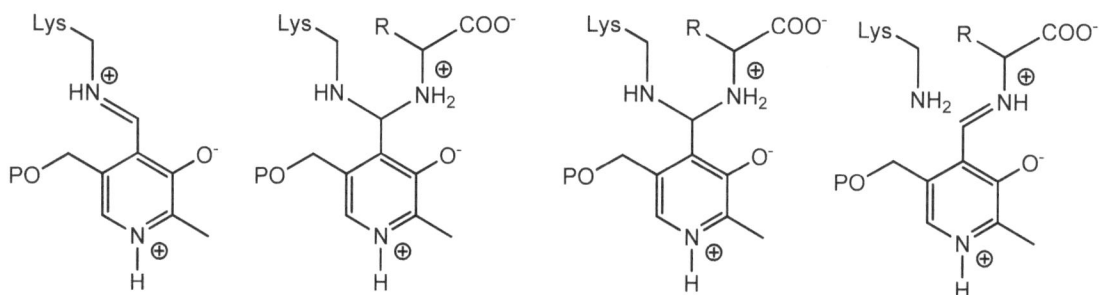

Fig. (**331**). Chemical structures of the Schiff base and the internal aldimine between HisC, PLP and L-histidinol phosphate.

The structure of a dimeric HisC from *Corinebacterium glutamicum* containing a N-terminal arm subunit (red and light blue), a large PLP-binding domain (blue and salmon) and a small C-terminal domain (dark blue and orange) were identified (Fig. **332**) [315].

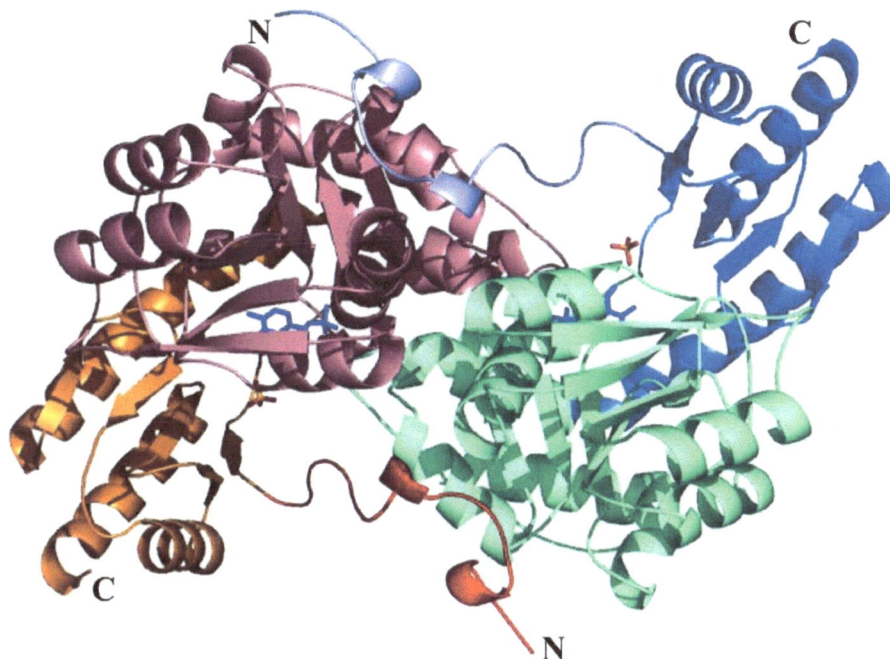

Fig. (332). Ribbon structure of a dimeric HisC from *Corinebacterium glutamicum* (PDB: 3CQ6).

12.13. Histidinol Phosphate Aminotransferase Inhibitors

Docking analysis using HisC complexed with PMP from *E. coli* (PDB id: 1fg7, 29% homology) were conducted and their binding energies and G-scores with favourable values calculated (Fig. **333**) [316].

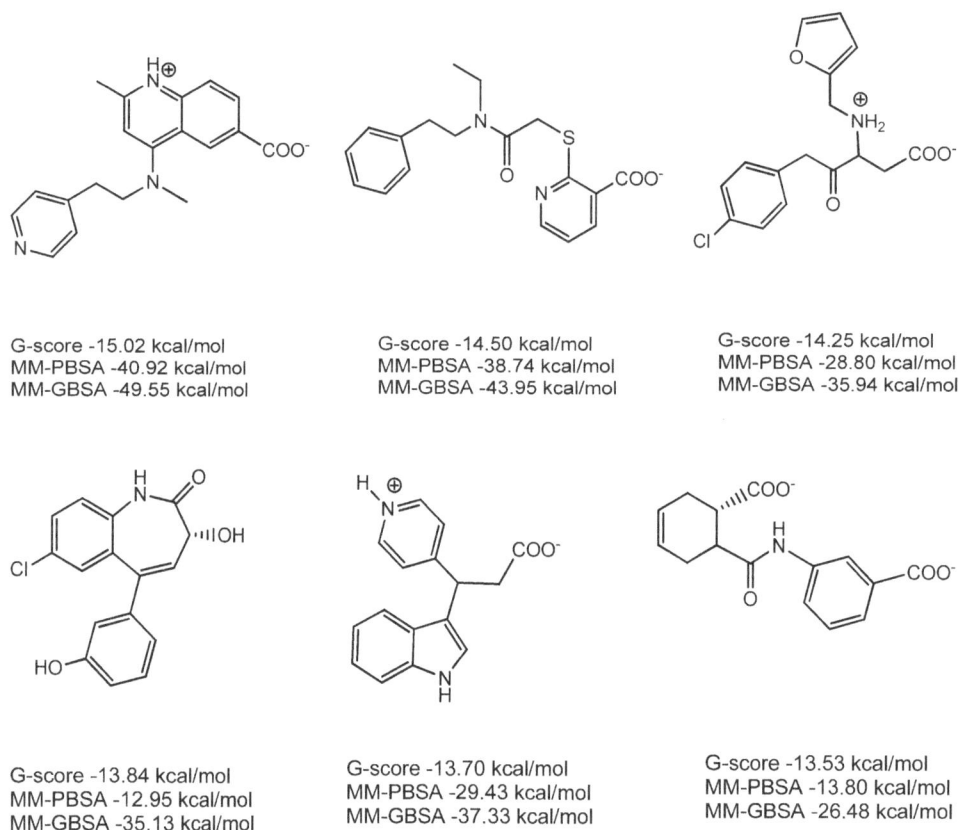

G-score -15.02 kcal/mol
MM-PBSA -40.92 kcal/mol
MM-GBSA -49.55 kcal/mol

G-score -14.50 kcal/mol
MM-PBSA -38.74 kcal/mol
MM-GBSA -43.95 kcal/mol

G-score -14.25 kcal/mol
MM-PBSA -28.80 kcal/mol
MM-GBSA -35.94 kcal/mol

G-score -13.84 kcal/mol
MM-PBSA -12.95 kcal/mol
MM-GBSA -35.13 kcal/mol

G-score -13.70 kcal/mol
MM-PBSA -29.43 kcal/mol
MM-GBSA -37.33 kcal/mol

G-score -13.53 kcal/mol
MM-PBSA -13.80 kcal/mol
MM-GBSA -26.48 kcal/mol

Fig. (333). Chemical structure and calculated binding energies and G-scores of potential HisC inhibitors.

12.14. Histidinol Phosphate Phosphatase (HPP)

The penultimate step in the histidine biosynthesis is catalysed by the enzyme histidinol phosphate phosphatase converting L-histidinol phosphate to L-histidinol, requiring metal ion for the catalysis (Fig. **334**).

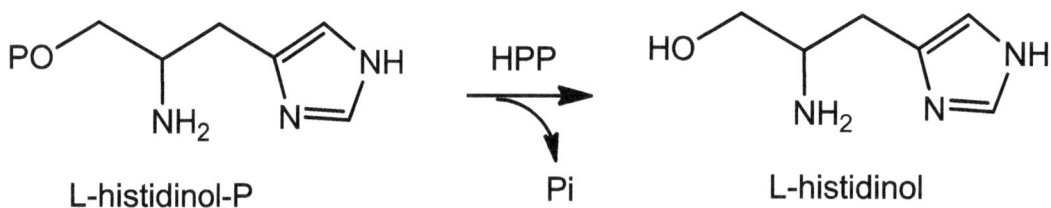

Fig. (334). The histidinol phosphate phosphatase reaction.

The proposed mechanism involves the binding of histidinol phosphate to the active site forming a bridge complex with metal ions. The oxygen not involved in the complex interacts with arginine residues (Arg160 and 197), resulting in the phosphorous-oxygen bond cleavage and consequently the phosphate group hydrolysis affording L-histidinol (Fig. **335**) [316].

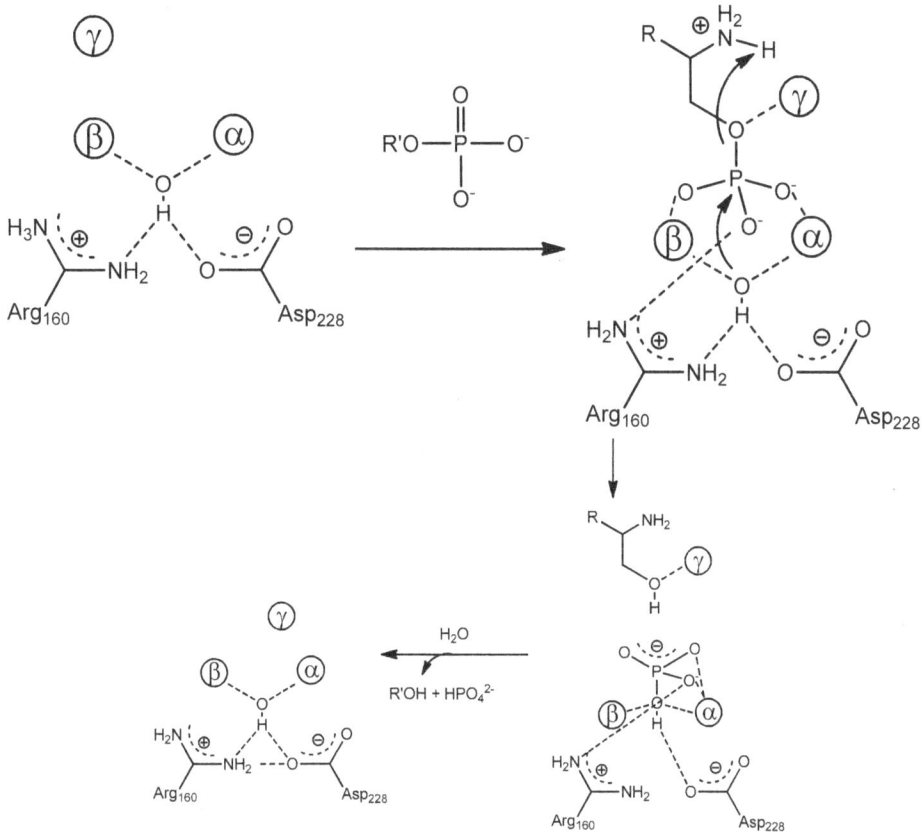

Fig. (335). Proposed mechanism the dephosphorylation process to convert L-histidinol phosphate to L-histidinol.

The ribbon diagram of histidinol phosphate phosphatase from *L. Lactis* is represented, with the α-helices in red, β-sheets in green, and at the active site a histidinol and phosphate interacting with Zn^{2+} ions as orange spheres, surrounded by the interacting residues (Fig. **336**) [316].

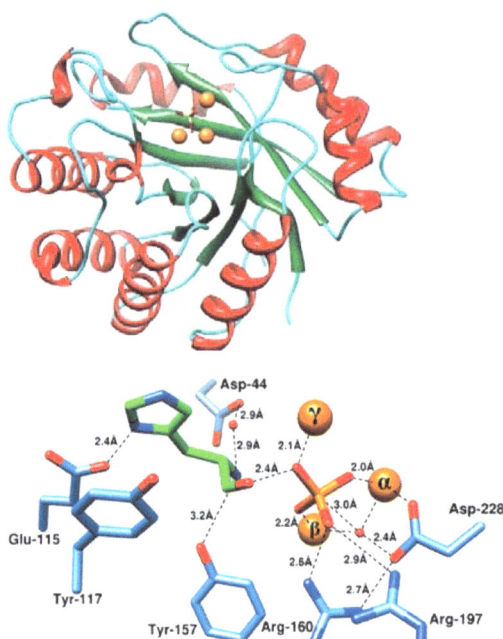

Fig. (336). The ribbon diagram of histidinol phosphate phosphatase from *L. Lactis* and histidinol and phosphate interacting with Zn^{2+} ions (PDB: 4GK8).

12.15. Histidinol Phosphate Phosphatase Inhibitors

High-throughput screening was performed to identify possible HPP inhibitors, being the carbazole derivative NSC311153 as the most potent displaying an IC_{50} value of 94.25 μM in *in vitro* assays (Fig. **337**) [317].

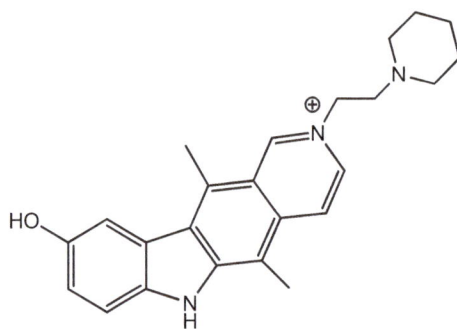

Fig. (337). Chemical structure of carbazole derivative NSC311153.

12.16. Histidinol Dehydrogenase (HISD)

The last step in the histidine biosynthesis is carried out by the metalloenzyme histidinol dehydrogenase which is a bifunctional enzyme converting L-histidinol to L-histidinaldeyde and subsequently to L-histidine, requiring Zn^{2+} and 2 NAD^+ equivalents (Fig. **338**).

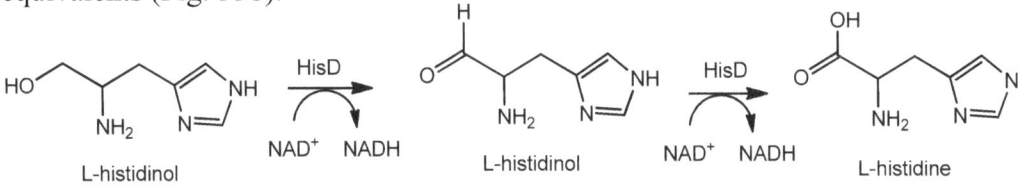

Fig. (338). The histidinol dehydrogenase reaction.

The mechanism requires two consecutive oxidation steps from L-histidinol to L-histidinaldehyde and to L-histidine with the participation of two NAD^+ molecules. Crystallographic analysis indicates that either the aldehyde or the alcohol, occurs at the active site (Fig. **339**) [318].

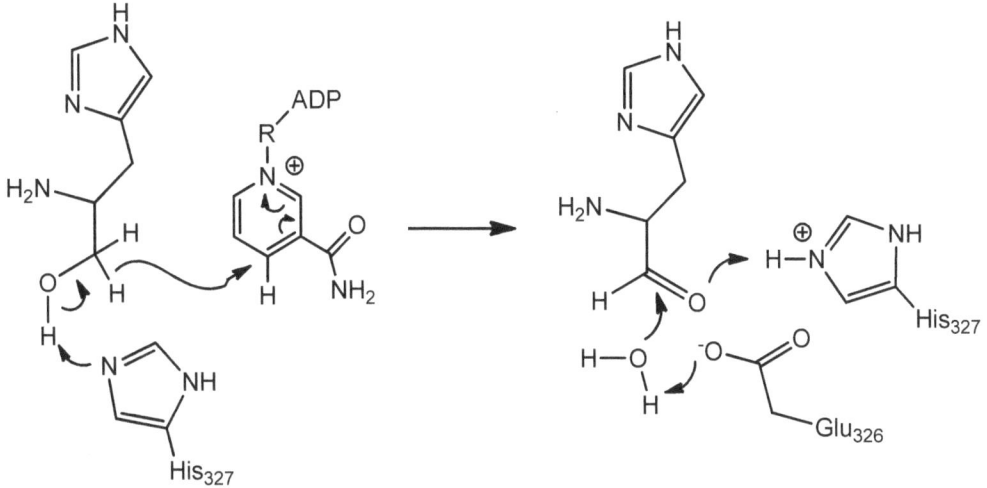

Fig. 339 cont.....

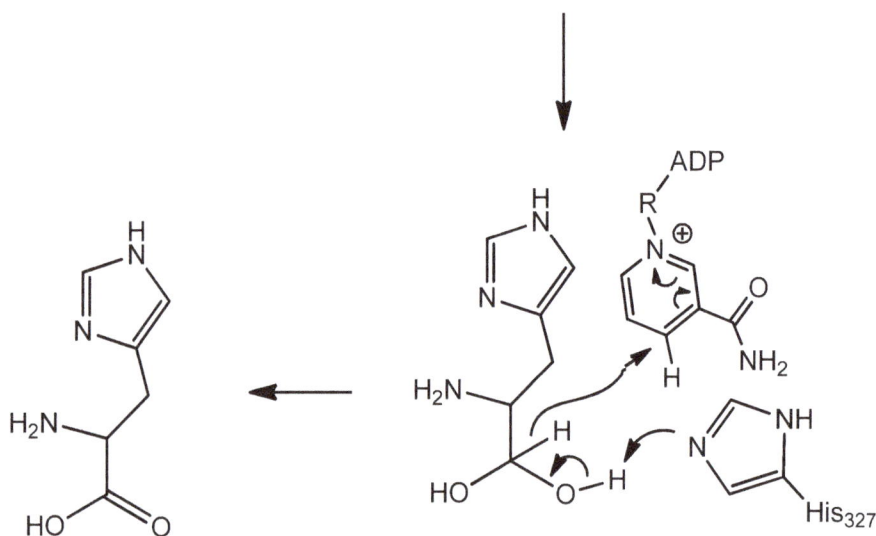

Fig. (339). Proposed reaction mechanism for L-histidine formation.

The crystal structure of HisD from *Mycobacterium tuberculosis* as homodimer presents domains I-IV, six stranded β-sheet with NAD⁺, L-histidinol (HOL) at the active site, and in a closer view the Zn^{2+} cation in deep pocket between domains I, II and IV, and the binding region of HOL with their surrounding residues (Fig. **340**) [319].

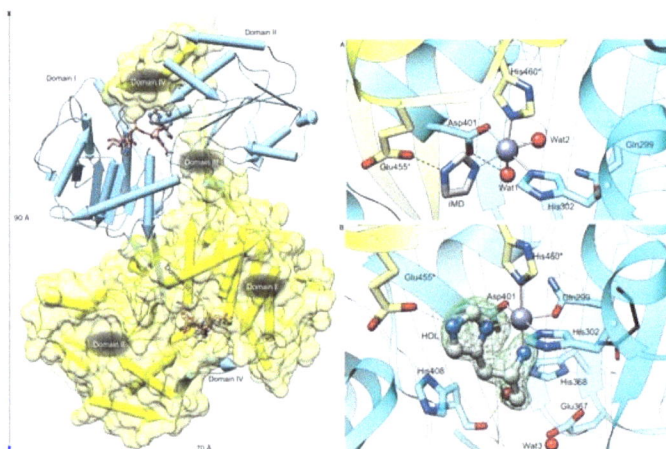

Fig. (340). Crystal structure of HisD from *Mycobacterium tuberculosis* and the binding region showing L-histidinol coordinated with Zn^{2+} and the residues interacting (PDB: 5VLC).

12.17. Histidinol Dehydrogenase Inhibitors

Potent HisD inhibitors were designed as herbicides with inhibition values at micro and nanomolar scale. An early study described the inhibitory capacity of histidine methyl ketone analogue having a K_i value of 5 µM, and based on these findings substituted benzyl ketones were designed and evaluated as histidinol dehydrogenase inhibitors. The results show potent inhibition at nano scale for some 3- and 4 substituents (Fig. **341**) [320].

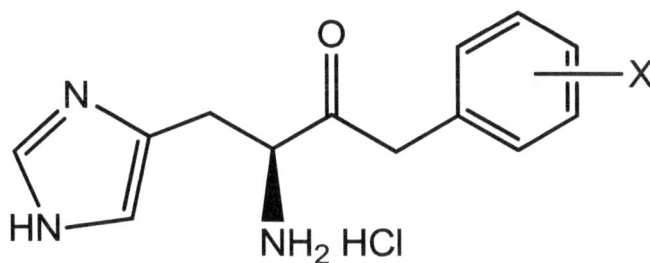

	IC_{50} µM	Ki nM
H	0.1	7.6
2-Br	1	
3-OH	0.3	
3-NH$_2$	0.05	7.4
3-Br	0.1	
4-OMe	1	
4-NH$_2$	0.5	
4-Cl	0.3	
4-Br	0.04	4.4
4-Ph	0.04	2.9
3-Br, 4-Br	0.05	

Fig. (341). Chemical structure and inhibition constants of histidine methyl ketone analogues.

Hydrazone derivatives of L-histidine were synthesized and evaluated as *Mt*HisD and their IC_{50} and inhibition constant (Ki) were measured, observing that naphtyl and 4-substituted with halogen (Cl, F) and electron withdrawing nitro group presented the highest inhibition values [321]. Other imidazole derivatives displaying IC_{50} values of significance are imidazole ethyl ester and imidazole pyrimidine carboxylic acid (Fig. **342**) [322].

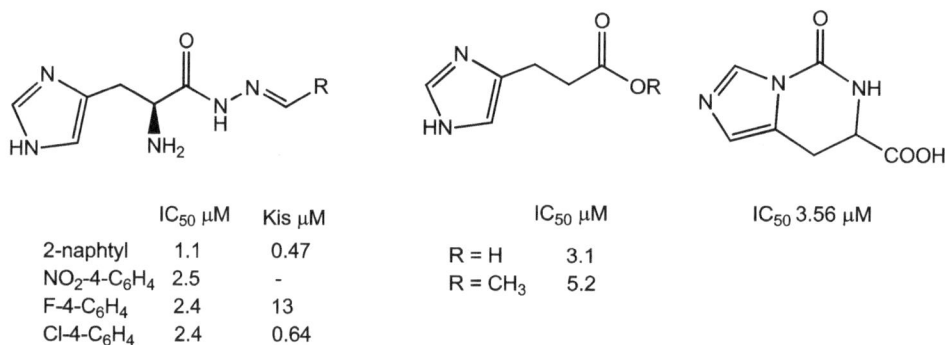

	IC_{50} μM	Kis μM
2-naphtyl	1.1	0.47
NO_2-4-C_6H_4	2.5	-
F-4-C_6H_4	2.4	13
Cl-4-C_6H_4	2.4	0.64

	IC_{50} μM
R = H	3.1
R = CH_3	5.2

IC_{50} 3.56 μM

Fig. (342). Chemical structure and inhibition constants of histidine hydrazone, imidazole ethyl ester and imidazole pyrimidine derivatives.

13. PROLINE BIOSYNTHESIS

Proline is a pyrrolidine heterocyclic amino acid with important implications in protein folding due his cyclic structure. Moreover, in cancer cells proline biosynthesis increase cell proliferation and indirectly supports biomass production. Its biosynthesis requires glutamine as starting material that is phosphorylated, and reduced to the aldehyde. The next step involves cyclization to the heterocyclic pyrrolidine-5-carboxylate and final reduction by pyrrolidine 5-carboxylase reductase enzyme to yield proline (Fig. **343**).

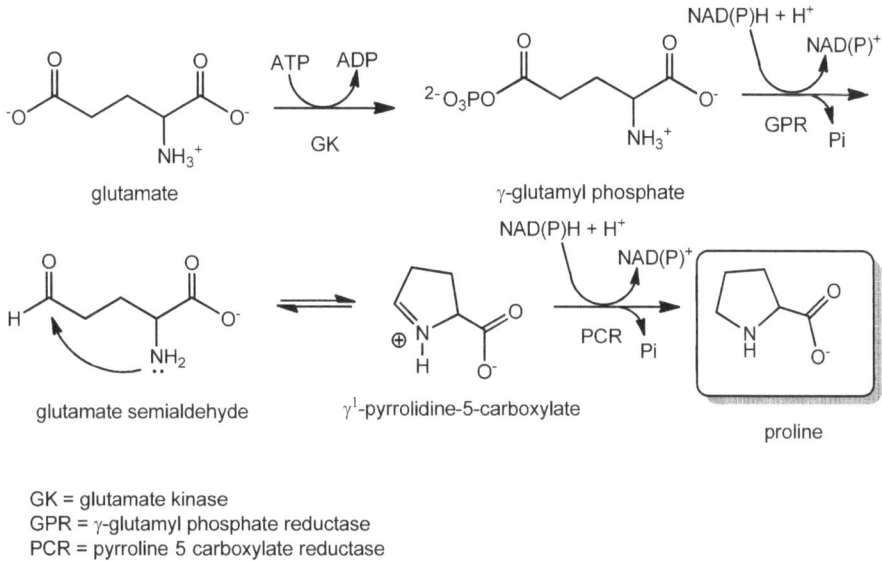

Fig. (343). The proline biosynthesis.

13.1. Glutamate 5-Kinase (G5K)

The first step in the proline synthesis is catalysed by the enzyme glutamate 5-kinase (G5-K), which converts glutamate to γ-glutamyl phosphate (Fig. **344**). In bacteria the process for producing L-proline requires a two-step sequence mediated by G5K and pyrrolidine-5-carboxylate reductase (GPR), while in higher eukaryotes the enzyme pyrroline-5-carboxylate synthase (P5CS) combines both activities.

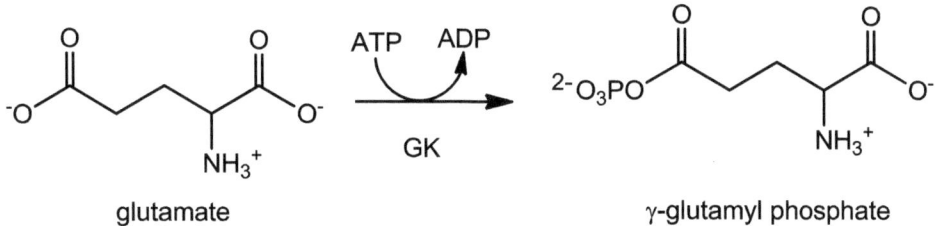

Fig. (344). The glutamate 5-kinase reaction.

The ribbon and surface representations of glutamate 5-kinase from *E. coli* were described, the first exhibiting a subunit containing the catalytic domains AAK and PUA. The surface representation shows the binding site with glutamate and ADP embedded in the pocket, and surrounded by the residues involved in the interactions (Fig. **345**) [323].

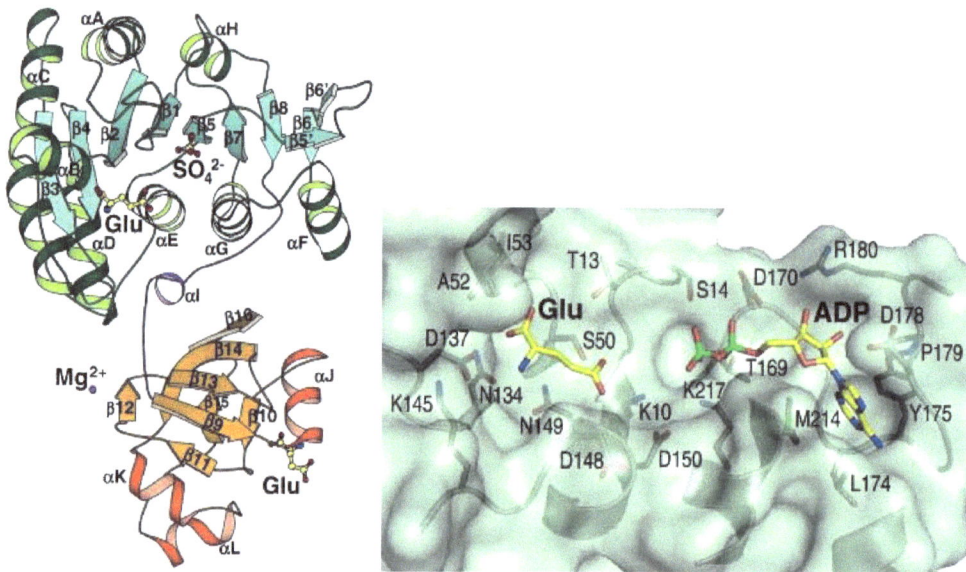

Fig. (345). Ribbon and surface structures of glutamate 5-kinase from *E. coli* (PDBB: 2J5T).

13.2. Glutamate 5-Kinase Inhibitors

Proline is a feedback inhibitor of bacterial glutamate kinase and plant pyrroline--carboxylate synthase enzymes. Different proline analogues were synthesized and tested as G5K, finding that the natural substrate and the unsaturated analogues were the most effective (Fig. **346**) [324].

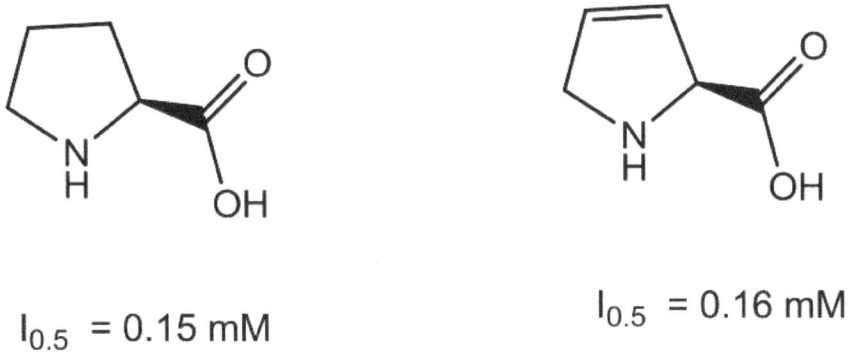

$I_{0.5} = 0.15$ mM

$I_{0.5} = 0.16$ mM

Fig. (346). Chemical structure and inhibition constants of L-proline and 3,4-dehydro-L-proline.

13.3. γ-Glutamyl Phosphate Reductase (GPR)

The second step of proline biosynthesis is mediated by GPR involving the conversion of γ-glutamyl phosphate into glutamate semialdehyde requiring NADPH as reducing cofactor (Fig. **347**).

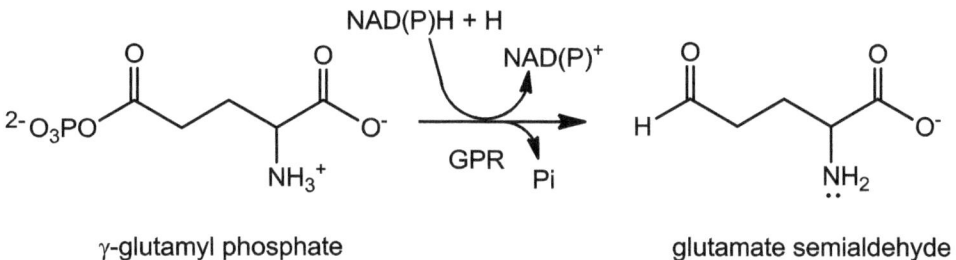

γ-glutamyl phosphate   glutamate semialdehyde

Fig. (347). The γ-glutamyl phosphate reductase reaction.

The ribbon representation of γ-glutamyl phosphate reductase from *Thermotoga maritima* monomer presenting three domains, indicating the NADPH binding pocket, a hinge region and Cys255 as a key residue for the catalysis (Fig. **348**) [325].

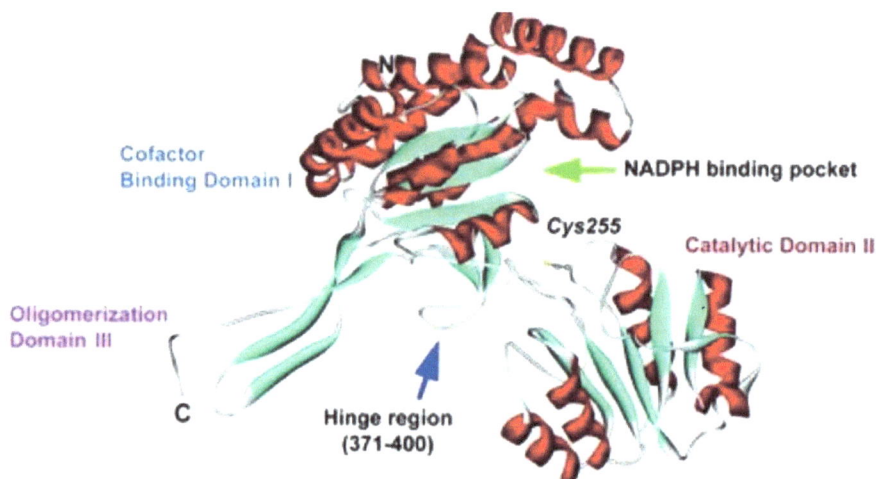

Fig. (348). Ribbon diagram of γ-glutamyl phosphate reductase from *Thermotoga maritime* (PDB: 1O20).

13.4. γ-Pyrroline-5-carboxylate reductase (P5CR)

The last step in the proline biosynthesis mediated by P5CR implies a redox transformation requiring NAD(P)H as reducing agent and unsaturated substrate Δ^1-pyrrolidine-5-carboxylate (P5C), which is reduced to L-proline and oxidized cofactor NADP$^+$ (Fig. **349**).

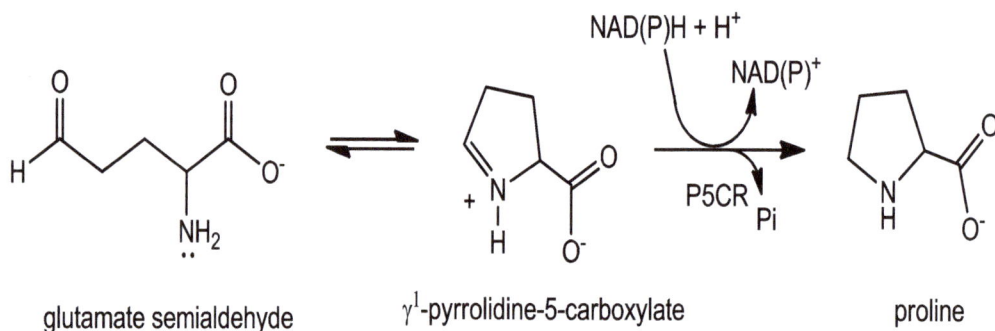

Fig. (349). The γ-pyrroline-5-carboxylate reductase reaction.

The monomer structure of human γ-pyrroline-5-carboxylate reductase shows two domains and the molecular surface diagram illustrates the binding site pocket, containing a substrate analogue and NADH (Fig. **350**) [326].

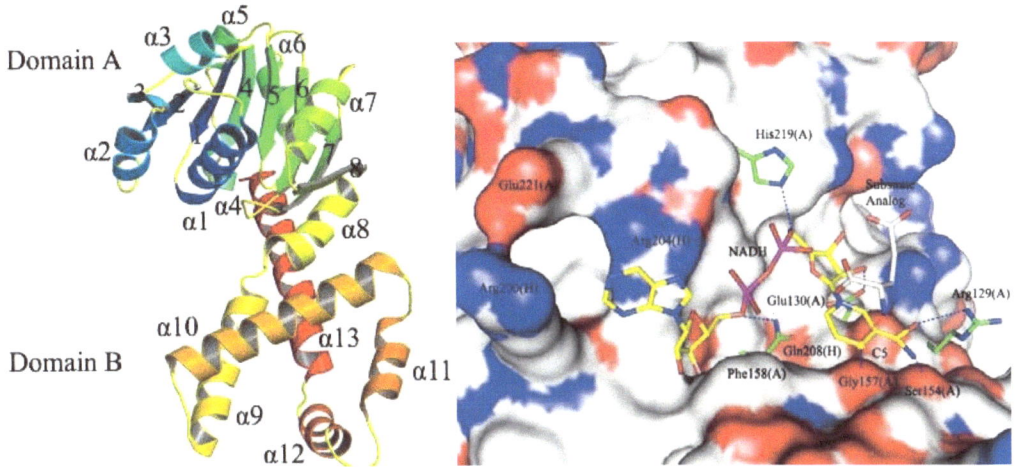

Fig. (350). Ribbon and surface representations of human γ-pyrroline-5-carboxylate reductase (PDB: 2GER).

14. Arginine Biosynthesis

Arginine is a guanidino amino acid involved in a number of essential steps such as the urea cycle, production of nitric oxide, synthesis of creatine, production of citrulline and ornithine, and in the collagen synthesis. In humans its biosynthesis is achieved in two step sequence, the first a condensing reaction between L-citrulline and L-aspartate mediated by the enzyme argininosuccinate synthase, to give N-(L-arginino) succinate, and the second catalysed by the enzyme argininosuccinate lyase, which performs a cleavage reaction, yielding L-arginine and fumarate. The scheme also illustrates the role of L-arginine as starting material in the formation of L-ornithine and L-citrulline, the later needed to restart the cycle (Fig. **351**).

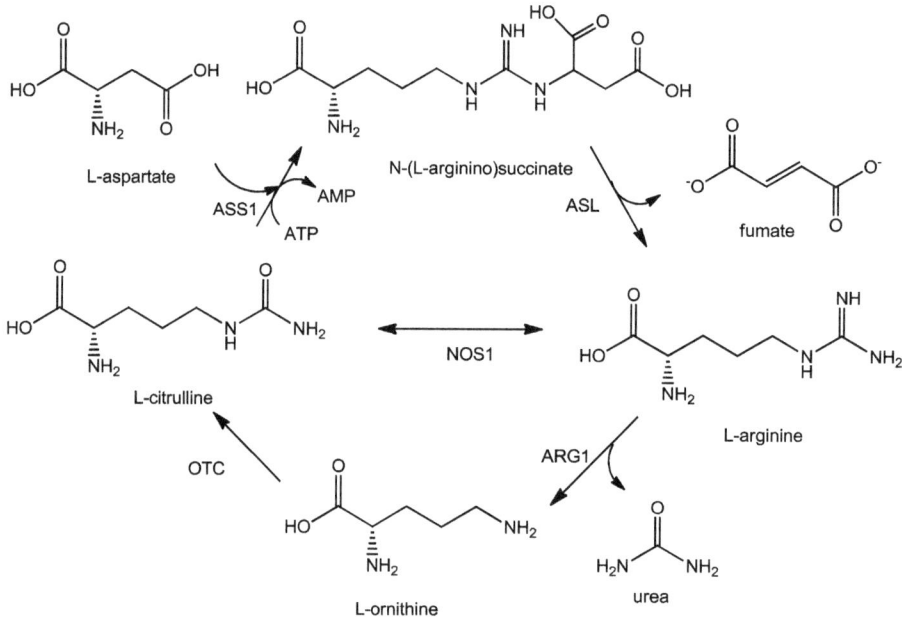

Fig. (351). The arginine biosynthesis.

ASS1 = argininosuccinate synthase

ASL = argininosuccinate lyase

ARG1 = argininase

OTC = ornithine carbamoyltransferase

NOS1 = nitric oxide synthase

14.1. Argininosuccinate Synthase (ASS1)

Is a rate limiting enzyme controlling the conversion of L-citrulline and L-aspartate to N-(L-arginino) succinate, and requires ATP molecule to complete the transformation (Fig. **352**).

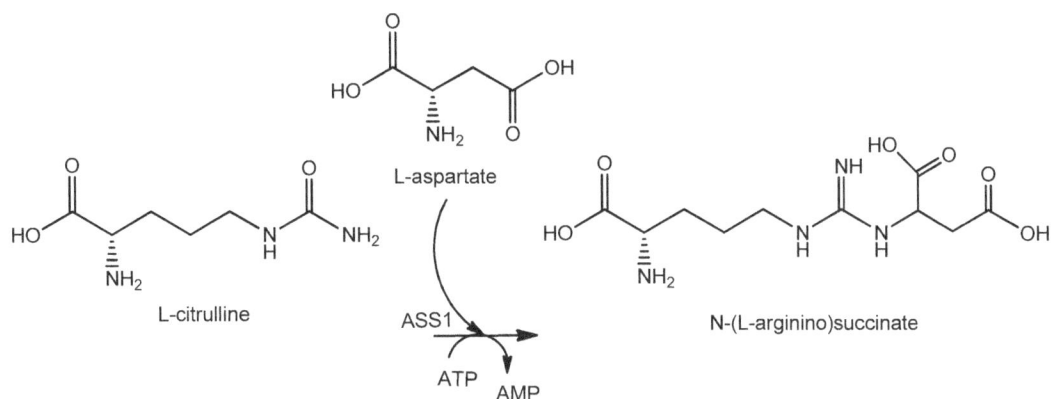

Fig. (352). The argininosuccinate synthase reaction.

A feasible mechanism proposes the phosphorylation reaction between citrulline and ATP to give citrulline adenylate and nucleophilic substitution from aspartate amino group to give the target molecule arginine succinate (Fig. **353**) [327].

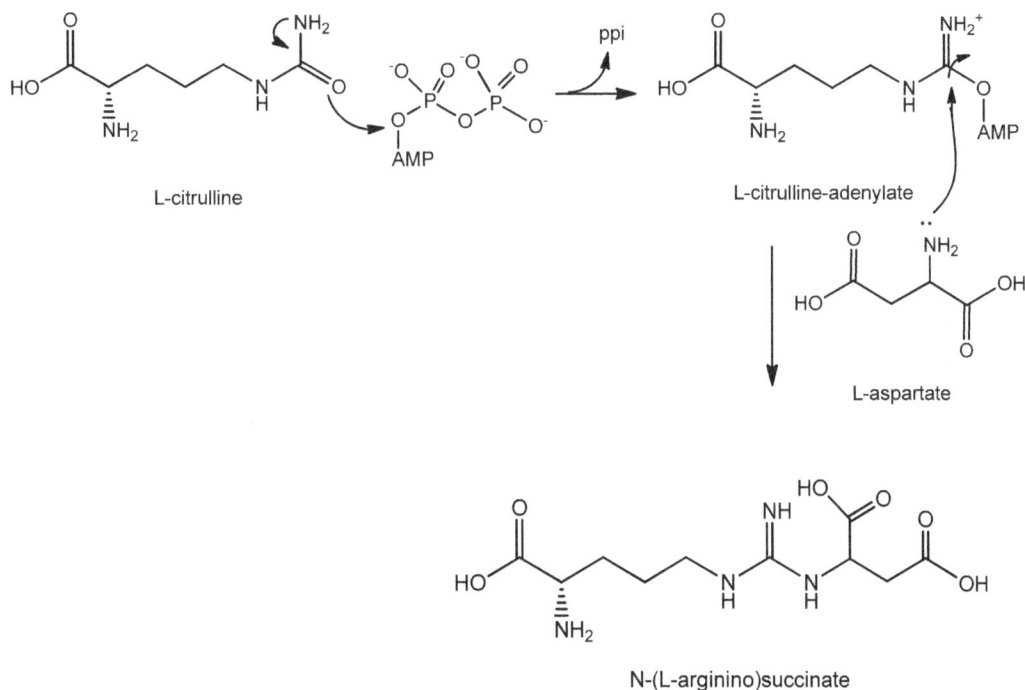

Fig. (353). Proposed mechanism for the biosynthesis of arginine succinate.

The crystal structure of human argininosuccinate synthase (hASS) was characterized and the active site identified (Fig. **354**). The substrates citrulline and aspartate were modelled into the electron density built into the active site [328].

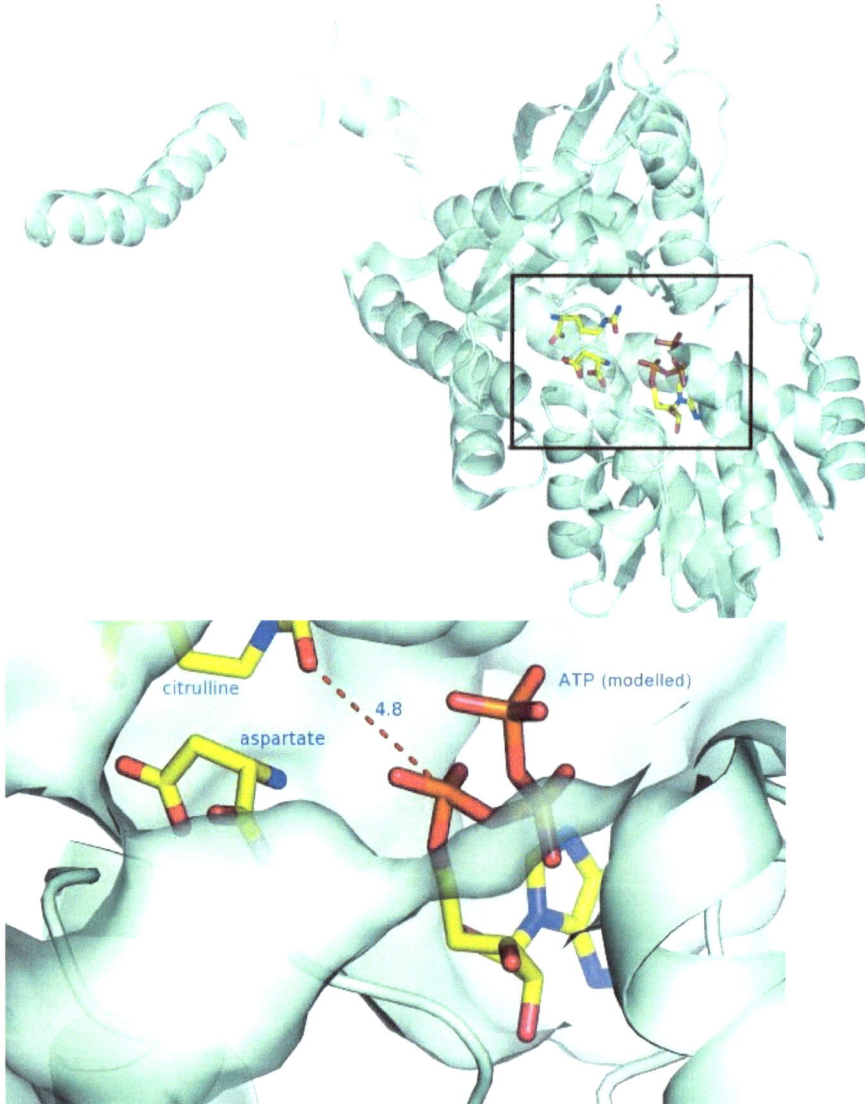

Fig. 354 cont.....

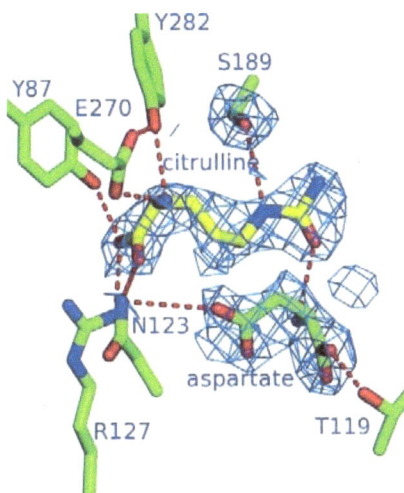

Fig. (354). Ribbon diagram human argininosuccinate synthase and electron density of citrulline and aspartate (PDB: 2NZ2)

14.2. Argininosuccinate Synthase Inhibitors

The fungal metabolite fumonisin B1 (Fig. **355**) produced by *Fusarium monilforme* was described as mycotoxin capable of altering sphingolipid metabolism because of the inhibition of ceramide synthase. Kinetic studies were conducted to evaluate *in vitro* inhibition of ASS1, observing mixed-type inhibition with K_i of 6 mM for FB1 [329].

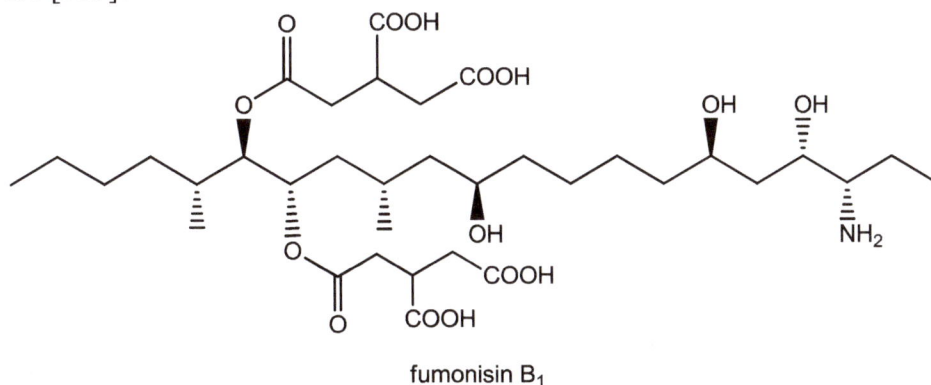

fumonisin B$_1$

Fig. (355). Chemical structure of fumonisin B1.

14.3. Argininosuccinate Lyase (ASL)

This enzyme participate in the reversible cleavage of N-(L-arginino)succinate producing L-arginine, highly implicated in cell growth and pathogenesis, and fumarate an intermediate in the citric acid cycle (Fig. **356**).

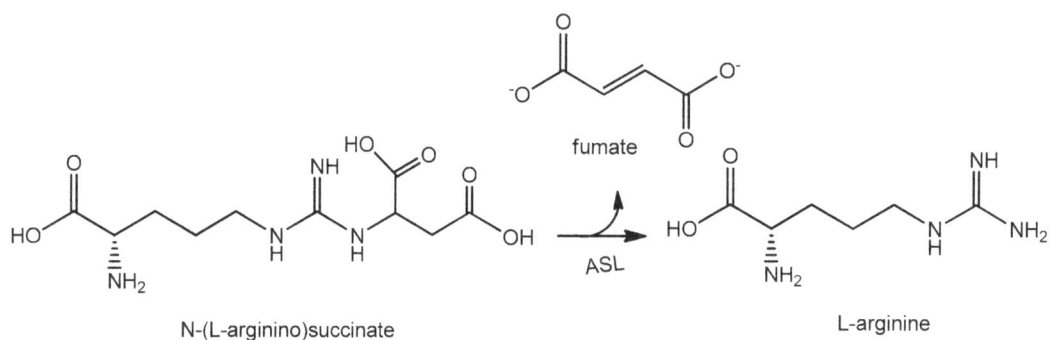

Fig. (356). The argininosuccinate lyase reaction.

The mechanism proposes an E1cB elimination-type proceeding with histidine residue acting as base to withdraw the hydrogen at C-9 position to generate the carbanion that establish resonance with the carboxylate, promoting the exit of fumarate group (Fig. **357**) [330].

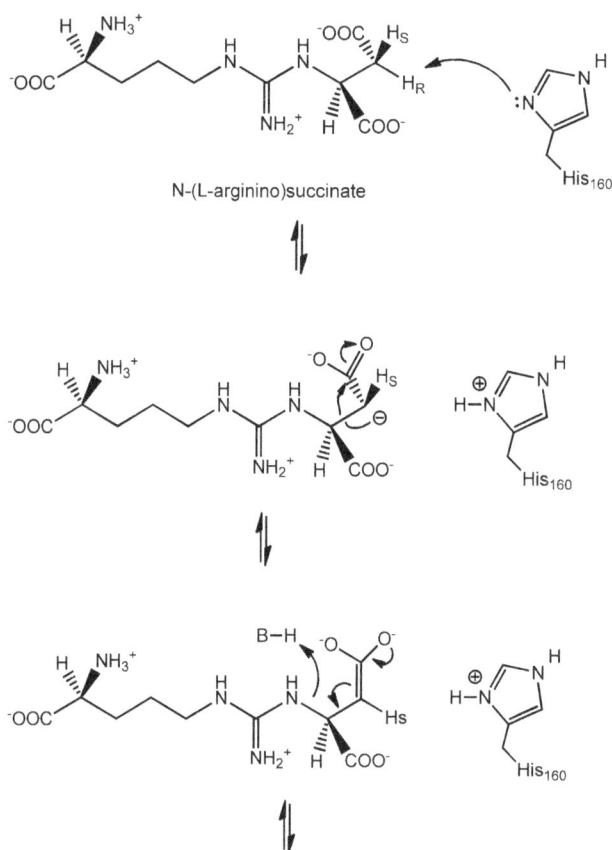

Fig. 357 cont.....

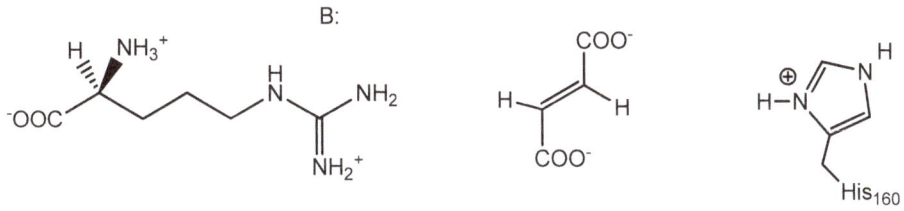

Fig. (357). Proposed mechanism for the biosynthesis of arginine and fumarate from N-(L-arginino)succinate.

The crystal structure of argininosuccinate lyase from *Mycobacterium tuberculosis* (*Mt*ASL) was characterized as tetramer containing the active site shown at the interface between monomer B and D. Each monomer is conformed by three domains, being the N-, the central and the C-domains (Fig. **358**) [331].

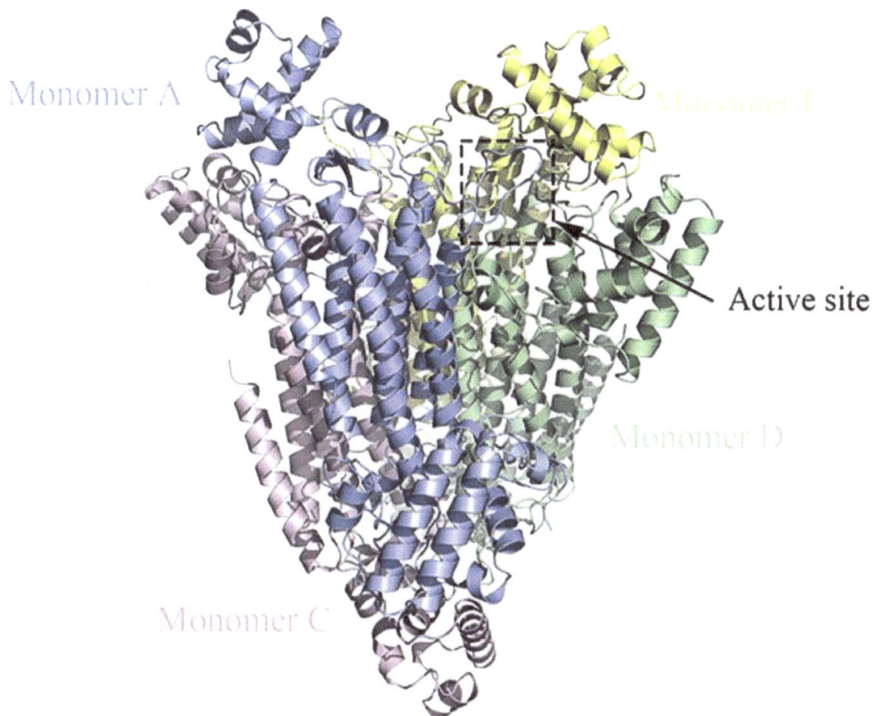

Fig. 358 cont.....

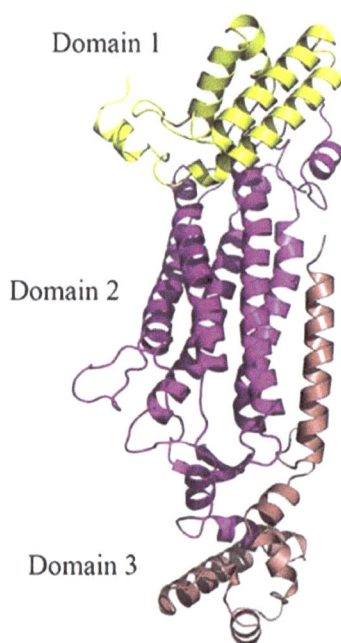

Fig. (358). The crystal structure of argininosuccinate lyase from *Mycobacterium tuberculosis* as tetramer and monomer (PDB: 6IGA).

14.4. Argininosuccinate Lyase Inhibitors

Arginine and nitric oxide have been identified as mediators in the development of tumours, and therefore the development of compounds capable for interrupting arginine biosynthesis result in a feasible strategy in the control of cancerous growth. However up to now only urea (Fig. **359**) has been evaluated as argininosuccinate lyase inhibitor showing competitive inhibition with a maximal around 4.15 mM of urea concentration [332].

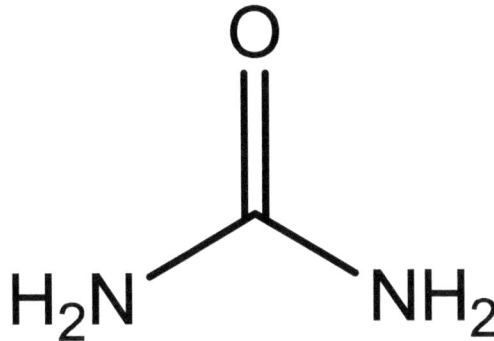

Fig. (359). Chemical structure of urea as ASL inhibitor.

15. Lysine Biosynthesis

Lysine is a basic amino acid not synthesized by humans, with importance in the production of carnitine and in combination with arginine the responsible of most of the advanced glycation products (AGES) observed during degenerative process such as diabetes, atherosclerosis, and neurodegenerative disease. In most of bacteria and plants lysine is synthesized from aspartate following the diaminopimelate (DAP) route consisting in the phosphorylation of carboxylate to give aspartate β-phosphate by the enzyme aspartate kinase (AK), then dephosphorylation and reduction providing aspartate semialdehyde under the catalysis of aspargyl semialdehyde dehydrogenase (ASD). Condensation of aspartate semialdehyde with pyruvate to produce 2,3-dihydrodipicolinate which is further reduced to Δ'-piperideine-2-6-dicarboxylate. Ring opening and esterification yields N-succinyl or acetyl α-amino-ε-ketopimelate which is submitted to transamination, followed by succinyl removal, epimerization and decarboxylation with the enzymes DapA-F to furnish Lysine (Fig. **360**).

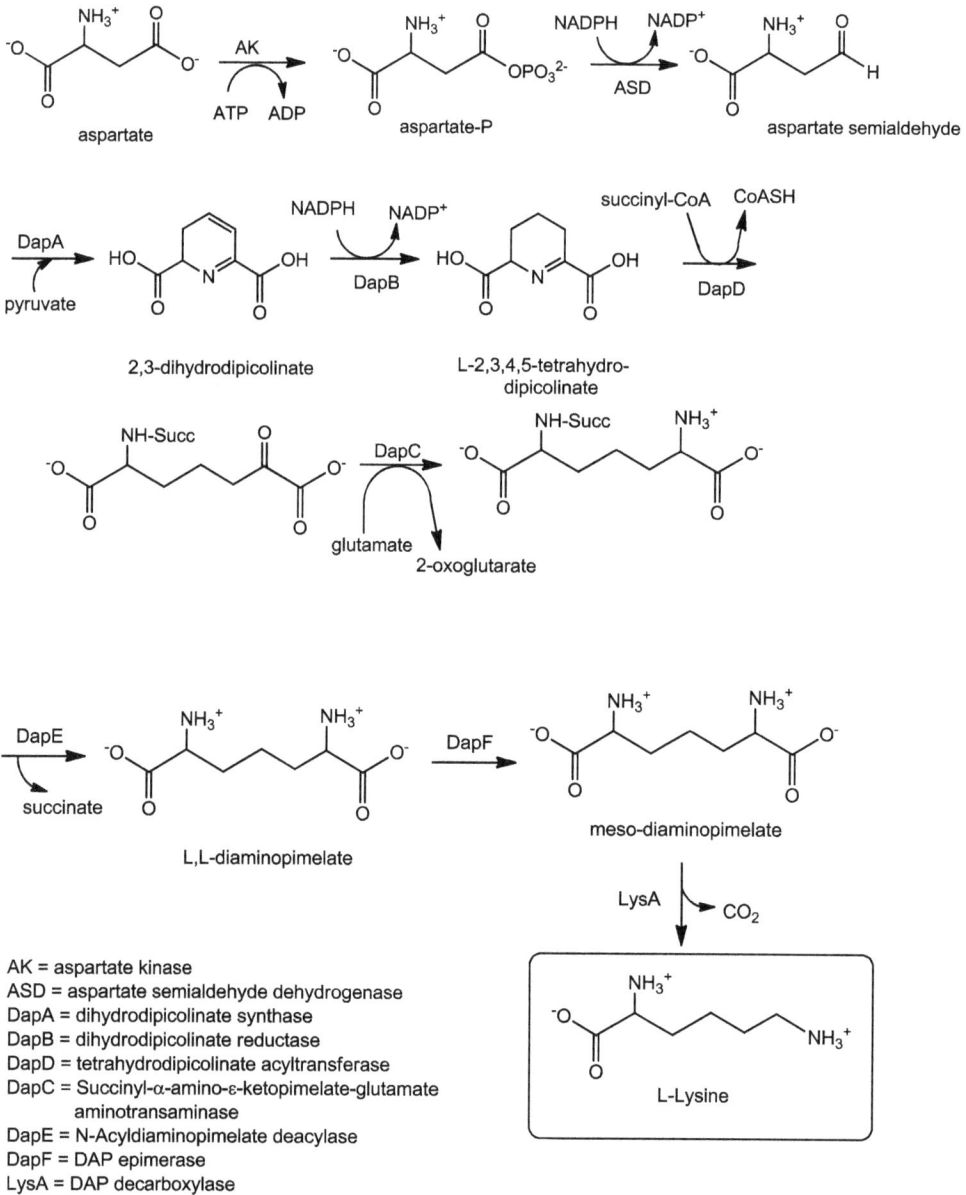

AK = aspartate kinase
ASD = aspartate semialdehyde dehydrogenase
DapA = dihydrodipicolinate synthase
DapB = dihydrodipicolinate reductase
DapD = tetrahydrodipicolinate acyltransferase
DapC = Succinyl-α-amino-ε-ketopimelate-glutamate
        aminotransaminase
DapE = N-Acyldiaminopimelate deacylase
DapF = DAP epimerase
LysA = DAP decarboxylase

Fig. (360). Biosynthesis of lysine from L-aspartate.

15.1. Dihydrodipicolinate Synthase (DHDPS)

Is an enzyme involved in the lysine biosynthesis in plant and microorganisms converting aspartate semialdehyde and pyruvate into (2S,4S)-2, 3-dihydrodipicolinate (HTPA) (Fig. **361**).

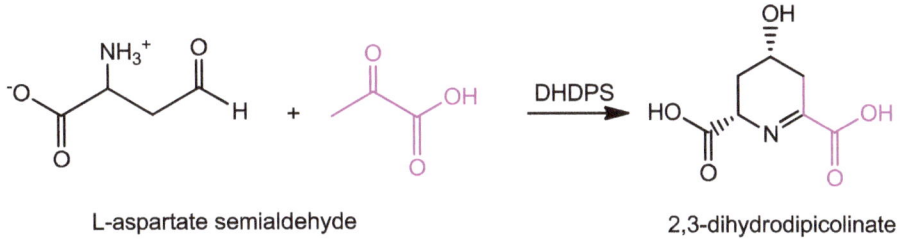

Fig. (361). The dihydrodipicolinate synthase reaction.

DHDPS from wild type E. *coli* is a homotetramer or dimer of dimers with the active site located at the inner face (Fig. **362**). At the active site threonine (T), tyrosine (Y), glycine (G), arginine (R) and isoleucine (I) are found in wild type and in mutant *E. coli* [333].

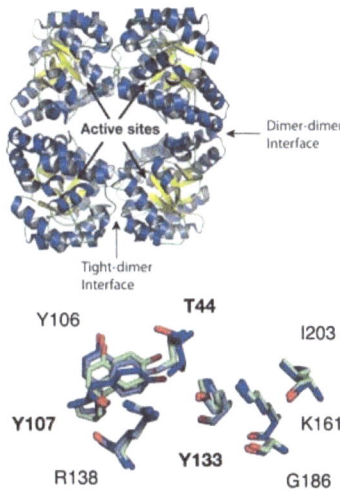

Fig. (362). Ribbon diagram of DHDPS from wild type E. *coli* (PDB: 1YXC).

The mechanism initiate with a nucleophilic attack from lys161 to pyruvate, producing an enamine, which reacts with (S)-aspartate-β-semialdehyde providing

(S)-ASA-K161 conjugate. The cyclisation proceed by amino attack to the iminium position and further cleavage of lys161 residue providing (2S,4S)2,3-dihydrodipicolinate (Fig. **363**) [334].

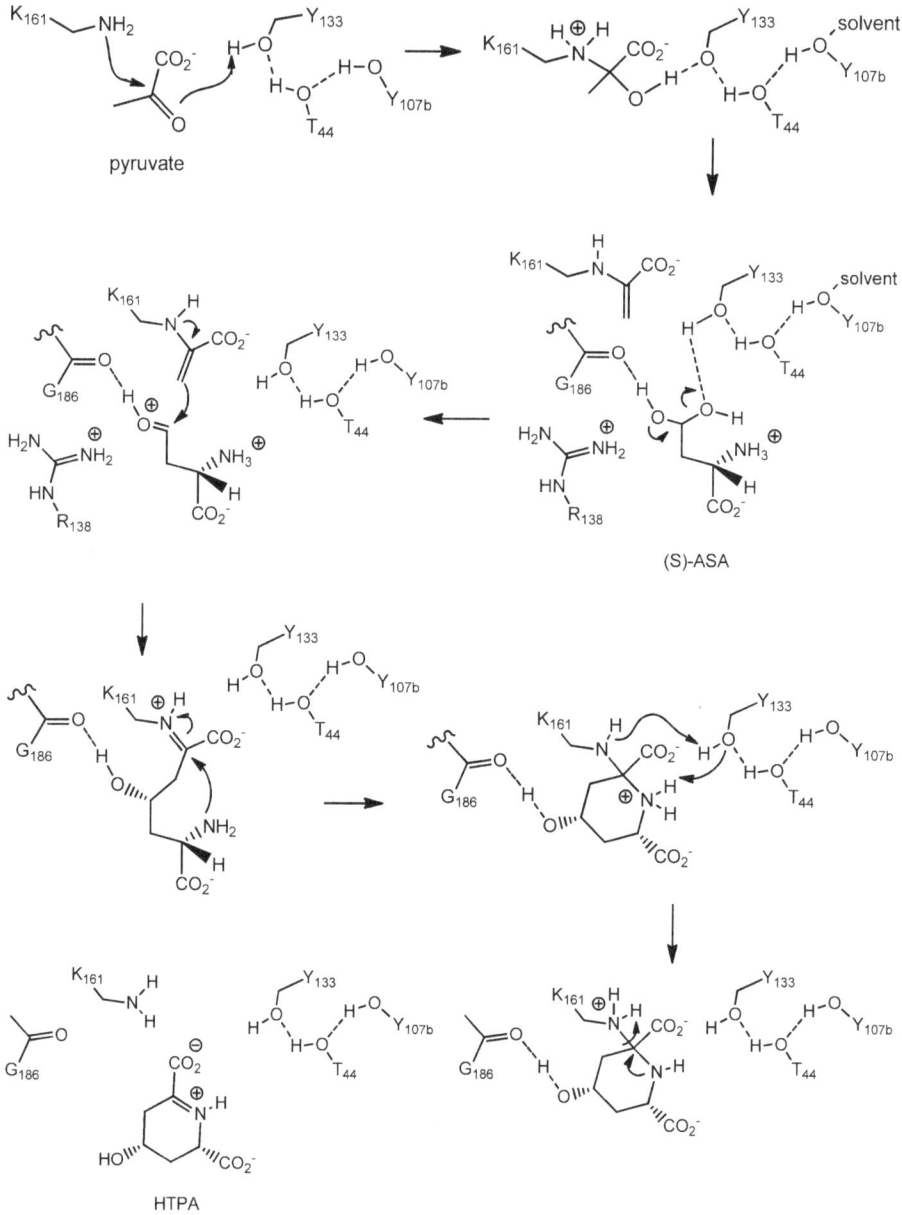

Fig. (363). Proposed reaction mechanism for the formation of (2S,4S)-2,3-dihydrodipicolinate (HTPA).

15.2. Dihydrodipicolinate Synthase Inhibitors

The synthesis and evaluation of (2E,5E)-4-oxoheptadienedionic acid 1a-b and (2E)-4-oxoheptenedioic acid 2a-b derivatives considered simplified HTPA mimics were described (Fig. **364**). The inhibition studies conclude that diester 1a was the most potent inhibitor, producing complete inactivation at 2 mM concentration. On the other hand, the diacid 1b produce lower potency, showing complete inactivation at 10 mM. As for the mono-alkene 2a and 2b they were the less potent inhibitors indicating poor binding to the active site [335].

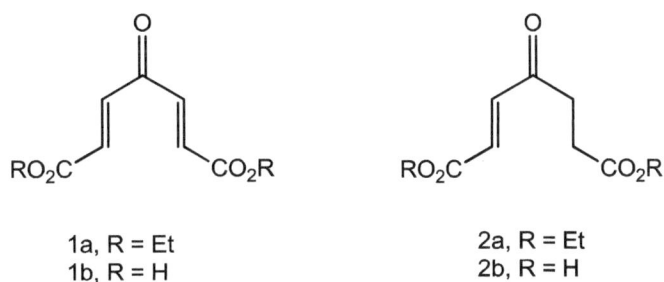

1a, R = Et
1b, R = H

2a, R = Et
2b, R = H

Fig. (364). Chemical structure of (2E,5E)-4-oxoheptadienedionic acid 1a-b and (2E)-4-oxoheptenedioic acid 2a-b.

Additionally, α-ketopimelic acid (Fig. **365**) was reported as DHDPS irreversible inhibitor, following a pseudo-first order and saturation kinetics with K_i value of 0.17 mM in the absence of substrate, although in the presence of pyruvate the enzyme activity is partially increased [336].

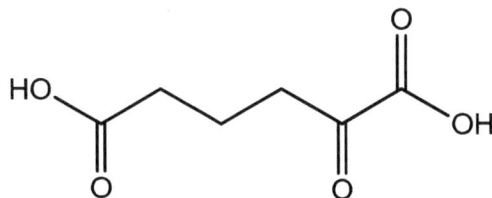

Fig. (365). Chemical structure of α-ketopimelic acid.

15.3. Dihydrodipicolinate Reductase (DHDPR)

DHDPR catalyses the second step in the lysine pathway which convert 2,3-dihydrodipicolinate (DHDP) to 2,3,4,5-tetrahydrodipicolinate (THDP), requiring NADPH as a reducing cofactor (Fig. **366**).

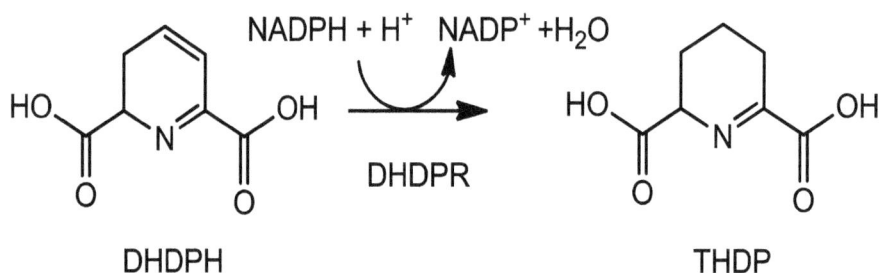

Fig. (366). The dihydrodipicolinate reductase reaction.

The three-dimensional structure of various DHDPR have been described, among them *E. coli*, S. aureus, M tuberculosis and more recently from *Bartonella henselae, Burkholderia thailandensis and Paenisporosarcina sp.* The structure superposition of apo (salmon) and DPA-bound (yellow) PaDHDPR the apo and DPA/NADH-bound (deep purple) forms of TmDHDPR revealed full domain similarities (Fig. **367**) [337].

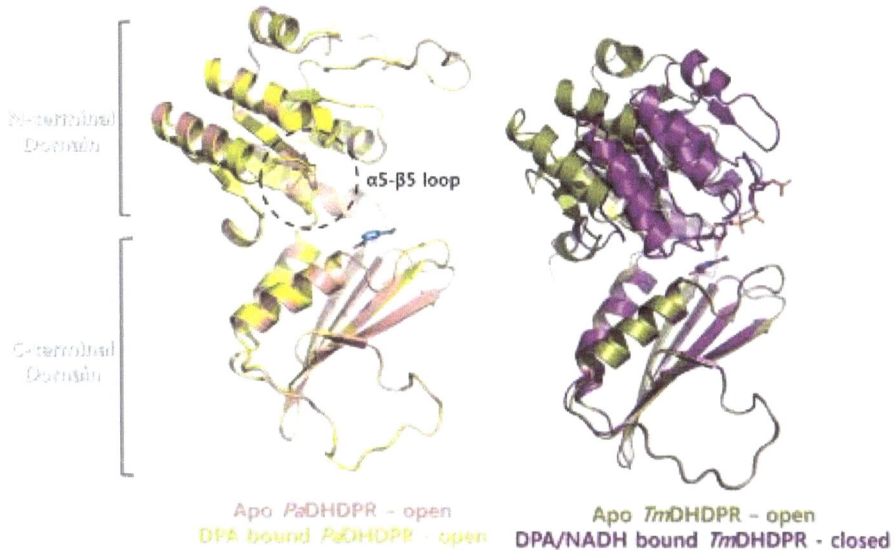

Apo *Pa*DHDPR – open
DPA bound *Pa*DHDPR – open

Apo *Tm*DHDPR – open
DPA/NADH bound *Tm*DHDPR - closed

Fig. (367). Ribbon representation of Dihydrodipicolinate reductase from *Mycobacterium tuberculosis* (PDB: 1C3V).

Despite the substantial information derived from crystallographic studies, only interactions between the enzyme DHPR and NADPH have been established, and conclusive structural analysis leading to understand the reduction step to convert the substrate DHDP into the product THDP remains to be done (Fig. **368**) [338].

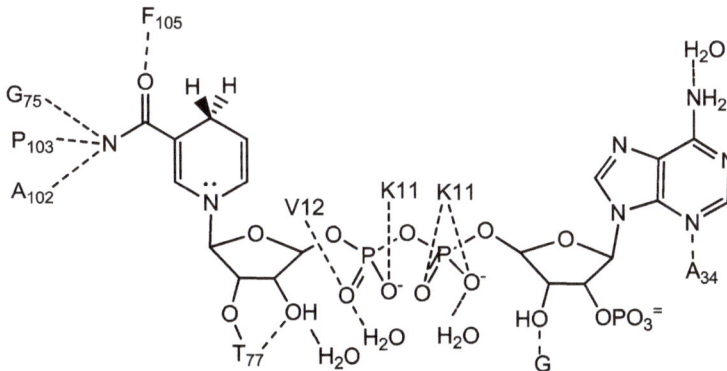

Fig. (368). Interactions between dihydrodipicolinate reductase residues with NADPH.

The reaction mechanism although not explained correspond to the general reduction process involving NADPH consisting in the hydride attack provided by NADPH and further protonation from a residue or water molecule (Fig. **369**)

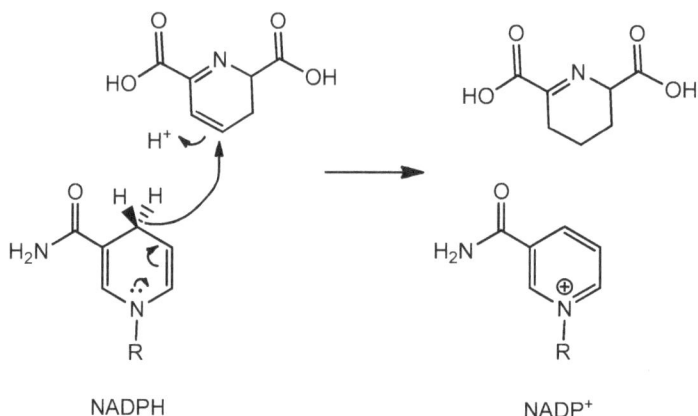

Fig. (369). Reduction step of NADPH to convert DHDP to THDP.

15.4. Dihydrodipicolinate Reductase Inhibitors

A series of sulphonamide derivatives were tested for DHDP reductase inhibition obtained from the Merck chemical collection, with the most potent inhibitors displaying K_i values from 7 to 48 μM (Fig. **370**) [339].

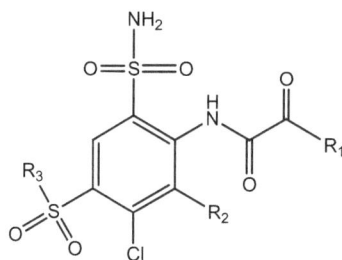

R_1 = OEt, R_2 = Cl, R_3 = NH_2 Ki= 22.6 μM
R_1 = OBu, R_2 = H, R_3 = NH_2 Ki= 35.4 μM
R_1 = Ph, R_2 = H, R_3 = NH_2 Ki= 42.3 μM
R_1 = Ph, R_2 = Cl, R_3 = NH_2 Ki= 48.4 μM
R_1 = Ph, R_2 = H, R_3 = iPr Ki= 7.20 μM

Fig. (370). Chemical structure and inhibition constants of sulphonamide derivatives.

15.5. Tetrahydrodipicolinate N-Succinyltranferase (THDP)

These enzyme catalyses the acylation of L-2-amino-6-oxopimelate, according to the scheme in a step sequence involving the imine hydrolysis of tetrahydrodipicolinate, to produce 2-N-succinil-6-oxopimelate (Fig. **371**).

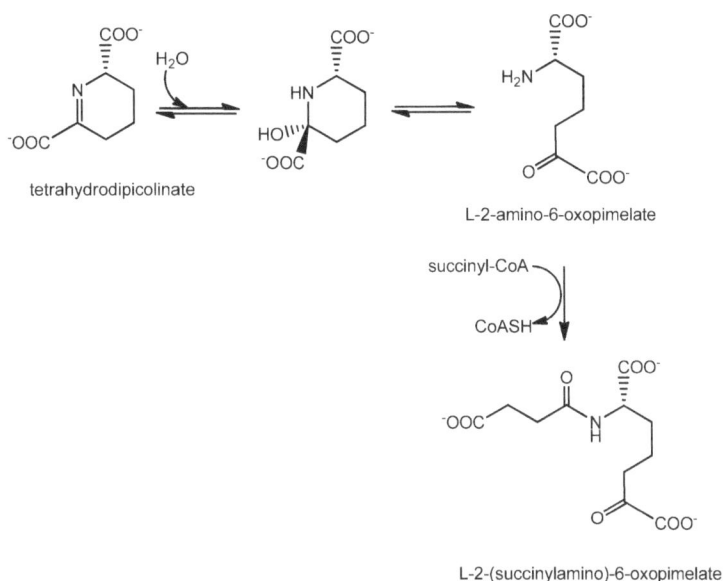

Fig. (371). The tetrahydrodipicolinate N-succinyltranferase reaction.

The structure of THDP in complex with pimelate and succinyl-CoA was determined and the binding interactions with the active site identified, showing the proximity of the succinyl carbonyl with Glu, Arg and two water molecules (Fig. **372**) [340].

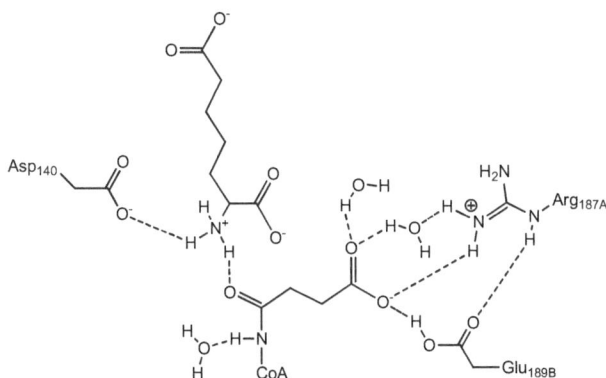

Fig. (372). Binding interactions of pimelate and succinyl-CoA with representative active site residues.

The crystal analysis of THDP from *Corynebacterium glutamicum* as monomer and trimer structure respectively shows the N-terminal helical domain (NTD), the left-handed β-helix domain (LβH), C-terminal domain (CTD) and succinyl-CoA and 2-aminopimelate binding sites (Fig. **373**) [341].

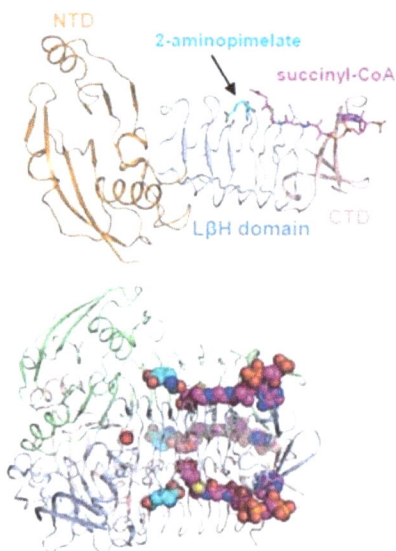

Fig. (373). Ribbon representation of THDP from *Corynebacterium glutamicum* as monomer and trimer (PDB 5E3P).

15.6. Tetrahydrodipicolinate N-Succinyltranferase Inhibitors

Structure analysis and gene deletion of THDP from *P. aeruginosa* reveal that the L isomer of 2-aminopimelate (Fig. **374**) is specific and acts as substrate, while the D isomer also binds to the same site behaving as weak inhibitor [342].

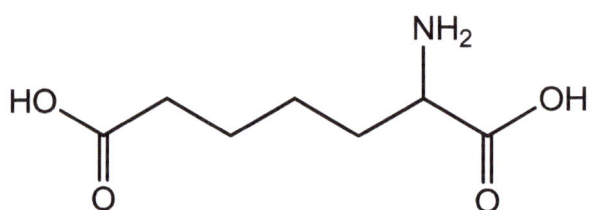

Fig. (374). Chemical structure of 2-aminopimelate.

15.7. Succinyl-A-Amino-E-Ketopimelate-Glutamate Aminotransferase (DAP-AT).

Represents the sixth step in the lysine biosynthesis corresponding to the transformation of N-succinyl-2-amino-6-keto-pimelate or N-succinyl-2-L-aino-6-oxoheptanedioate to N-succinyl-diaminopimelate or N-succinyl-L,L-2, 6-diaminoheptanedioate, and is mediated by the enzyme N-succiny--diaminopimelate aminotransferase using as coupling amine donor glutamate which is converted to 2-oxoglutarate (Fig. **375**).

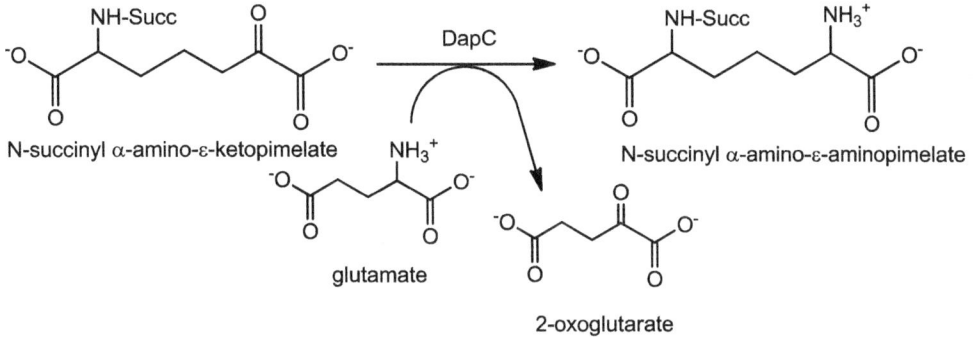

Fig. (375). The succinyl-α-amino-ε-ketopimelate-glutamate aminotransferase reaction.

The tridimensional structure of DAP-AT homodimer from *Mycobacterium tuberculosis* has been described, observing at the active site the presence of two chlorine ions, nine water molecules, glycerol molecule (not observed), and the residues interacting with the internal aldimide glutamate-PLP and N-succinyl-2-amino-6-ketopimelate-PLP complex (Fig. **376**) [343].

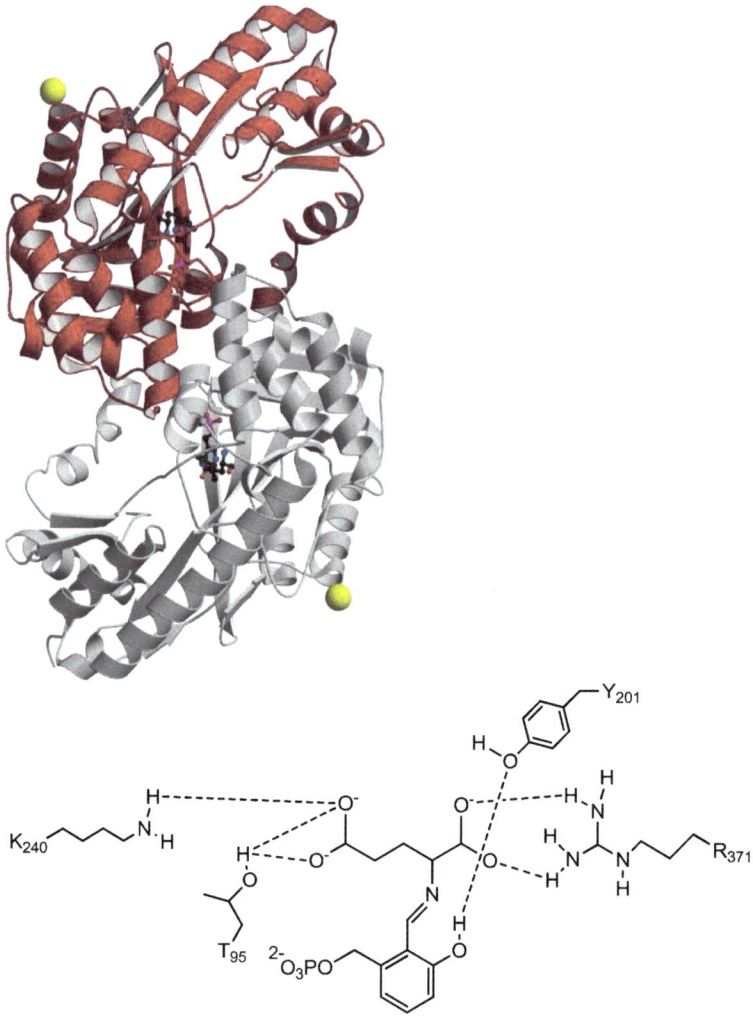

Fig. 376 cont.....

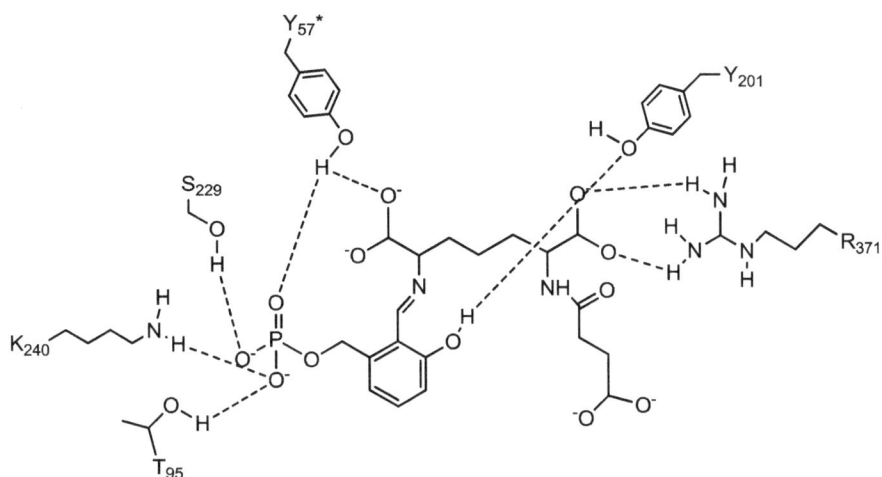

Fig. (376). Tridimensional structure of DapC homodimer from *Mycobacterium tuberculosis* and interaction map of internal aldimide glutamate-PLP and N-succinyl-2-amino-6-ketopimelate-PLP complex (PDB: 2O0R).

15.8. Succinyl-A-Amino-E-Ketopimelate-Glutamate Aminotransferase Inhibitors

Different hydrazine pimelate derivatives were synthesized and evaluated as DAP-AT from wild type *Escherichia coli* ATCC 9637, being most potent 2-(-(succinylamino))-6-hydrazoheptane-1,7-dioic acid and 2-(N-Cbz-amio-6-hydrazinoheptane-1,7-dioic acid (Fig. **377**) [344].

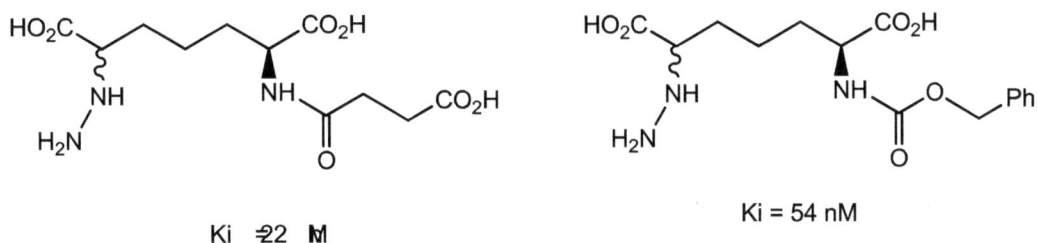

$Ki = 22$ M

$Ki = 54$ nM

Fig. (377). Chemical structure and inhibitions constants of hydrazine pimelate derivatives.

15.9. N-Acyldiaminopimelate Deacylase (N-Acyldap)

The step consisting in the hydrolysis of N-succinyl-LL-diaminopimelic acid to L,L-diaminopimelic acid is mediated by the enzyme N-acyldiaminopimelate deacylase classified as a dinuclear Zn^{2+} metalloprotease (Fig. **378**).

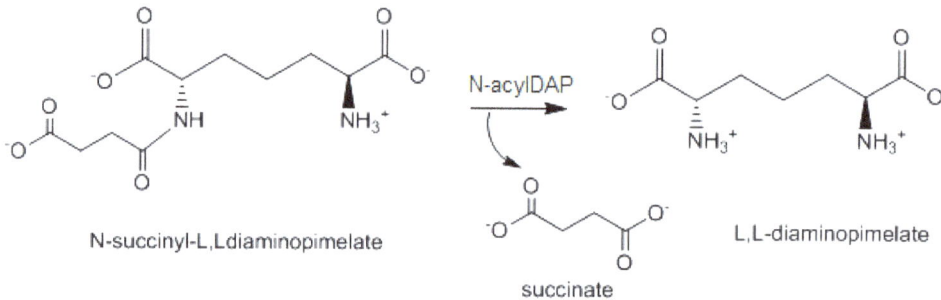

Fig. (378). The N-acyldiaminopimelate deacylase reaction.

The crystal structure of N-succinyl-L,L-diaminopimelic acid desuccinylase from *Haemophilus influenzae* (HiDapE) was described, providing unique information about conformational forms (open and close). Also, the catalytic domain bearing the substrates succinic acid, L-diaminopimelic acid, and histidine194.b residue, forming an interaction with the succinic amide, coordinated zinc-H_2O is represented (Fig. **379**).

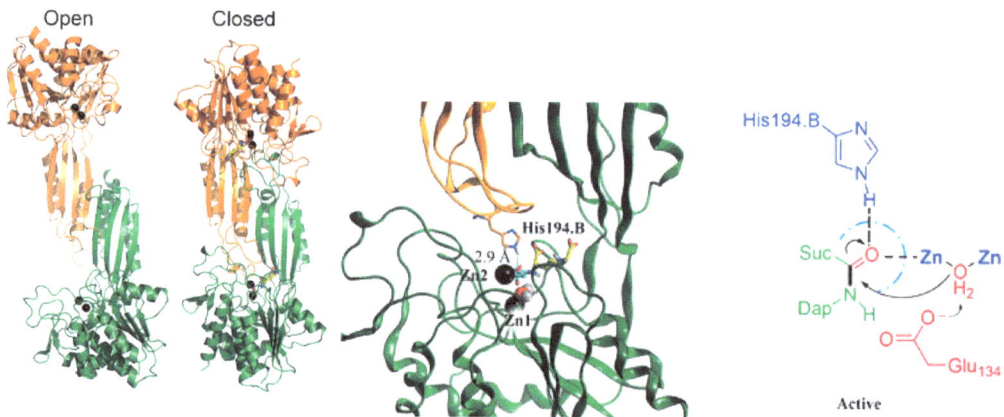

Fig. (379). N-succinyl-L,L-diaminopimelic acid desuccinylase from *Haemophilus influenzae* (PDB: 5VO3).

Also a detailed mechanism for the transformation of N-succinyl-L-L-diaminopimelic acid to L,L-diaminopimelic acid is provided according to the intermediates illustrated in (Fig. **380**) [345].

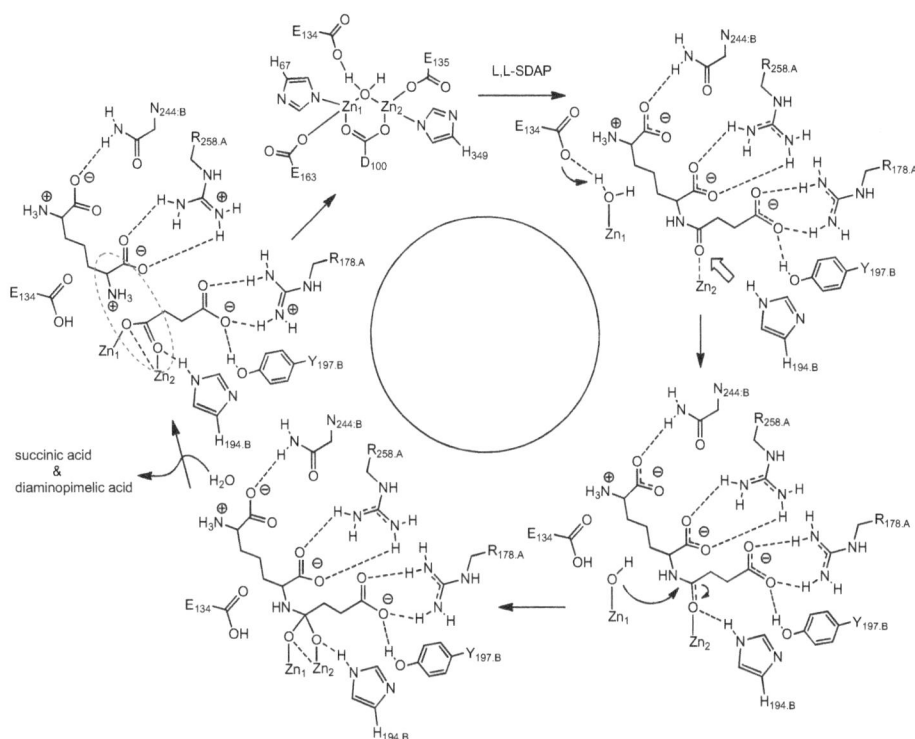

Fig. (380). Proposed mechanism for the formation of L,L-diaminopimelic acid.

15.10. N-Acyldiaminopimelate Deacylase Inhibitors

Several classes of compounds has been tested as N-acylDap inhibitors against *Escherichia coli* including thiols, hydroxamates, carboxylic acids, boronic acids and phosphates, being the most competitive the anti-hypertensive L-captopril and L-penicillamine (Fig. **381**) [346].

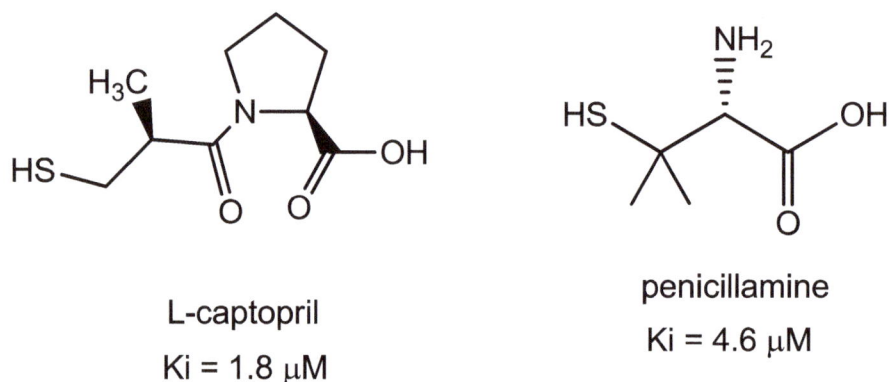

L-captopril

Ki = 1.8 μM

penicillamine

Ki = 4.6 μM

Fig. (381). Chemical structure and inhibition constants of L-captopril and L-penicillamine.

15.11. Diaminopimelate Epimerase (DAPE)

This enzyme is responsible for the interconversion of LL-DAP to the meso-DAP (Fig. **382**) being the last precursor in the L-lysine biosynthetic pathway.

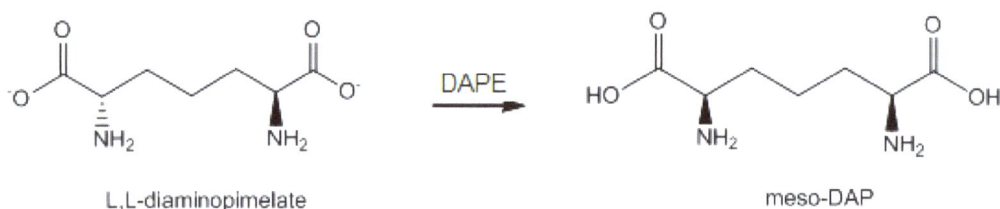

L,L-diaminopimelate

meso-DAP

Fig. (382). The diaminopimelate epimerase reaction.

Structural studies have been carried out to explain the interconversion of LL-DAP to the meso form suggesting a two-base mechanism involving cysteine residues, one acting as base to accept a proton at the α-position to generate an enolate and the second cysteine having the function of reprotonate the enolate once the inversion process occurs (Fig. **383**) [347].

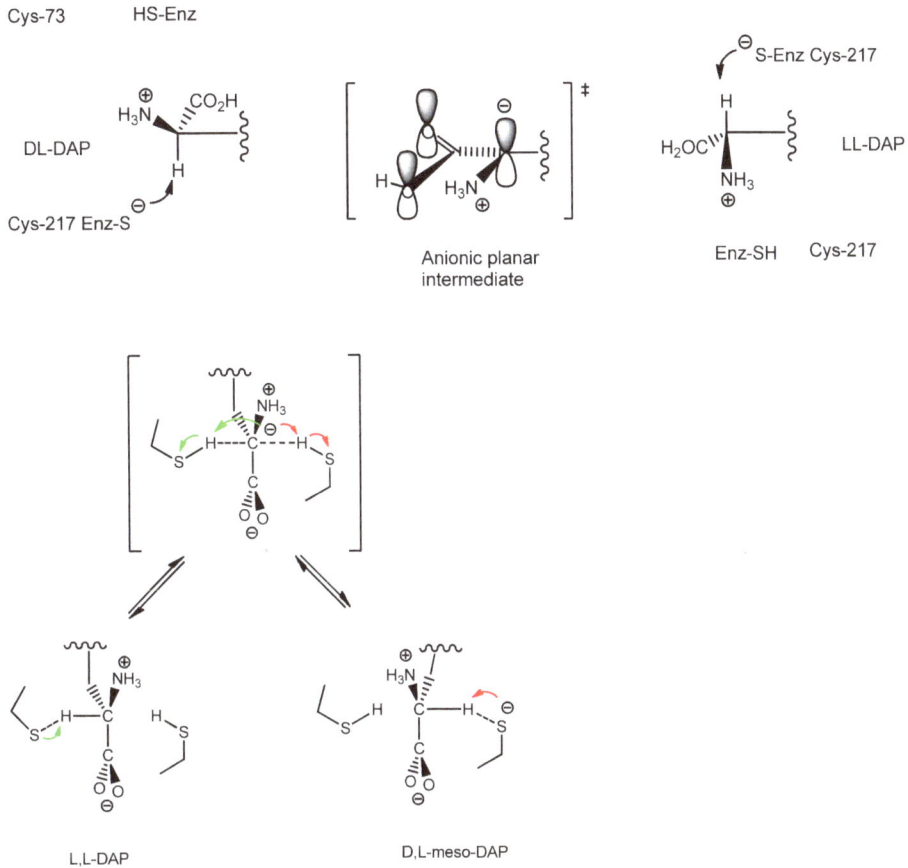

Fig. (383). Proposed mechanism for the epimerization of LL-DAP to meso-DAP.

The crystal structure of DAPE from various microorganism has been determined, among them *Mycobacterium tuberculosis* (PDB 3FVE), *Corynebacterium glutamicum* (PDB 5H2G), *Escherichia coli* (PDB 6D13) and *Hemophilus influenziae* (PDB 1BWZ). The overall structure of *Corynebacterium glutamicum* *Cg*DapF monomer allowed to identify the NTD and CTD regions, as well as the D,L-DAP molecule attached to the enzyme. Also the hydrogen bound interactions between D,L-DAP and the residues involved in the catalysis are represented (Fig. **384**) [348].

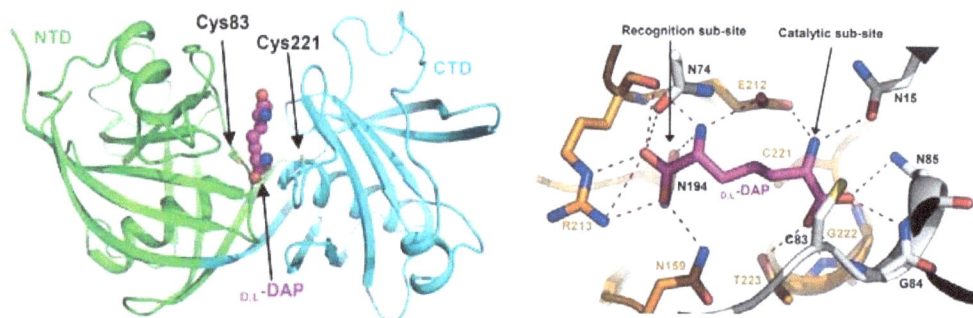

Fig. (384). Ribbon representation of DapF monomer from *Corynebacterium glutamicum* and recognition and catalytic subsites for D,L-DAP (PDB: 5H2G).

15.12. Diaminopimelate Epimerase Inhibitors

Aziridine DAP analogues (Fig. **385**) have been synthesized and evaluated as DAP epimerase inhibitors using *Escherichia coli* mutant BL21(DE3), observing moderate inhibition values with IC_{50} of 2.88 mM [349].

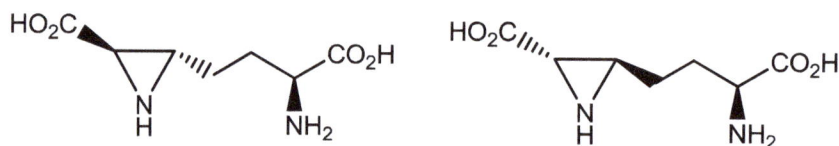

Fig. (385). Chemical structure and inhibition values of aziridine DAP analogues.

15.13. Diaminopimelate Decarboxylase (DAPDC)

The last step in the L-lysine biosynthetic pathway is mediated by DAP decarboxylase which converts meso-diaminopimelate to L-lysine (Fig. **386**).

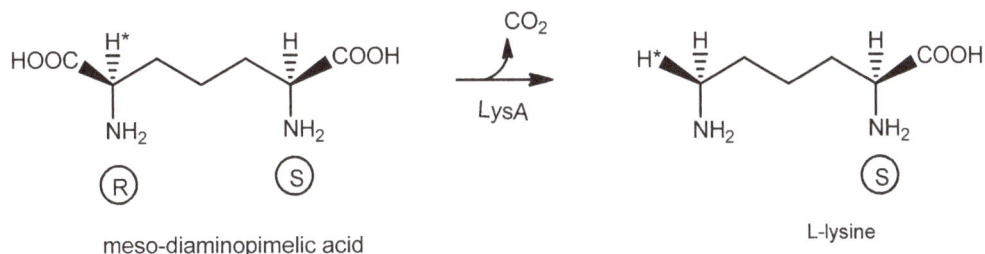

Fig. (386). The diaminopimelate decarboxylase reaction.

The crystal structure of Diaminopimelate decarboxylase from *Corynebacterium glutamicum* (*cg*LysA) in complex with pyridoxal phosphate and L-lysine was described. In the cofactor binding mode it is possible to see the residues participating in the binding pocket, their interactions with PLP as red dots and three water molecules as red spheres (Fig. **387**) [350].

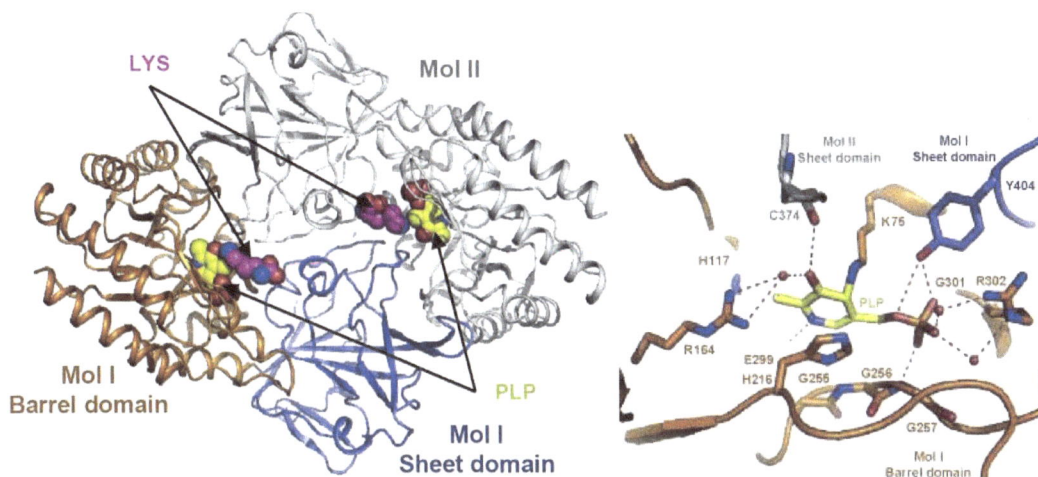

Fig. (387). The crystal structure of Diaminopimelate decarboxylase from *Corynebacterium glutamicum* in complex with pyridoxal phosphate and L-lysine (PDB: 5X7M).

A mechanism explaining the decarboxylation process uses mirror plane structures, showing the initial PLP-enzyme iminium complex, followed by nucleophilic attack from the amino acid to give a PLP-amino acid iminium complex conserving the quiral center of the amino acid. Then a loss of hydrogen from the tetrahedral carbon promotes resonance until reaching a quinoid form and assuming planar geometry. Finally the pyridinium intermediate is hydrolysed to the amino acid L-Lysine along with PLP cofactor (Fig. **388**) [350].

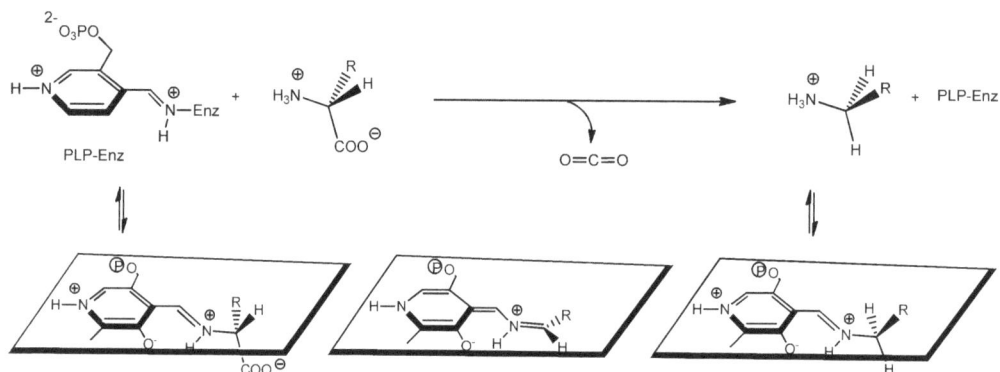

Fig. (388). Proposed mechanism for conversion of meso-diaminopimelate to L-lysine.

15.14. Diaminopimelate Decarboxylase Inhibitors

Diaminopimelate analogues were synthesized and evaluated as inhibitors of growth against *Escherichia coli, Bacilus subtilus, B. cereus, B. sphaericus* and *T. vulgaris*. The Ki values correspond to inhibitory concentration against *T. vulgaris* at mM concentrations (Fig. **389**) [351].

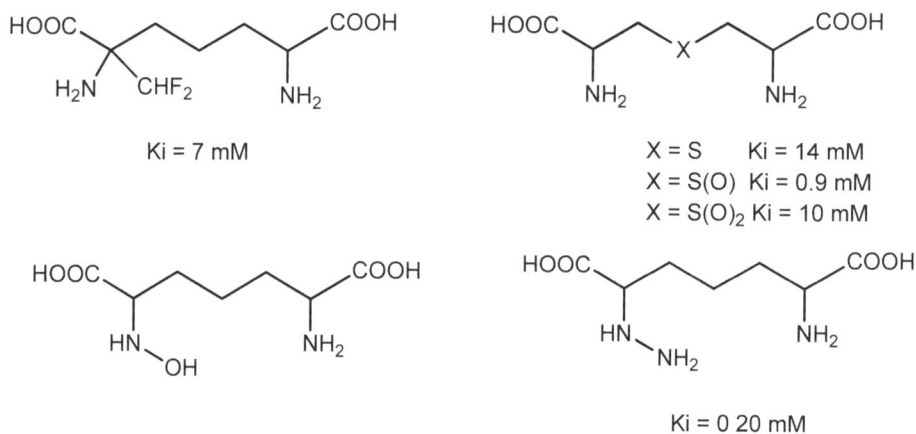

Fig. (389). Chemical structure and inhibition constants of Diaminopimelate analogs.

15.15. Lysine Biosynthesis From A-Aminoadipate.

In yeast and higher fungi the biosynthesis of lysine comes from the α-aminoadipate pathway (AAA) starting from α-ketoglutarate as a precursor, which is converted sequentially in homocitrate, homoaconitate, homoisocitrate, and α-ketoadipate, being transformed by transamination to α-aminoadipate. This key intermediate is reduced in the cytoplasm to α-aminoadipate-δ-semialdehyde and condensed with glutamate to produce L-saccharopine which undergoes oxidative deamination to yield lysine as shown in Fig. (**390**) [352].

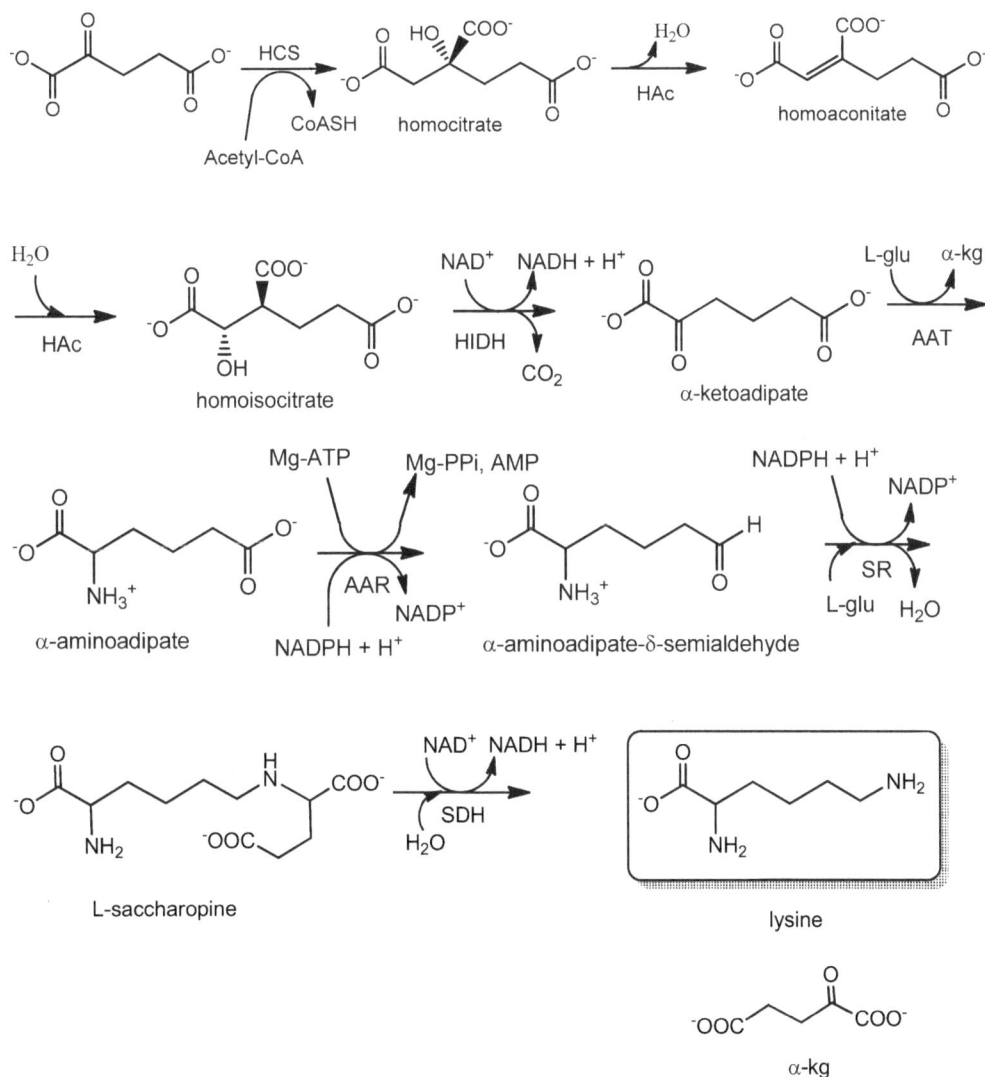

Fig. 390 cont.....

HCS = homocitrate synthase
HAc = homoaconitase
HIDH = homoisocitrate dehydrogenase
AAT = aminoadipate aminotransferase
AAR = aminoadipate reductase
SR = saccharopine reductase
SDH = saccharopine dehydrogenase

Fig. (390). Biosynthesis of lysine from α-aminoadipate.

16. ASPARTATE, ASPARAGINE AND GLUTAMATE BIOSYNTHESIS

Aspartate is a carboxylic amino acid considered a precursor of the amino acids asparagine, glutamate, threonine, methionine, lysine and isoleucine and several nucleotides. It is synthesized from a transamination reaction between oxaloacetate and glutamate, catalysed by the enzyme aspartate transaminase (Fig. **391**).

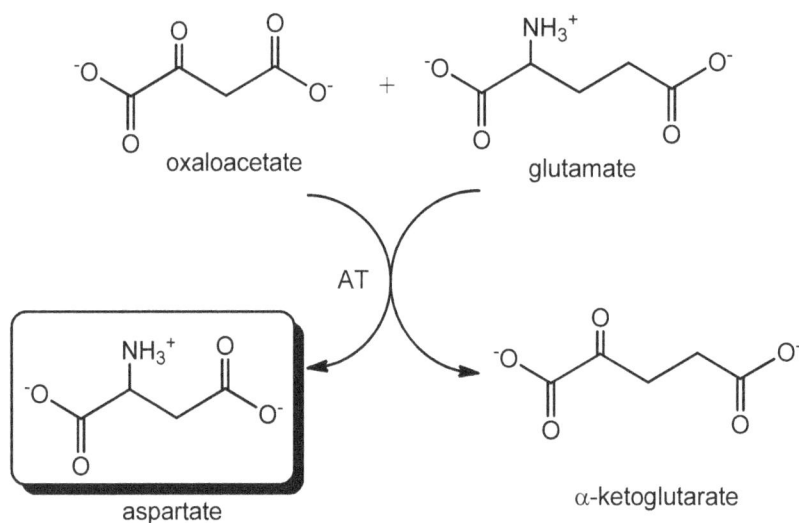

Fig. 391 cont.....

AT = aspartate transaminase
AS = asparagine synthetase

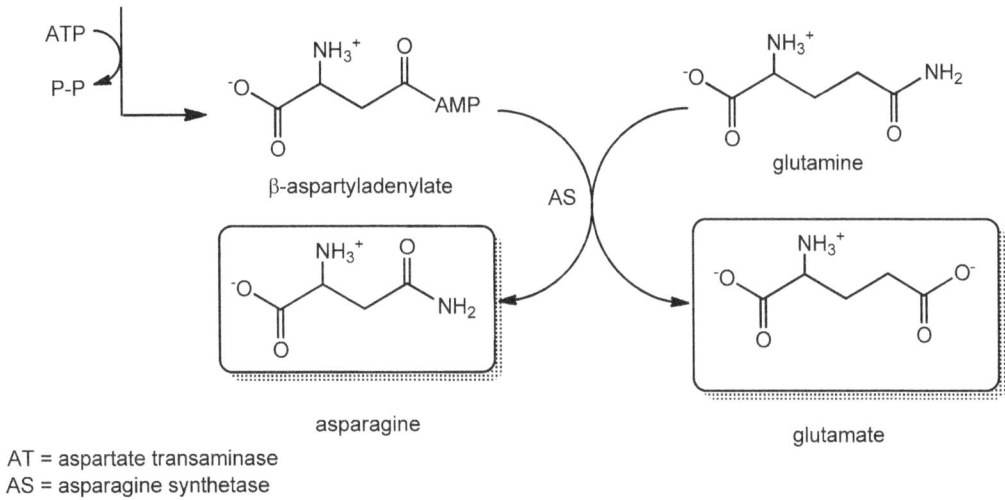

Fig. (391). Biosynthetic pathway for aspartate and glutamate formation.

16.1. Aspartate Aminotransferase (AT)

Aspartate aminotransferase is a PLP-dependent enzyme responsible for the reversible conversion of glutamate to α-keto glutarate and from oxaloacetate to aspartate *Via* ping-pong bi-bi cycle (Fig. **392**) [353].

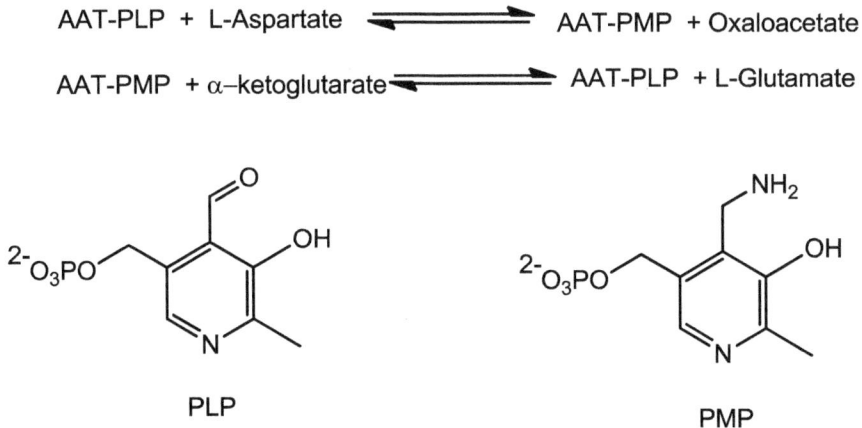

AAT-PLP + L-Aspartate ⇌ AAT-PMP + Oxaloacetate

AAT-PMP + α–ketoglutarate ⇌ AAT-PLP + L-Glutamate

PLP

PMP

Fig. (392). The aspartate aminotransferase reaction.

From the substantial information about aminotransaminase mechanisms general steps occurring for the transamination process involves intermediates gem-diamine 1 and 2, external aldimine, quinoid, ketamine, and carbinolamine 1 and 2 taking part of the transformation (Fig. **393**).

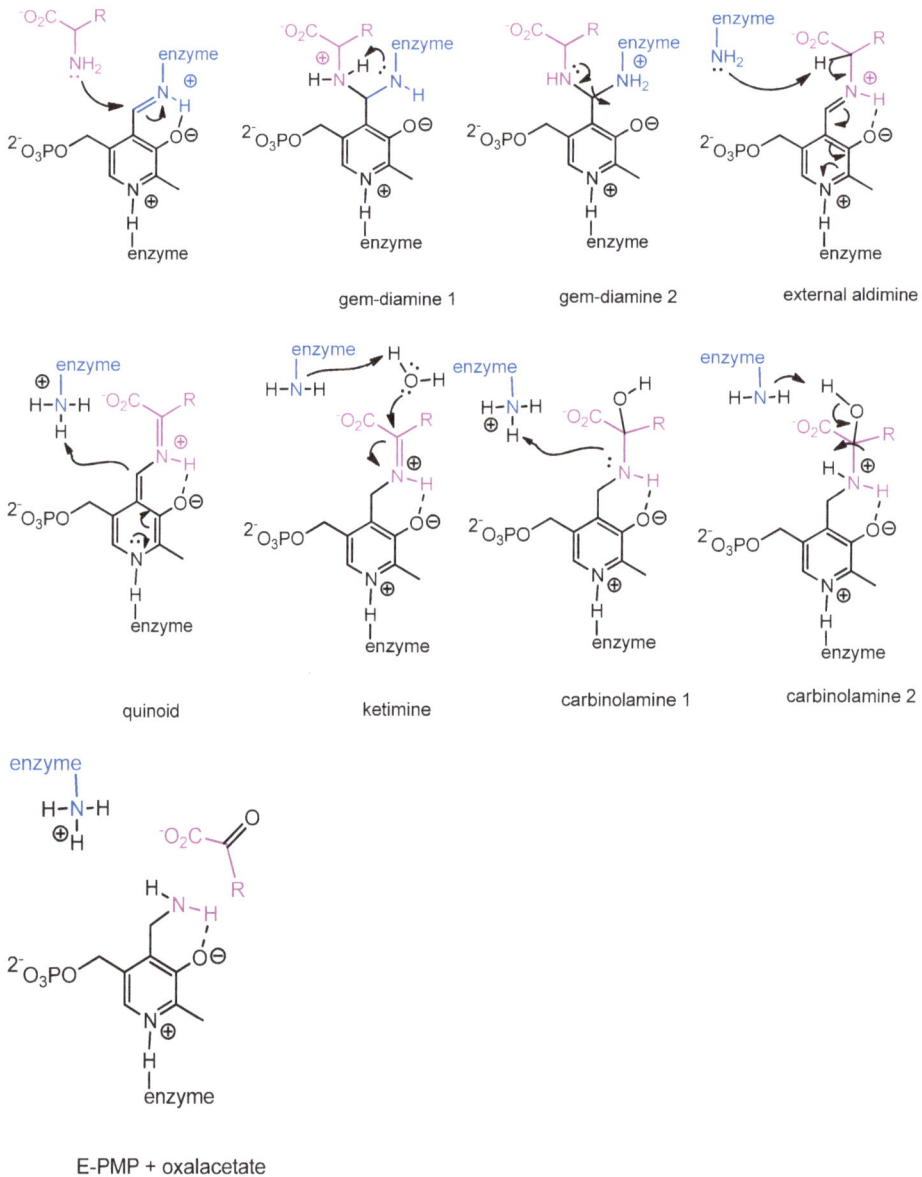

Fig. (393). Intermediates participating in the aspartate aminotransferase reaction.

Neutron crystallography was successfully applied to determine the hydrogen/deuterium positions in the enzyme aspartate aminotransferase (AT) from recombinant porcine, observing an homodimer with two active sites, the first catalysing the external aldimide (PLA) and the second the internal aldimine (PLP) (Fig. **394**) [354].

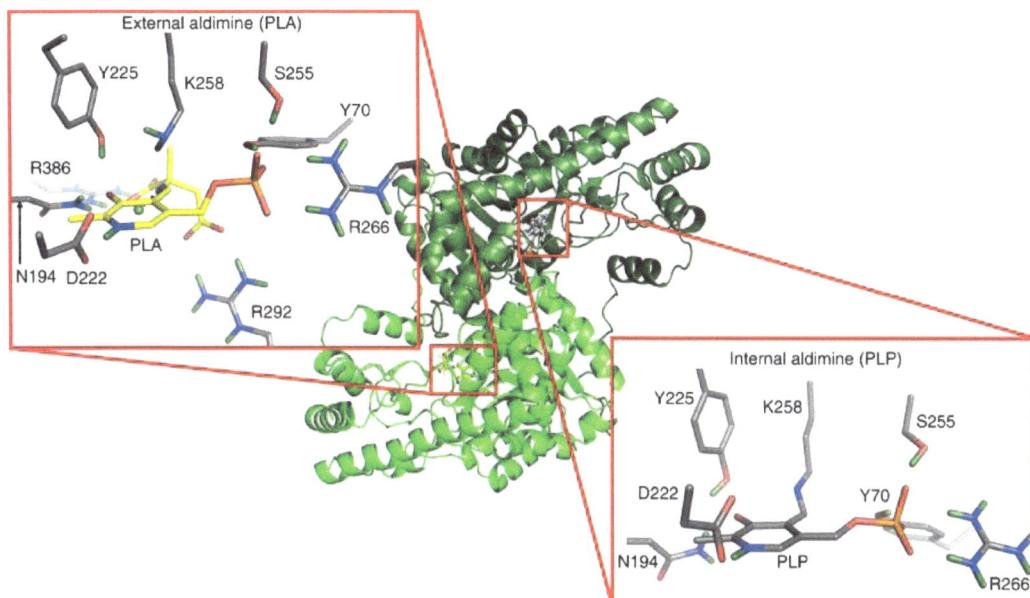

Fig. (394). Neutron crystallography showing the ribbon structure with two active sites containing PLP, PLA the interacting residues.

16.2. Aspartate Aminotransferase Inhibitors

Amino-oxyacetate is a well-known inhibitor of pyridoxal phosphate required in hepatocyte AT, producing 93% inhibition after 10 minutes [269, 174]. Another inhibitors of arginine aminotransferases reported are L-cycloserine (seromycin) decreasing enzyme activity at 50 μM over 90%, β-chloro-L-alanine and L---amino-4-methoxy-trans-but-3-enoic acid (Fig. **395**) [355, 356].

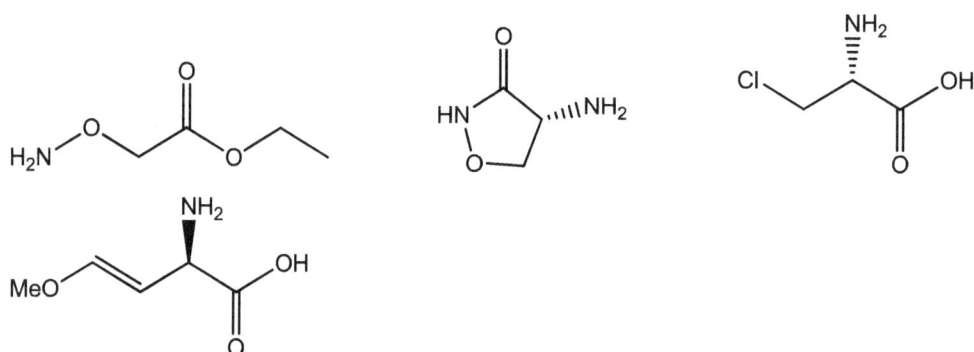

Fig. (395). Chemical structure of amino-oxyacetate, L-cycloserine, β-chloro-L-alanine and L-2-amin--4-methoxy-trans-but-3-enoic acid.

16.3. Asparagine Synthetase (AS)

This enzyme is associated to cellular stress and tumour biology, and catalyses the conversion of β-aspartyladenylate to asparagine, and from glutamine to glutamate (Fig. **396**). In mammals the AS uses glutamine as nitrogen source (AS-B), and some bacteria uses ammonia (AS-A) [357].

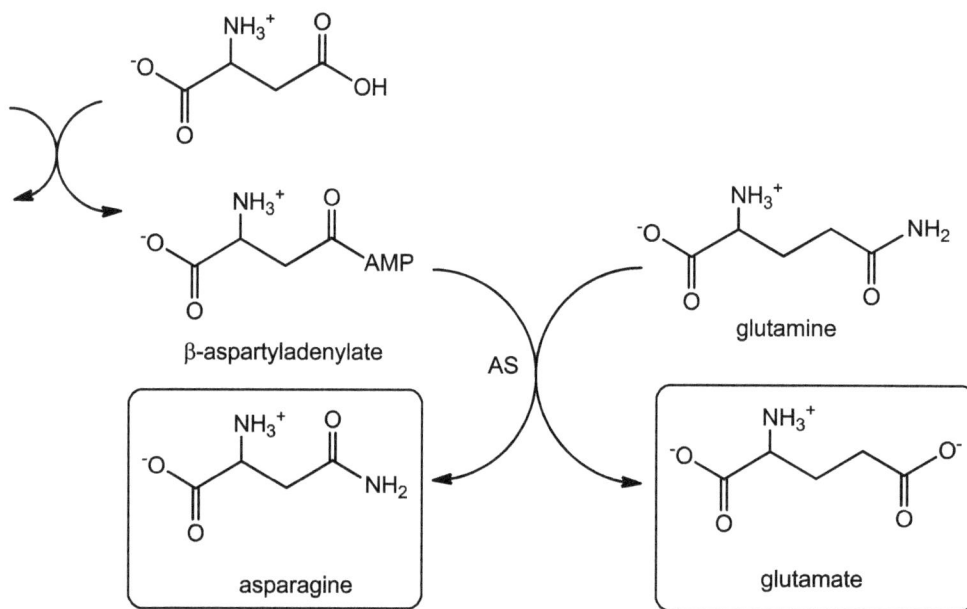

Fig. (396). The asparagine synthetase reaction.

The crystal structure of *archaedal* asparagine synthetase has been solved as dimeric structure and compared with other AsnA structures, presenting seven-stranded anti-parallel β-sheets surrounded by α-helices (Fig. **397**) [358].

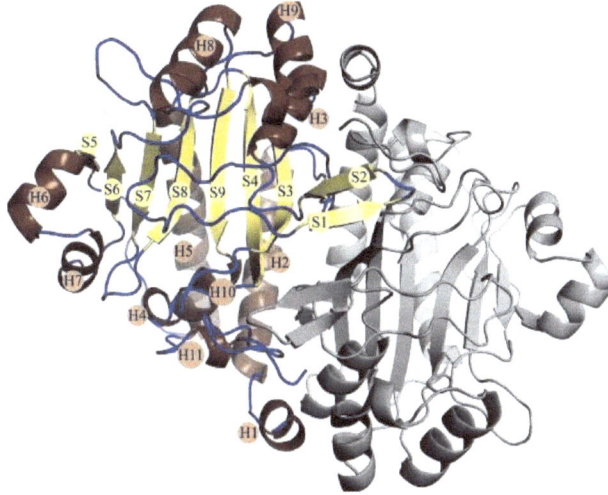

Fig. (397). Crystal structure of *archaedal* asparagine synthetase (PDB: 3P8T).

The active site of asparagine synthetase identifies aspartate as natural substrate, which interacts through the α-carboxylate group with Arg222, Lys80, and Asp195 residues establishing hydrogen bond contacts with water molecules. Also the Asp amino group interacts with Asp118, Ser75 and Gln116, while the β-carboxylate with Arg99, Gly262, and Ser218 with participation of water molecules (Fig. **398**) [359].

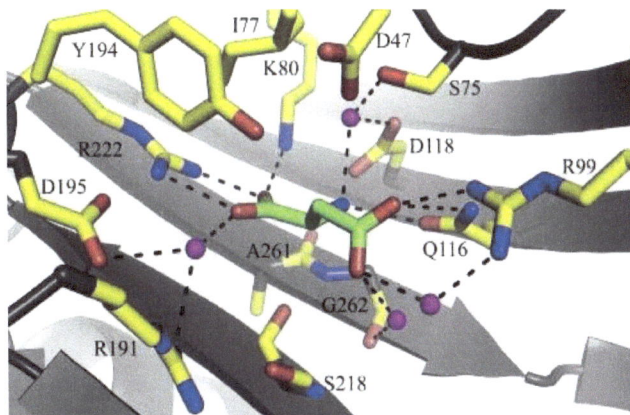

Fig. (398). The active site of asparagine synthetase showing interaction of asparagine with residues and water molecules.

16.4. Asparagine Synthetase Inhibitors

The overexpression of asparagine synthetase in human T-cells could lead to lymphoblastic leukemia and therefore the design of AS inhibitors have been used as rational strategy to control cell proliferation. The synthesis of N-acylsulfonamide analogues was described, and the measurement of inhibitory activity was determined based on the PPi production. From the six analogues evaluated the sulphonamide 6 displayed potent inhibition with K_i of 728 nM (Fig. 399) [360].

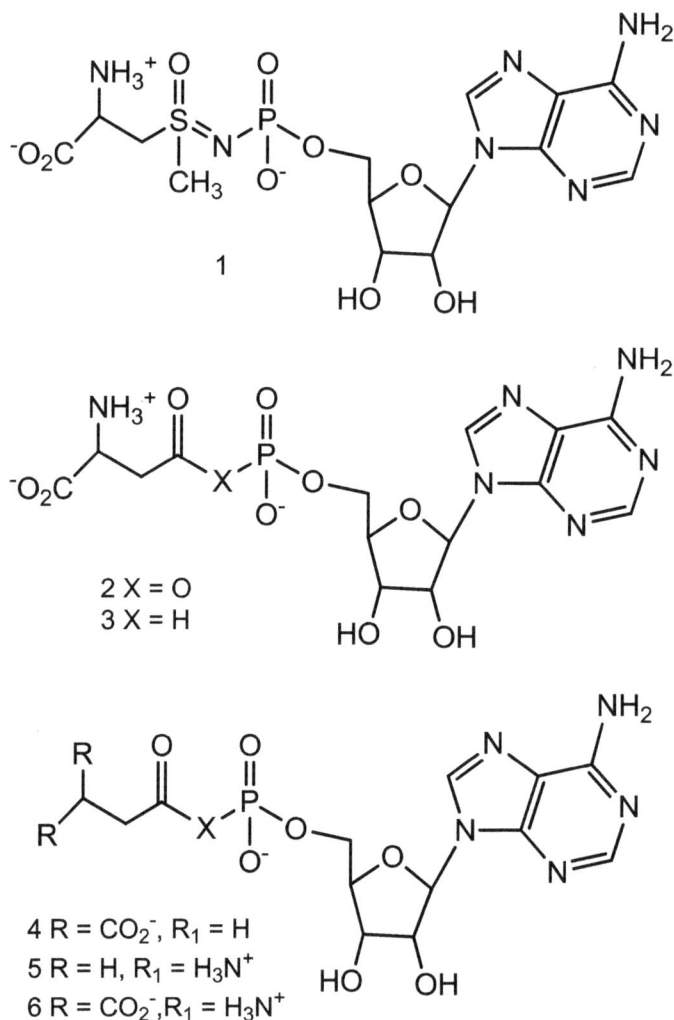

Fig. (399). Chemical structure and inhibition constants of N-acylsulfonamide analogues.

17. GLUTAMINE AND GLUTATHIONE BIOSYNTHESIS

Glutamine is a non-essential amino acid produced in sufficient amounts under normal circumstances and conditionally essential under stressing conditions. It is strongly associated with the good performance of the immune system and intestinal health. It is synthesized from glutamate by two steps sequence the first involving a phosphorylation and the second amination catalysed by glutamine synthetase (Fig. **400**).

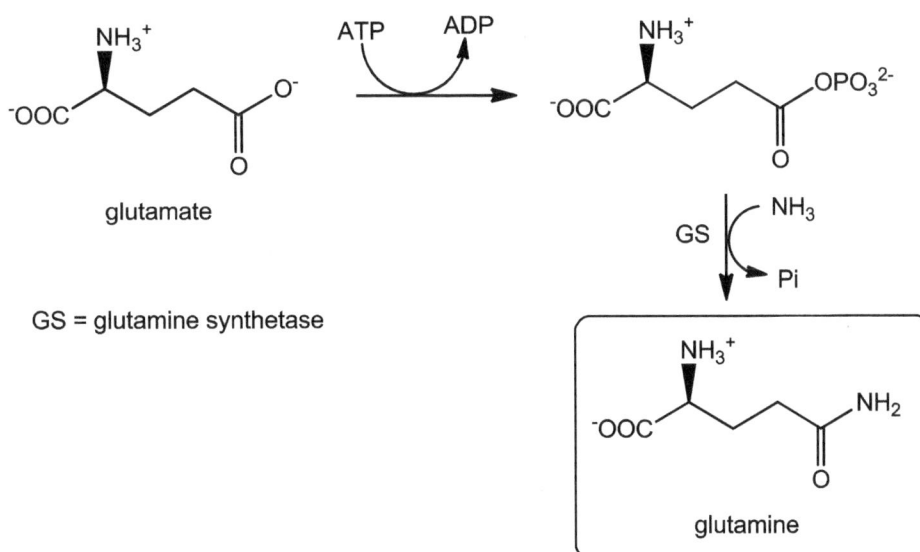

Fig. (400). Glutamine biosynthesis.

The synthesis of glutathione is an ATP dependent reaction involving the condensation between L-cysteine with L-glutamate catalysed by the enzyme glutamate-cysteine ligase (GSL) to form γ-glutamyl-L-cysteine. The next steps consisting in the condensation of the dipeptide with glycine also requiring an ATP molecule, and catalysed by the enzyme glutathione synthase (GS), providing reduced glutathione (GSH), which is finally oxidized by the catalysis of glutathione disulfide reductase to produce oxidized glutathione (GSSG) as illustrated in Fig. (**401**).

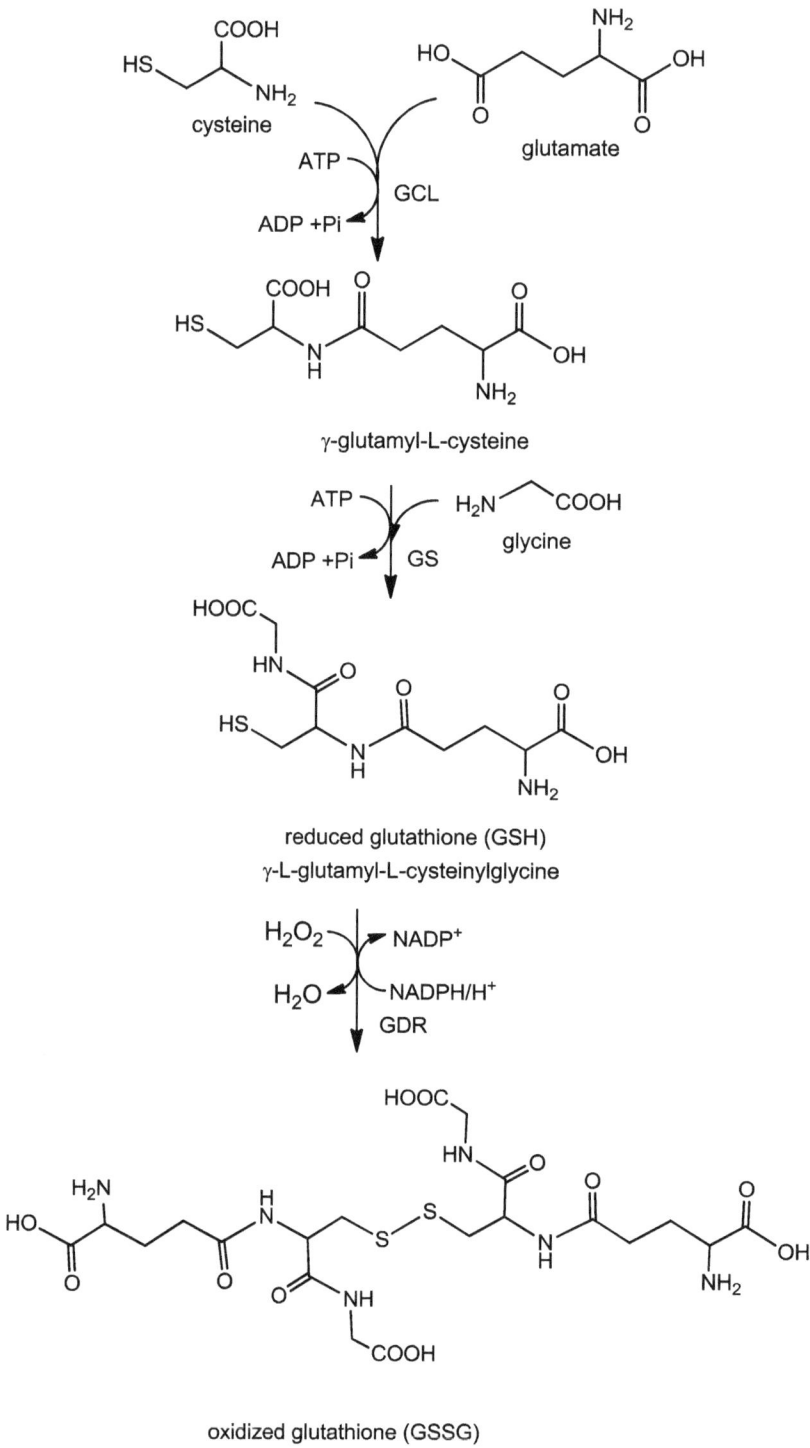

γ-glutamyl-L-cysteine

reduced glutathione (GSH)
γ-L-glutamyl-L-cysteinylglycine

oxidized glutathione (GSSG)

Fig. 401 cont.....

GCL = glutamate-cysteine ligase

GS = glutathione synthase

GDR = glutathione disulfide reductase

Fig. (401). Biosynthetic pathway for the formation of glutathione.

17.1. Glutamate Cysteine Ligase (GCL)

This rate-limiting enzyme is the most abundant in the biosynthetic pathway of the cellular antioxidant glutathione GSH that produce γ-glutamyl-L-cysteine by condensation between L-cysteine and glutamate, in the presence of ATP (Fig. **402**) [361].

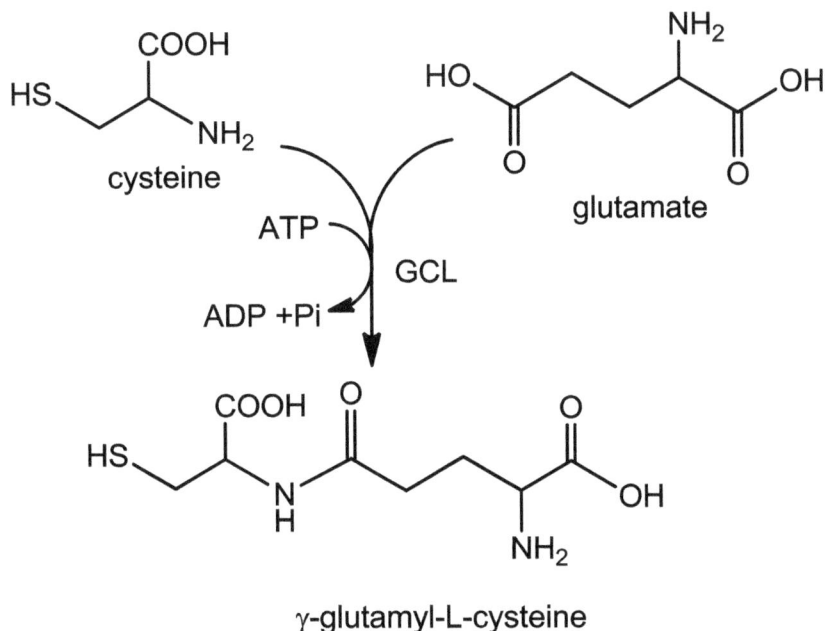

Fig. (402). The glutamate cysteine ligase reaction.

The binding site of glutamate cysteine ligase from *Saccharomyces cerevisiae* has been identified with L-glutamate interacting through the α-carboxylate with Arg313 and Tyr362, and the Mg^{2+} cofactor. Also in the ScGCL/Glu/Mg^{2+}/ADP complex it is possible to identify the ADP molecule with the adenine ring on the outer edge of the active site pocket and the phosphate group near the γ-carboxylate of glutamate (Fig. **403**) [362].

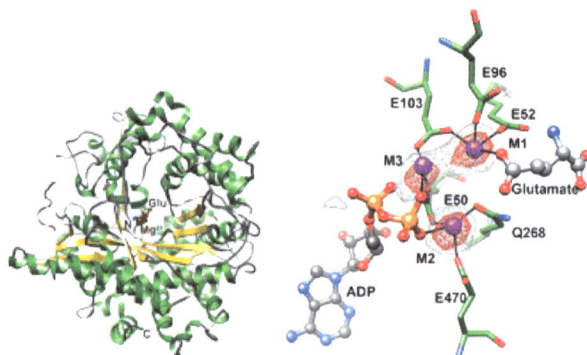

Fig. (403). The crystal structure and binding site of glutamate cysteine ligase form *Saccharomyces cerevisiae* (PDB: 3IG5).

17.2. Glutamate Cysteine Ligase Inhibitors

The inhibition of this enzyme is receiving attention in the cancer therapy due recent findings suggesting that glutathione (GSH) protects tumoral cells against reactive oxygen species and therefore the inhibition of the enzymes involved in the synthesis of this important ROS quencher becomes a valuable strategy.

L-buthionine sulfoximine (BSO) is a potent GCL inhibitor in rat, mouse and *Drosophila* with K_i values in the range of 110-150 μM. The absence of ATP or Mg^{2+} suppress BSO inhibition, suggesting that phosphorylation is required for its inhibition (Fig. **404**) [363].

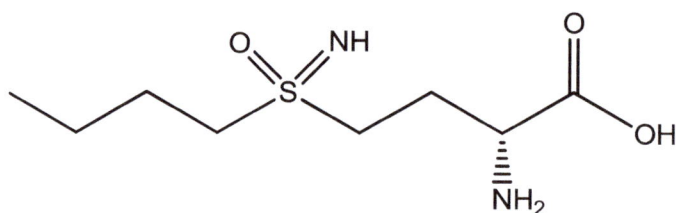

Fig. (404). Chemical structure of L-buthionine sulfoximine.

17.3. Glutathione Synthetase (GS)

The last step in the synthesis of reduced glutathione, a tripeptide with strategic implications in the inactivation of reactive oxygen species (ROS) takes place by conjugation of γ-glutamyl-L-cysteine with glycine catalysed by glutathione synthase (Fig. **405**).

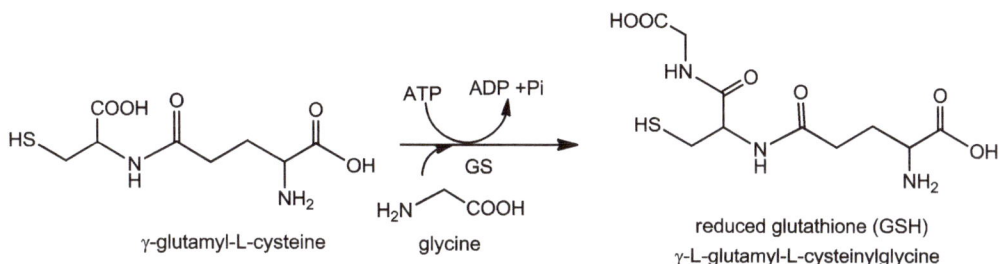

Fig. (405). The glutathione synthetase reaction.

Human and yeast GS in complex with ADP, Mg^{2+} ions, sulphate, and glutathione has been described. The γ-glutamyl moiety interacts with Arg267 and presents hydrogen bonding with Ser151, Asn216, Gln220, and Arg267 (Fig. **406**) [363, 364].

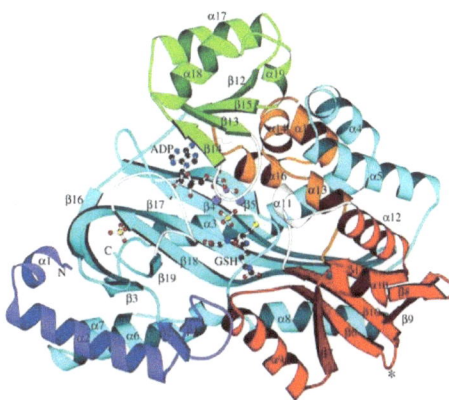

Fig. (406). A ribbon representation of human glutathione synthetase (PDB: 2HGS).

The proposed mechanism consisted in the phosphorylation of the C-terminal of γ-glutamatyl cysteine by ATP molecule, producing glutamatecysteine phosphate, which undergoes a nucleophilic substitution by the amino group of glycine, providing glutathione (Fig. **407**) [364, 365].

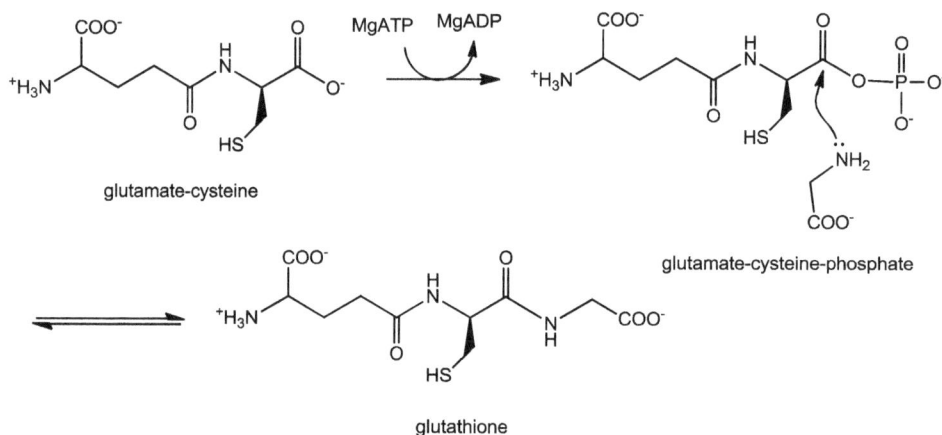

Fig. (407). The proposed mechanism for glutathione formation.

17.4. Glutathione Disulphide Reductase (GDR)

It is a NAD dependent enzyme involved in the oxidation of reduced form (GSH) to oxidized glutathione (GSSG), with important implications as redox agent (Fig. **408**).

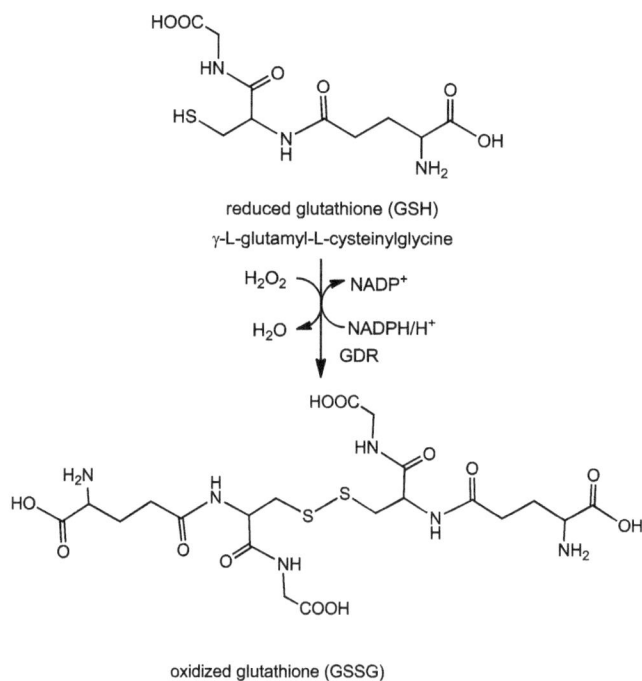

Fig. (408). The glutathione disulphide reductase reaction.

A simplified half reaction mechanism for the glutathione reductase is illustrated in (Fig. **409**), involving a hydride transfer from NADPH to FAD, which cleaves a cysteine disulfide bond. Then a histidine residue working as donor/acceptor takes a proton from reduced cysteine and the resulting anion attacks an oxidized glutathione molecule (GSSG) resulting in the reduced glutathione formation (GSH) [366].

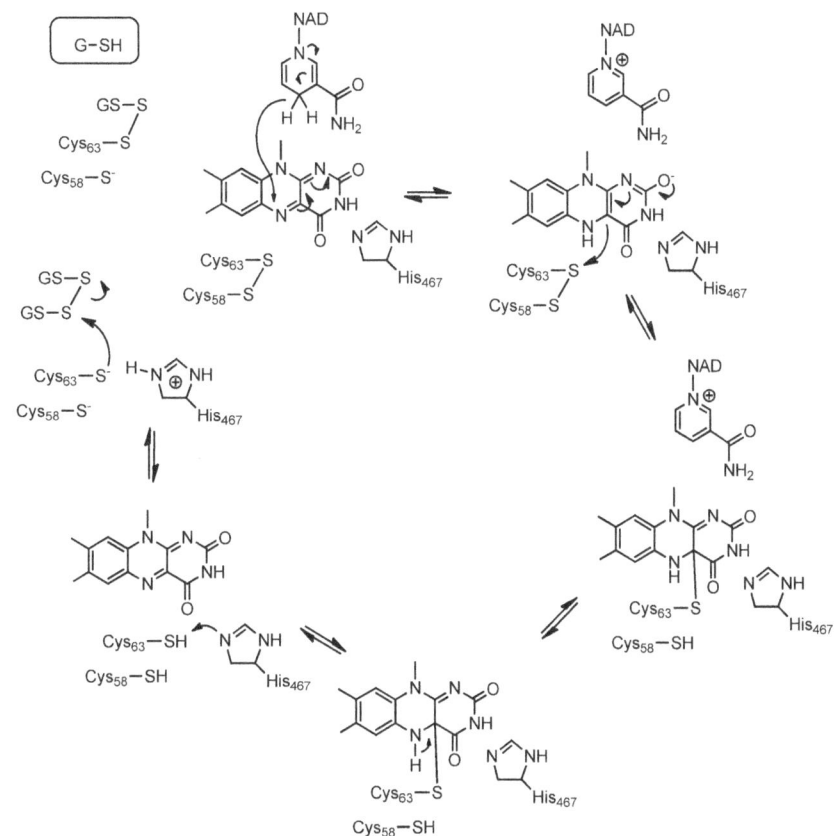

Fig. (409). Proposed reaction mechanism for the transformation of oxidized (GSSG) to reduced glutathione (GSH).

The crystal structure of the dimeric flavoenzyme glutathione reductase from human and *E. coli* have been established, presenting 52% homology and three domains: the FAD domain, the NADP domain and the interface domain (Fig. **410**) [367, 368].

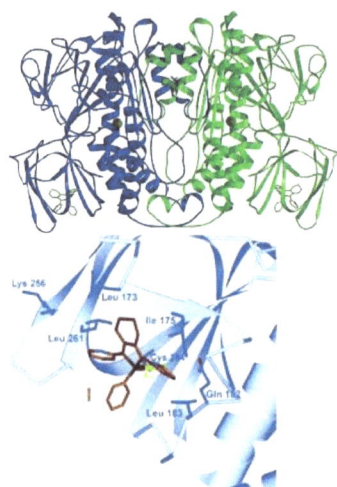

Fig. (410). The crystal structure of human glutathione reductase and binding site (PDB: 2AAQ).

17.5. Glutathione Reductase Inhibitors

As it was mentioned, glutathione reductase plays a key role in redox balance, protein-nucleotide metabolism, and cellular defence. However during cancer proliferation glutathione provides protection against reactive oxygenated species (ROS) and therefore the use of inhibitors has been implemented as a promising therapy. Some inhibitors have been identified such as [1-phenyl-2,5-d-(2-pyridyl)phosphole]AuCl, having 2-pyridyl or 2-thienyl substituents on the phosphole ring and gold(I) (AuCl) or platinum(II) (PtCl$_2$), displaying K$_i$ = 0.46 ± 0.06 μM for the Au phosphole complex and 19.7 ± 0.48 μM for the Pt phosphole complex (Fig. **411**) [369].

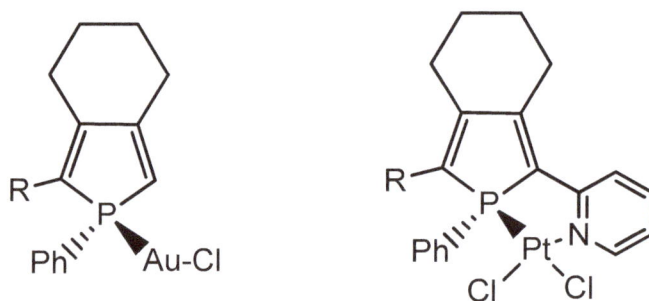

Fig. (411). Chemical structure of phosphole rings with gold(I) and platinum(II) complex.

The alkylating agents N,N-bis(2-chloroethyl)-N-nitrosourea (carmustine) and lomustine, the first used as a reference GR inhibitor with $IC_{50} = 647$ μM against yeast, although with limited used in therapy because of his side effects on DNA synthesis. Other potentially useful inhibitors are the dithiocarbamate derivatives having IC_{50} values of 50 and 56 μM, respectively (Fig. **412**) [370].

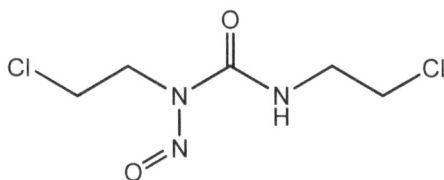

Carmustine (BCNU)   Lomustine (CCNU

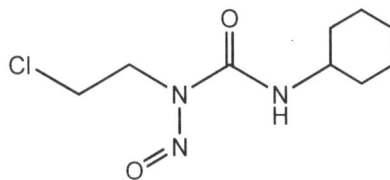

Fig. (412). Chemical structure of alkylating agent's carmustine and lomustine.

References

[1]   Flamholz A, Noor E, Bar-Even A, Liebermeister W, Milo R. Glycolytic strategy as a tradeoff between energy yield and protein cost. Proc Natl Acad Sci USA 2013; 110(24): 10039-44.
[http://dx.doi.org/ 10.1073/pnas.1215283110] [PMID: 23630264]

[2]   Liberti MV, Locasale JW. The warburg effect: How does it benefit cancer cells? Trends Biochem Sci 2016; 41(3): 211-8.
[http://dx.doi.org/10.1016/j.tibs.2015.12.001] [PMID: 26778478]

[3]   Roy S, Vega MV, Harmer NJ. Carbohydrate kinases: A conserved mechanism across differing folds catalyst 2019; 9(29): 1-19.
[http://dx.doi.org/10.3390/catal9010029]

[4]   Nishimasu H, Fushinobu S, Shoun H, Wakagi T. Crystal structures of an ATP-dependent hexokinase with broad substrate specificity from the hyperthermophilic archaeon Sulfolobus tokodaii. J Biol Chem 2007; 282(13): 9923-31.
[http://dx.doi.org/10.1074/jbc.M610678200] [PMID: 17229727]

[5]   Nancolas B, Guo L, Zhou R, *et al.* The anti-tumour agent lonidamine is a potent inhibitor of the mitochondrial pyruvate carrier and plasma membrane monocarboxylate transporters. Biochem J 2016; 473(7): 929-36.
[http://dx.doi.org/10.1042/BJ20151120] [PMID: 26831515]

[6]   Ihrlund LS, Hernlund E, Khan O, Shoshan MC. 3-Bromopyruvate as inhibitor of tumour cell energy metabolism and chemopotentiator of platinum drugs. Mol Oncol 2008; 2(1): 94-101.
[http://dx.doi.org/10.1016/j.molonc.2008.01.003] [PMID: 19383331]

[7]   Huang CC, Wang S-Y, Lin L-L, *et al.* Glycolytic inhibitor 2-deoxyglucose simultaneously targets cancer and endothelial cells to suppress neuroblastoma growth in mice. Dis Model Mech 2015; 8(10): 1247-54.
[http://dx.doi.org/10.1242/dmm.021667] [PMID: 26398947]

[8]   Davies C, Muirhead H, Chirgwin J. The structure of human phosphoglucose isomerase complexed with a transition-state analogue. Acta Crystallogr D Biol Crystallogr 2003; 59(Pt 6): 1111-3.
[http://dx.doi.org/10.1107/S0907444903007352] [PMID: 12777791]

[9]   Read J, Pearce J, Li X, Muirhead H, Chirgwin J, Davies C. The crystal structure of human phosphoglucose isomerase at 1.6 A resolution: implications for catalytic mechanism, cytokine activity and haemolytic anaemia. J Mol Biol 2001; 309(2): 447-63.
[http://dx.doi.org/10.1006/jmbi.2001.4680] [PMID: 11371164]

[10]   Arsenieva D, Appavu BL, Mazock GH, Jeffery CJ. Crystal structure of phosphoglucose isomerase from *Trypanosoma brucei* complexed with glucose-6-phosphate at 1.6 A resolution. Proteins 2009; 74(1): 72-80.
[http://dx.doi.org/10.1002/prot.22133] [PMID: 18561188]

[11]   Hardré R, Salmon L, Opperdoes FR. Competitive inhibition of *Trypanosoma brucei* phosphoglucose isomerase by D-arabinose-5-phosphate derivatives. J Enzyme Inhib 2000; 15(5): 509-15.
[http://dx.doi.org/10.3109/14756360009040706] [PMID: 11030090]

[12]   Murillo-López J, Zinovjev K, Pereira H, *et al.* Studying the phosphoryl transfer mechanism of the E. coli phosphofructokinase-2: from X-ray structure to quantum mechanics simulations. Chem Sci (Camb) 2019; 10: 2882-92.
[http://dx.doi.org/10.1039/C9SC00094A] [PMID: 30996866]

[13]   Kloos M, Brüser A, Kirchberger J, Schöneberg T, Sträter N. Crystallization and preliminary crystallographic analysis of human muscle phosphofructokinase, the main regulator of glycolysis. Acta Crystallogr F Struct Biol Commun 2014; 70(Pt 5): 578-82.
[http://dx.doi.org/10.1107/S2053230X14008723] [PMID: 24817713]

[14] Banaszak K, Mechin I, Obmolova G, *et al.* The crystal structures of eukaryotic phosphofructokinases from baker's yeast and rabbit skeletal muscle. J Mol Biol 2011; 407(2): 284-97.
[http://dx.doi.org/10.1016/j.jmb.2011.01.019] [PMID: 21241708]

[15] Brimacombe KR, Walsh MJ, Liu L, *et al.* Identification of ML251, a potent inhibitor of *T. brucei* and *T. cruzi* Phosphofructokinase. ACS Med Chem Lett 2013; 5(1): 12-7.
[http://dx.doi.org/10.1021/ml400259d] [PMID: 24900769]

[16] Lorentzen E, Siebers B, Hensel R, Pohl E. Mechanism of the schiff base forming fructose-1,-
-bisphosphate aldolase: Structural analysis of reaction intermediates. Biochemistry 2005; 44(11): 4222-9.
[http://dx.doi.org/10.1021/bi048192o] [PMID: 15766250]

[17] Dalby A, Dauter Z, Littlechild JA. Crystal structure of human muscle aldolase complexed with fructose 1,6-bisphosphate: Mechanistic implications. Protein Sci 1999; 8(2): 291-7.
[http://dx.doi.org/10.1110/ps.8.2.291] [PMID: 10048322]

[18] Dax C, Duffieux F, Chabot N, *et al.* Selective irreversible inhibition of fructose 1,6-bisphosphate aldolase from *Trypanosoma brucei*. J Med Chem 2006; 49(5): 1499-502.
[http://dx.doi.org/10.1021/jm050237b] [PMID: 16509566]

[19] Lewis DJ, Lowe G. Phosphoglycollohydroxamic Acid: an Inhibitor of Class I and I1 Aldolases and Triosephosphate Isomerase A Potential Antibacterial and Antifungal agent. JCS Chem. Comm 1973; pp. 713-5.

[20] Hernandez-Alcántara G, Torres-Larios A, Enríquez-Flores S, *et al.* Structural and functional perturbation of giardia lamblia triosephosphate isomerase by modification of a non-catalytic Non-Conserved region. PLoS One 2013; 8: e69031.

[21] Grüning N-M, Du D, Keller MA, Luisi BF, Ralser M. Inhibition of triosephosphate isomerase by phosphoenolpyruvate in the feedback-regulation of glycolysis. Open Biol 2014; 4130232
[http://dx.doi.org/10.1098/rsob.130232] [PMID: 24598263]

[22] Didierjean C, Corbier C, Fatih M, *et al.* Crystal structure of two ternary complexes of phosphorylating glyceraldehyde-3-phosphate dehydrogenase from *Bacillus stearothermophilus* with NAD and D-glyceraldehyde 3-phosphate. J Biol Chem 2003; 278(15): 12968-76.
[http://dx.doi.org/10.1074/jbc.M211040200] [PMID: 12569100]

[23] Fermani S, Sparla F, Falini G, *et al.* Molecular mechanism of thioredoxin regulation in photosynthetic A2B2-glyceraldehyde-3-phosphate dehydrogenase. Proc Natl Acad Sci USA 2007; 104(26): 11109-14.
[http://dx.doi.org/10.1073/pnas.0611636104] [PMID: 17573533]

[24] Qvit N, Joshi AU, Cunningham AD, Ferreira JCB, Mochly-Rosen D. Glyceraldehyde-3-Phosphate Dehydrogenase (GAPDH). J Biol Chem 2016; 291(26): 13608-21.
[http://dx.doi.org/10.1074/jbc.M115.711630] [PMID: 27129213]

[25] Wilson M, Lauth N, Perie J, Callens M, Opperdoes FR. Inhibition of glyceraldehyde-3-Phosphate dehydrogenase by phosphorylated epoxides and α-Enones. Biochemistry 1994; 33(1): 214-0.

[26] Bruno S, Uliassi E, Zaffagnini M, *et al.* Molecular basis for covalent inhibition of glyceraldehyde--
-phosphate dehydrogenase by a 2-phenoxy-1,4-naphthoquinone small molecule. Chem Biol Drug Des 2017; 90(2): 225-35.
[http://dx.doi.org/10.1111/cbdd.12941] [PMID: 28079302]

[27] Ulanovskaya OA, Cui J, Kron SJ, Kozmin SA. A Pairwise Chemical Genetic Screen Identifies New Inhibitors of Glucose Transport Chemistry & Biology 2011.

[28] Sawyer GM, Monzingo AF, Poteet EC, O'Brien DA, Robertus JD. X-ray analysis of phosphoglycerate kinase 2, a sperm-specific isoform from Mus Musculus. Proteins 2008; 71: 1134-44.

[29]   Gondeau C, Chaloin L, Lallemand P, *et al.* Molecular basis for the lack of enantioselectivity of human 3-phosphoglycerate kinase. Nucleic Acids Res 2008; 36(11): 3620-9.
[http://dx.doi.org/10.1093/nar/gkn212] [PMID: 18463139]

[30]   Kotsikorou E, Sahota G, Oldfield E. Bisphosphonate inhibition of phosphoglycerate kinase: quantitative structure-activity relationship and pharmacophore modeling investigation. J Med Chem 2006; 49(23): 6692-703.
[http://dx.doi.org/10.1021/jm0604833] [PMID: 17154500]

[31]   Wang Y, Wei Z, Bian Q, *et al.* Crystal structure of human bisphosphoglycerate mutase. J Biol Chem 2004; 279(37): 39132-8.
[http://dx.doi.org/10.1074/jbc.M405982200] [PMID: 15258155]

[32]   Hirano M. Neurobiology of Disease Ed S, Gilman Section Myopathies and Neuromuscular Disorders Elsevier Academic Press 2007; 947-56.

[33]   Wang Y, Wei Z, Liu L, *et al.* Crystal structure of human B-type phosphoglycerate mutase bound with citrate. Biochem Biophys Res Commun 2005; 331(4): 1207-15.
[http://dx.doi.org/10.1016/j.bbrc.2005.03.243] [PMID: 15883004]

[34]   Wang P, Jiang L, Cao Y, Ye D, Zhou L. The design and synthesis of *N*-Xanthone benzenesulfonamides as novel phosphoglycerate mutase 1 (PGAM1) inhibitors. Molecules 2018; 23(6): 1396.
[http://dx.doi.org/10.3390/molecules23061396] [PMID: 29890679]

[35]   Nowicki MW, Kuaprasert B, Mc Nae IW, *et al.* Crystal structures of *Leishmania Mexicana* phosphoglycerate mutase suggest a one-metal mechanism and a new enzyme subclass. J Mol Biol 2009; 394: 533-43.

[36]   Li X, Tang S, Wang Q-Q, *et al.* Identification of Epigallocatechin-3- Gallate as an inhibitor of phosphoglycerate mutase 1. Front Pharmacol 2017; 8: 325.
[http://dx.doi.org/10.3389/fphar.2017.00325] [PMID: 28611670]

[37]   Kang HJ, Jung S-K, Kim SJ, Chung SJ. Structure of human α-enolase (hENO1), a multifunctional glycolytic enzyme. Acta Crystallogr D Biol Crystallogr 2008; 64(Pt 6): 651-7.
[http://dx.doi.org/10.1107/S0907444908008561] [PMID: 18560153]

[38]   Qin J, Chai G, Brewer JM, Lovelace LL, Lebioda L. Fluoride inhibition of enolase: Crystal structure and thermodynamics. Biochemistry 2006; 45(3): 793-800.
[http://dx.doi.org/10.1021/bi051558s] [PMID: 16411755]

[39]   Leonard PG, Satani N, Maxwell D, *et al.* SF2312 is a natural phosphonate inhibitor of Enolase Nature chemical biology 2016; 12: 1053-58.

[40]   Pietkiewicz J, Gamian A, Staniszewska M, Danielewicz R. Inhibition of human muscle-specific enolase by methylglyoxal and irreversible formation of advanced glycation end products. J Enzyme Inhib Med Chem 2009; 24(2): 356-64.
[http://dx.doi.org/10.1080/14756360802187679] [PMID: 18830874]

[41]   Valentini G, Chiarelli LR, Fortin R, *et al.* Structure and function of Human Erythrocyte Pyruvate Kinase. J Biol Chem 2002; 277(26): 23807-14.

[42]   aDombrauckas JD, Santarsiero BD, Mesecar AD. Structural basis for tumor pyruvate kinase m2 allosteric regulation and catalysis. Biochemistry 2005; 44(27): 9417-29.bSrivastava D, Razzaghi M, Henz MT, Dey M. Structural investigation of a dimeric variant of pyruvate kinase muscle isoform 2. Biochemistry 2017; 56(50): 6517-20.

[43]   Vander Heiden MG, Christofk HR, Schuman E, *et al.* Identification of small molecule inhibitors of

pyruvate kinase M2. Biochem Pharmacol 2010; 79(8): 1118-24.
[http://dx.doi.org/10.1016/j.bcp.2009.12.003] [PMID: 20005212]

[44]   Feksa LR, Cornelio AR, Vargas CR, *et al.* Alanine prevents the inhibition of pyruvate kinase activity caused by tryptophan in cerebral cortex of rats. Metab Brain Dis 2003; 18(2): 129-37.
[http://dx.doi.org/10.1023/A:1023811019023] [PMID: 12822831]

[45]   Anastasiou D, Poulogiannis G, Asara JM, *et al.* Inhibition of pyruvate kinase m2 by reactive oxygen species contributes to cellular antioxidant responses. Science 2011; 334(6060): 1278-83.

[46]   Alves-Filho JC, Pålsson-McDermott EM. Pyruvate kinase m2: A potential target for regulating inflammation. Front Immunol 2016; 7: 145.

[47]   Chen J, Xie J, Jiang Z, Wang B, Wang Y, Hu X. Shikonin and its analogs inhibit cancer cell glycolysis by targeting tumor pyruvate kinase-M2. Oncogene 2011; 30(42): 4297-306.

[48]   Karpusas M, Branchaud B, Remington SJ. Proposed mechanism for the condensation reaction of citrate synthase: 19 å structure of the ternary complex with oxaloacetate and carboxymethyl coenzyme A Biochemistry 1990; 29: 2213-19.

[49]   Russell RJM, Fergusson JMC, Hough DW, Danson MJ, Taylor GL. The crystal structure of citrate synthase from the hyperthermophilic archaeon pyrococcus furious at 19 å resolution Biochemistry 1997; 36: 9983-4.

[50]   Russell RJM, Gerike U, Danson MJ, Hough DW, Taylor GL. Structural adaptations of the cold-active citrate synthase from an antarctic bacterium. Structure 1998; 6(3): 351-61.
[http://dx.doi.org/10.1016/S0969-2126(98)00037-9] [PMID: 9551556]

[51]   Bayer E, Bauer B, Eggerer H. Evidence from inhibitor studies for conformational changes of citrate synthase. Eur J Biochem 1981; 120(1): 155-60.
[http://dx.doi.org/10.1111/j.1432-1033.1981.tb05683.x] [PMID: 7308213]

[52]   Lloyd SJ, Lauble H, Prasad GS, Stout CD. The mechanism of aconitase: 1.8 A resolution crystal structure of the S642a:citrate complex. Protein Sci 1999; 8(12): 2655-62.
[http://dx.doi.org/10.1110/ps.8.12.2655] [PMID: 10631981]

[53]   (a) Dupuy J, Volbeda A, Carpentier P, Darnault C, Moulis J-M, Fontecilla-Camps JC. Crystal structure of human iron regulatory protein 1 as cytosolic aconitase. Structure 2006; 14(1): 129-39.
[http://dx.doi.org/10.1016/j.str.2005.09.009] (b) Lu Y. Assembly and transfer of iron-sulfur clusters in the plastid. Front Plant Sci 2018; 9: 336.
[http://dx.doi.org/10.3389/fpls.2018.00336] [PMID: 29662496]

[54]   Gupta KJ, Shah JK, Brotman Y, *et al.* Inhibition of aconitase by nitric oxide leads to induction of the alternative oxidase and to a shift of metabolism towards biosynthesis of amino acids. J Exp Bot 2012; 63(4): 1773-84.
[http://dx.doi.org/10.1093/jxb/ers053] [PMID: 22371326]

[55]   Lauble H, Kennedy MC, Emptage MH, Beinert H, Stout CD. The reaction of fluorocitrate with aconitase and the crystal structure of the enzyme-inhibitor complex. Proc Natl Acad Sci USA 1996; 93(24): 13699-703.
[http://dx.doi.org/10.1073/pnas.93.24.13699] [PMID: 8942997]

[56]   Ighodaro OM, Adeosun AM, Akinloye OA. Alloxan-induced diabetes, a common model for evaluating the glycemic-control potential of therapeutic compounds and plants extracts in experimental studies. Medicina (Kaunas) 2017; 53(6): 365-74.
[http://dx.doi.org/10.1016/j.medici.2018.02.001] [PMID: 29548636]

[57]   Boquist L, Ericsson I. Inhibition by alloxan of mitochondrial aconitase and other enzymes associated with the citric acid cycle. FEBS 1984; 178(2): 245-8.

[58]   Lampa M, Arlt H, He T, *et al.* Glutaminase is essential for the growth of triple-negative breast cancer cells with a deregulated glutamine metabolism pathway and its suppression synergizes with mTOR inhibition. PLoS One 2017; 12(9): e0185092.

[http://dx.doi.org/10.1371/journal.pone.0185092] [PMID: 28950000]

[59]   DeBerardinis RJ, Mancuso A, Daikhin E, *et al.* Beyond aerobic glycolysis: transformed cells can engage in glutamine metabolism that exceeds the requirement for protein and nucleotide synthesis. Proc Natl Acad Sci USA 2007; 104(49): 19345-50.
[http://dx.doi.org/10.1073/pnas.0709747104] [PMID: 18032601]

[60]   Taylor AB, Hu G, Hart PJ, McAlister-Henn L. Allosteric motions in structures of yeast NAD+-specific isocitrate dehydrogenase. J Biol Chem 2008; 283(16): 10872-80.
[http://dx.doi.org/10.1074/jbc.M708719200] [PMID: 18256028]

[61]   Yaoi T, Miyazaki K, Oshima T. Substrate recognition of isocitrate dehydrogenase and 3-isopropylmalate dehydrogenase from Thermus thermophilus HB8. J Biochem 1997; 121(1): 77-81.
[http://dx.doi.org/10.1093/oxfordjournals.jbchem.a021573] [PMID: 9058195]

[62]   Jones S, Ahmet J, Ayton K, *et al.* Discovery and optimization of allosteric inhibitors of mutant isocitrate dehydrogenase 1 (r132h idh1) displaying activity in human acute myeloid leukemia cells. J Med Chem 2016; 59(24): 11120-37.
[http://dx.doi.org/10.1021/acs.jmedchem.6b01320] [PMID: 28002956]

[63]   Tretter L, Adam-Vizi V. Alpha-ketoglutarate dehydrogenase: a target and generator of oxidative stress. Philos Trans R Soc Lond B Biol Sci 2005; 360(1464): 2335-45.
[http://dx.doi.org/10.1098/rstb.2005.1764] [PMID: 16321804]

[64]   McLain AL, Szweda PA, Szweda LI. α-Ketoglutarate dehydrogenase: a mitochondrial redox sensor. Free Radic Res 2011; 45(1): 29-36.
[http://dx.doi.org/10.3109/10715762.2010.534163] [PMID: 21110783]

[65]   Nemeria NS, Ambrus A, Patel H, *et al.* Human 2-Oxoglutarate dehydrogenase complex e1 component forms a thiamin-derived radical by aerobic oxidation of the enamine intermediate Journal Biological Chemistry 2014; 289: 29859-73.

[66]   Frank RA, Price AJ, Northrop FD, Perham RN, Luisi BF. Crystal structure of the E1 component of the *Escherichia coli* 2-oxoglutarate dehydrogenase multienzyme complex. J Mol Biol 2007; 368(3): 639-51.
[http://dx.doi.org/10.1016/j.jmb.2007.01.080] [PMID: 17367808]

[67]   Brito-Arias M. The chemistry of oxidative stress and glycoxidation as risk factors for developing degenerative disease. Ann Rev Research 2019; 4(5): 1-14.

[68]   Stuart SD, Schauble A, Gupta S, *et al.* A strategically designed small molecule attacks alpha-ketoglutarate dehydrogenase in tumor cells through a redox process. Cancer Metab 2014; 2(1): 4-15.
[http://dx.doi.org/10.1186/2049-3002-2-4] [PMID: 24612826]

[69]   Atlante S, Visintin A, Marini E, *et al.* α-ketoglutarate dehydrogenase inhibition counteracts breast cancer-associated lung metastasis Cell Death and Disease 2018; 9(7): 2-18.

[70]   Wolodko WT, Fraser ME, James MNG, Bridger WA. The crystal structure of succinyl-CoA synthetase from *Escherichia coli* at 2.5-A resolution. J Biol Chem 1994; 269(14): 10883-90.
[PMID: 8144675]

[71]   Huang J, Malhi M, Deneke J, Fraser ME. Structure of GTP-specific succinyl-CoA synthetase in complex with CoA. Acta Crystallogr F Struct Biol Commun 2015; 71(Pt 8): 1067-71.
[http://dx.doi.org/10.1107/S2053230X15011188] [PMID: 26249701]

[72]   Hunger-Glaser I, Brun R, Linder M, Seebeck T. Inhibition of succinyl CoA synthetase histidine-phosphorylation in *Trypanosoma brucei* by an inhibitor of bacterial two-component systems. Mol Biochem Parasitol 1999; 100(1): 53-9.
[http://dx.doi.org/10.1016/S0166-6851(99)00032-8] [PMID: 10376993]

[73]   Boquist L, Ericsson I. Inhibition by streptozotocin of the activity of succinyl-CoA synthetase *in vitro* and *in vivo* FEBS Lett 1985; 196(2): 341-3.

[74]   Hederstedt L, Rutberg L. Succinate dehydrogenase--a comparative review. Microbiol Rev 1981; 45(4): 542-55.
        [http://dx.doi.org/10.1128/MMBR.45.4.542-555.1981] [PMID: 6799760]

[75]   Rustin P, Munnich A, Rötig A. Succinate dehydrogenase and human diseases: New insights into a well-known enzyme. Eur J Hum Genet 2002; 10(5): 289-91.
        [http://dx.doi.org/10.1038/sj.ejhg.5200793] [PMID: 12082502]

[76]   Yankovskaya V, Horsefield R, Törnroth S, *et al.* Architecture of succinate dehydrogenase and reactive oxygen species generation. Science 2003; 299(5607): 700-4.
        [http://dx.doi.org/10.1126/science.1079605] [PMID: 12560550]

[77]   Wojtczak L, Wojtczak AB, Ernster L. The inhibition of succinate dehydrogenase by oxalacetate. Biochim Biophys Acta 1969; 191(1): 10-21.
        [http://dx.doi.org/10.1016/0005-2744(69)90310-6] [PMID: 5823490]

[78]   Anastacio MM, Kanter EM, Keith AD, Schuessler RB, Nichols CG, Lawton JS. Inhibition of succinate dehydrogenase by diazoxide is independent of the KATP channel subunit SUR 1. J Am Coll Surg 2013; 216: 1144-9.
        [http://dx.doi.org/10.1016/j.jamcollsurg.2013.01.048] [PMID: 23535164]

[79]   Masgras I, Sanchez-Martin C, Colombo G, Rasola A. The chaperone trap1 as a modulator of the mitochondrial adaptations in cancer cells. Front Oncol 2017; 8: 58.

[80]   Guzzo G, Sciacovelli M, Bernardi P, Rasola A. Inhibition of succinate dehydrogenase by the mitochondrial chaperone TRAP1 has anti-oxidant and anti-apoptotic effects on tumor cells. Oncotarget 2014; 5(23): 11897-908.
        [http://dx.doi.org/10.18632/oncotarget.2472] [PMID: 25564869]

[81]   Feliciano PR, Drennan CL, Nonato MC. Crystal structure of an Fe-S cluster-containing fumarate hydratase enzyme from Leishmania major reveals a unique protein fold. PNAS 2016; 113(35): 9804-9.
        [http://dx.doi.org/10.1073/pnas.1605031113]

[82]   Kasbekar M, Fischer G, Mott BT, *et al.* Selective small molecule inhibitor of the *Mycobacterium tuberculosis* fumarate hydratase reveals an allosteric regulatory site. Proc Natl Acad Sci USA 2016; 113(27): 7503-8.
        [http://dx.doi.org/10.1073/pnas.1600630113] [PMID: 27325754]

[83]   Takeuchi T, Schumacker PT, Kozmin SA. Identification of fumarate hydratase inhibitors with nutrient-dependent cytotoxicity. J Am Chem Soc 2015; 137(2): 564-7.
        [http://dx.doi.org/10.1021/ja5101257] [PMID: 25469852]

[84]   González JM, Marti-Arbona R, Chen JCH, Broom-Peltz B, Unkefer CJ. Conformational changes on substrate binding revealed by structures of Methylobacterium extorquens malate dehydrogenase. Acta Crystallogr F Struct Biol Commun 2018; 74(Pt 10): 610-6.
        [http://dx.doi.org/10.1107/S2053230X18011809] [PMID: 30279311]

[85]   Irimia A, Vellieux FMD, Madern D, *et al.* The 2.9A resolution crystal structure of malate dehydrogenase from Archaeoglobus fulgidus: mechanisms of oligomerisation and thermal stabilisation. J Mol Biol 2004; 335(1): 343-56.
        [http://dx.doi.org/10.1016/j.jmb.2003.10.054] [PMID: 14659762]

[86]   Lee K, Ban HS, Naik R, *et al.* Identification of malate dehydrogenase 2 as a target protein of the HIF-1 inhibitor LW6 using chemical probes. Angew Chem Int Ed Engl 2013; 52(39): 10286-9.
        [http://dx.doi.org/10.1002/anie.201304987] [PMID: 23934700]

[87]   Eleftheriadis T, Pissas G, Antoniadi G, Liakopoulos V, Stefanidis I. Malate dehydrogenase-2 inhibitor LW6 promotes metabolic adaptations and reduces proliferation and apoptosis in activated human T-cells. Exp Ther Med 2015; 10(5): 1959-66.
        [http://dx.doi.org/10.3892/etm.2015.2763] [PMID: 26640580]

[88]   Dall'aglio P, Arthur CJ, Williams C, *et al.* Analysis of *Streptomyces coelicolor* phosphopantetheinyl

transferase, AcpS, reveals the basis for relaxed substrate specificity. Biochemistry 2011; 50(25): 5704-17.
[http://dx.doi.org/10.1021/bi2003668] [PMID: 21595442]

[89]   Keatinge-Clay AT, Shelat AA, Savage DF, *et al.* Catalysis, specificity, and ACP docking site of *Streptomyces coelicolor* malonyl-CoA:ACP transacylase. Structure 2003; 11(2): 147-54.
[http://dx.doi.org/10.1016/S0969-2126(03)00004-2] [PMID: 12575934]

[90]   Yu X, Hiromasa Y, Tsen H, Stoops JK, Roche TE, Zhou ZH. Structures of the human pyruvate dehydrogenase complex cores: a highly conserved catalytic center with flexible N-terminal domains. Structure 2008; 16(1): 104-14.
[http://dx.doi.org/10.1016/j.str.2007.10.024] [PMID: 18184588]

[91]   Kato M, Li J, Chuang JL, Chuang DT. Distinct structural mechanisms for inhibition of pyruvate dehydrogenase kinase isoforms by AZD7545, dichloroacetate, and radicicol. Structure 2007; 15(8): 992-1004.
[http://dx.doi.org/10.1016/j.str.2007.07.001] [PMID: 17683942]

[92]   Patel MS, Korotchkina LG. The Biochemistry of the pyruvate dehydrogenase Complex. Biochem Mol Biol Educ 2003; 31: 5-15.
[http://dx.doi.org/10.1002/bmb.2003.494031010156]

[93]   Milne JLS, Wu X, Borgnia MJ, *et al.* Molecular structure of a 9-MDa icosahedral pyruvate dehydrogenase subcomplex containing the E2 and E3 enzymes using cryoelectron microscopy. J Biol Chem 2006; 281(7): 4364-70.
[http://dx.doi.org/10.1074/jbc.M504363200] [PMID: 16308322]

[94]   Izard T, Aevarsson A, Allen MD, *et al.* Principles of quasi-equivalence and Euclidean geometry govern the assembly of cubic and dodecahedral cores of pyruvate dehydrogenase complexes. Proc Natl Acad Sci USA 1999; 96(4): 1240-5.
[http://dx.doi.org/10.1073/pnas.96.4.1240] [PMID: 9990008]

[95]   Zhou Y, Zhang S, He H, *et al.* Design and synthesis of highly selective pyruvate dehydrogenase complex E1 inhibitors as bactericides Bioorganic &. Med Chem 2016; 26: 84-95.
[http://dx.doi.org/10.1016/j.bmc.2017.11.021]

[96]   Adamson SR, Stevenson KJ. Stevenson inhibition of pyruvate dehydrogenase multienzyme complex from Escherichia coli with a bifunctional arsenoxide: selective inactivation of lipoamide dehydrogenase1 Biochemistry 1981; 20: 3418-24.

[97]   Behrens M, Hüwel S, Galla H-J, Humpf H-U. Blood-brain barrier effects of the fusarium mycotoxins deoxynivalenol, 3 acetyldeoxynivalenol, and moniliformin and their transfer to the brain PLOS ONE 2015; 23: 1-20.

[98]   Larson SB, Day JS, Nguyen C, Cudney R, McPherson A. Structure of pig heart citrate synthase at 1.78 A resolution. Acta Crystallogr Sect F Struct Biol Cryst Commun 2009; 65(Pt 5): 430-4.
[http://dx.doi.org/10.1107/S1744309109008343] [PMID: 19407370]

[99]   Russell RJM, Hough DW, Danson MJ, Taylor GL. The crystal structure of citrate synthase from the thermophilic archaeon, Thermoplasma acidophilum. Structure 1994; 2(12): 1157-67.
[http://dx.doi.org/10.1016/S0969-2126(94)00118-9] [PMID: 7704526]

[100] Karpusas M, Branchaud B, Remington SJ. Proposed mechanism for the condensation reaction of citrate synthase: 19-a structure of the ternary complex with oxaloacetate and carboxymethyl coenzyme A Biochemistry 1990; 29: 2213-19.

[101] Lee SH, Son HF, Kim K-J. Structural insights into the inhibition properties of archaedon citrate synthase from Metallosphaera sedula. PLoS One 2019; 1: 1-16.

[102] Sun T, Hayakawa K, Bateman KS, Fraser ME. Identification of the citrate-binding site of human ATP-Citrate lyase using X-ray Crystallography 2010; 285: 27418-28.

[103] Zaidi N, Swinnen JV, Smans K. ATP-Citrate Lyase: A key player in cancer metabolism. Cancer Res

2012; 72(15): 3709-14.
[http://dx.doi.org/10.1158/0008-5472.CAN-11-4112]

[104] Khwairakpam AD, Shyamananda MS, Salio BL, *et al.* ATP Citrate Lyase (ACLY): A promising target for cancer prevention and treatment current drug targets 2015; 16: 156-63.

[105] Wang D, Yin L, Wei J, Yang Z, Jiang G. ATP citrate lyase is increase in human breast cancer, depletion of which promotes apoptosis. Tumour Biol 2017; 39(4): 1-10.
[http://dx.doi.org/10.1177/1010428317698338]

[106] Wei J, Leit S, Kuai J, *et al.* An allosteric mechanism for potent inhibition of human ATP-citrate lyase. Nature 2019; 568(7753): 566-70.
[http://dx.doi.org/10.1038/s41586-019-1094-6] [PMID: 30944472]

[107] Li JJ, Wang H, Tino JA, *et al.* 2-hydroxy-N-arylbenzenesulfonamides as ATP-citrate lyase inhibitors. Bioorg Med Chem Lett 2007; 17(11): 3208-11.
[http://dx.doi.org/10.1016/j.bmcl.2007.03.017] [PMID: 17383874]

[108] Qiu X, Janson CA, Konstantinidis AK, *et al.* Crystal Structure of –ketoacyl-Acyl carrier protein synthase III. J Biol Chem 1999; 274: 36465-71.
[http://dx.doi.org/10.1074/jbc.274.51.36465] [PMID: 10593943]

[109] Witkowski A, Joshi AK, Smith S. Mechanism of the β-ketoacyl synthase reaction catalyzed by the animal fatty acid synthase. Biochemistry 2002; 41(35): 10877-87.

[110] (a) Christensen CE, Kragelund BB, von Wettstein-Knowles P, Henriksen A. Structure of the human β-ketoacyl [ACP] synthase from the mitochondrial type II fatty acid synthase. Protein Sci 2007; 16(2): 261-72.
[http://dx.doi.org/10.1110/ps.062473707] [PMID: 17242430] (b) Moche M, Schneider G, Edwards P, Dehesh K, Lindqvist Y. Structure of the complex between the antibiotic cerulenin and its target, β-ketoacyl-acyl carrier protein synthase. J Biol Chem 1999; 274(10): 6031-4.
[http://dx.doi.org/10.1074/jbc.274.10.6031] [PMID: 10037680]

[111] Hardwicke MA, Rendina AR, Williams SP, *et al.* A human fatty acid synthase inhibitor binds β-ketoacyl reductase in the keto-substrate site. Nat Chem Biol 2014; 10(9): 774-9.
[http://dx.doi.org/10.1038/nchembio.1603] [PMID: 25086508]

[112] Vázquez MJ, Leavens W, Liu R, *et al.* Discovery of GSK837149A, an inhibitor of human fatty acid synthase targeting the β-ketoacyl reductase reaction. FEBS J 2008; 275(7): 1556-67.
[http://dx.doi.org/10.1111/j.1742-4658.2008.06314.x] [PMID: 18312417]

[113] Wang H, Klein MG, Zou H, *et al.* Crystal structure of human stearoyl-coenzyme A desaturase in complex with substrate. Nat Struct Mol Biol 2015; 22(7): 581-5.
[http://dx.doi.org/10.1038/nsmb.3049] [PMID: 26098317]

[114] Lyn RK, Singaravelu R, Kargman S, *et al.* Stearoyl-CoA desaturase inhibition blocks formation of hepatitis C virus-induced specialized membranes. Sci Rep 2015; 4: 4549.

[115] Uto Y. Recent progress in the discovery and development of stearoyl CoA desaturase inhibitors. Chem Phys Lipids 2016; 197: 3-12.
[http://dx.doi.org/10.1016/j.chemphyslip.2015.08.018] [PMID: 26344107]

[116] Peláez R, Pariente A, Pérez-Sala A, Larráyoz M. Sterculic Acid: The Mechanisms of Action beyond Stearoyl-CoA Desaturase Inhibition and Therapeutic Opportunities in Human Disease Cells 2020; 9: 2-20.

[117] Zhang Z, Sun S, Kodumuru V, *et al.* Discovery of piperazin-1-ylpyridazine-based potent and selective stearoyl-CoA desaturase-1 inhibitors for the treatment of obesity and metabolic syndrome. J Med Chem 2013; 56(2): 568-83.
[http://dx.doi.org/10.1021/jm301661h] [PMID: 23245208]

[118] Hao P, Alaraj IQ, Dulayymi JR, Baird MS, Liu J, Liu Q. sterculic acid and its analogues are potent inhibitors of toxoplasma gondii. Korean J Parasitol 2016; 54(2): 139-45.

[http://dx.doi.org/10.3347/kjp.2016.54.2.139] [PMID: 27180571]

[119]   Yeh JI, Kettering R, Saxl R, *et al*. Structural characterizations of glycerol kinase: unraveling phosphorylation-induced long-range activation. Biochemistry 2009; 48(2): 346-56.
[http://dx.doi.org/10.1021/bi8009407] [PMID: 19102629]

[120]   Schnick C, Polley SD, Fivelman QL, *et al*. Structure and non-essential function of glycerol kinase in Plasmodium falciparum blood stages. Mol Microbiol 2009; 71(2): 533-45.
[http://dx.doi.org/10.1111/j.1365-2958.2008.06544.x] [PMID: 19040641]

[121]   Balogun EO, Inaoka DK, Shiba T, *et al*. Molecular basis for the reverse reaction of African human trypanosomes glycerol kinase. Mol Microbiol 2014; 94(6): 1315-29.

[122]   Tisdale MJ, Threadgill MD. (+/-)2,3-Dihydroxypropyl dichloroacetate, an inhibitor of glycerol kinase. Cancer Biochem Biophys 1984; 7(3): 253-9.
[PMID: 6091865]

[123]   Monroy G, Kelker HC, Pullman ME. Partial purification and properties of an acyl coenzyme A:sn-glycerol 3-phosphate acyltransferase from rat liver mitochondria. J Biol Chem 1973; 248(8): 2845-52.
[PMID: 4697393]

[124]   Zhang Y-M, Rock CO. Thematic review series: Glycerolipids. Acyltransferases in bacterial glycerophospholipid synthesis. J Lipid Res 2008; 49(9): 1867-74.
[http://dx.doi.org/10.1194/jlr.R800005-JLR200] [PMID: 18369234]

[125]   Yao J, Rock CO. Phosphatidic acid synthesis in bacteria. Biochim Biophys Acta 2013; 1831(3): 495-502.
[http://dx.doi.org/10.1016/j.bbalip.2012.08.018] [PMID: 22981714]

[126]   Turnbull AP, Rafferty JB, Sedelnikova SE, *et al*. Analysis of the structure, substrate specificity, and mechanism of squash glycerol-3-phosphate (1)-acyltransferase. Structure 2001; 9(5): 347-53.
[http://dx.doi.org/10.1016/S0969-2126(01)00595-0] [PMID: 11377195]

[127]   Li Z, Tang Y, Wu Y, *et al*. Structural insights into the committed step of bacterial phospholipid biosynthesis. Nat Commun 2017; 8(1): 1691.
[http://dx.doi.org/10.1038/s41467-017-01821-9] [PMID: 29167463]

[128]   Wydysh EA, Medghalchi SM, Vadlamudi A, Townsend CA. Design and synthesis of small molecule glycerol 3-phosphate acyltransferase inhibitors. J Med Chem 2009; 52(10): 3317-27.
[http://dx.doi.org/10.1021/jm900251a] [PMID: 19388675]

[129]   Kuhajda FP, Aja S, Tu Y, *et al*. Pharmacological glycerol-3-phosphate acyltransferase inhibition decreases food intake and adiposity and increases insulin sensitivity in diet-induced obesity. Am J Physiol Regul Integr Comp Physiol 2011; 301(1): R116-30.
[http://dx.doi.org/10.1152/ajpregu.00147.2011] [PMID: 21490364]

[130]   Yamashita A, Hayashi Y, Matsumoto N, *et al*. Glycerophosphate/Acylglycerophosphate Acyltransferases Biology 2014; 3: 801-30.

[131]   Shindou H, Shimizu T. Acyl-CoA:lysophospholipid acyltransferases. J Biol Chem 2009; 284(1): 1-5.
[http://dx.doi.org/10.1074/jbc.R800046200] [PMID: 18718904]

[132]   Currie E, Schulze A, Zechner R, Walther TC, Farese RV Jr. Cellular fatty acid metabolism and cancer. Cell Metab 2013; 18(2): 153-61.
[http://dx.doi.org/10.1016/j.cmet.2013.05.017] [PMID: 23791484]

[133]   Coon M, Ball A, Pound J, *et al*. Inhibition of lysophosphatidic acid acyltransferase β disrupts proliferative and survival signals in normal cells and induces apoptosis of tumor cells. Mol Cancer Ther 2003; 2(10): 1067-78.
[PMID: 14578472]

[134]   Takeuchi K, Reue K. Biochemistry, physiology, and genetics of GPAT, AGPAT, and lipin enzymes in triglyceride synthesis Am J Physiol Endorinol Metab 2009; 296: E1195-209.

[135] Carman GM, Han G-S. Phosphatidic acid phosphatase, a key enzyme in the regulation of lipid synthesis. J Biol Chem 2009; 284(5): 2593-7.

[136] Jin H-H, Jiang J-G. Phosphatidic acid phosphatase and diacylglycerol acyltransferase: potential targets for metabolic engineering of microorganism oil. J Agric Food Chem 2015; 63(12): 3067-77.
[http://dx.doi.org/10.1021/jf505975k] [PMID: 25672855]

[137] Elabbadi N, Day ChP, Virden R, Yeaman SJ. Regulation of phosphatidic acid phosphohydrolase 1 by fatty acids. Lipids 2002; 37: 69-73.

[138] Jin Y, McFie PJ, Banman SL, Brandt C, Stone SJ. Diacylglycerol acyltransferase-2 (DGAT2) and monoacylglycerol acyltransferase-2 (MGAT2) interact to promote triacylglycerol synthesis. J Biol Chem 2014; 289(41): 28237-48.
[http://dx.doi.org/10.1074/jbc.M114.571190] [PMID: 25164810]

[139] Caldo KMP, Acedo JA, Panigrahi R, Vederas JC, Weselake RJ, Lemieux MJ. Diacylglycerol acyltransferase 1 is regulated by its n-terminal domain in response to allosteric effectors. Plant Physiology 2017; 175: 667-80.

[140] DeVita RJ, Pinto S. Current status of the research and development of diacylglycerol O-acyltransferase 1 (DGAT1) inhibitors. J Med Chem 2013; 56(24): 9820-5.
[http://dx.doi.org/10.1021/jm4007033] [PMID: 23919406]

[141] Schirch V, Szebenyi DME. Serine hydroxymethyltransferase revisited. Curr Opin Chem Biol 2005; 9(5): 482-7.
[http://dx.doi.org/10.1016/j.cbpa.2005.08.017] [PMID: 16125438]

[142] Vivoli M, Angelucci F, Ilari A, *et al.* Role of a conserved active site cation−π interaction in *Escherichia coli* Serine hydroxymethyltransferase. Biochemistry 2009; 48(50): 12034-46.

[143] Angelucci F, Morea V, Angelaccio S, Saccoccia F, Contestabile R, Ilari A. The crystal structure of archaeal serine hydroxymethyltransferase reveals idiosyncratic features likely required to withstand high temperatures. Proteins 2014; 82(12): 3437-49.
[http://dx.doi.org/10.1002/prot.24697] [PMID: 25257552]

[144] Trivedi V, Gupta A, Jala VR, *et al.* Crystal structure of binary and ternary complexes of serine hydroxymethyltransferase from *Bacillus stearothermophilus*: insights into the catalytic mechanism. J Biol Chem 2002; 277(19): 17161-9.
[http://dx.doi.org/10.1074/jbc.M111976200] [PMID: 11877399]

[145] Scarsdale JN, Kazanina G, Radaev S, Schirch V, Wright HT. Crystal structure of rabbit cytosolic serine hydroxymethyltransferase at 2.8 A resolution: mechanistic implications. Biochemistry 1999; 38(26): 8347-58.
[http://dx.doi.org/10.1021/bi9904151] [PMID: 10387080]

[146] Ducker GS, Ghergurovich JM, Mainolfi N, *et al.* Human SHMT inhibitors reveal defective glycine import as a targetable metabolic vulnerability of diffuse large B-cell lymphoma. Proc Natl Acad Sci USA 2017; 114(43): 11404-9.
[http://dx.doi.org/10.1073/pnas.1706617114] [PMID: 29073064]

[147] Yu X, Wang X, Engel PC. The specificity and kinetic mechanism of branched-chain amino acid aminotransferase from Escherichia coli studied with a new improved coupled assay procedure and the enzyme's potential for biocatalysis. FEBS J 2014; 281(1): 391-400.
[http://dx.doi.org/10.1111/febs.12609] [PMID: 24206068]

[148] Yoshikane Y, Yokochi N, Yamasaki M, *et al.* Crystal structure of pyridoxamine-pyruvate aminotransferase from Mesorhizobium loti MAFF303099. J Biol Chem 2008; 283(2): 1120-7.
[http://dx.doi.org/10.1074/jbc.M708061200] [PMID: 17989071]

[149] Whalen WA, Wang M-D, Berg CM. β-Chloro-L-alanine inhibition of the *Escherichia coli* alanine-valine transaminase. J Bacteriol 1985; 164(3): 1350-2.
[http://dx.doi.org/10.1128/JB.164.3.1350-1352.1985] [PMID: 3934143]

[150] Wolfe RR. Branched-chain amino acids and muscle protein synthesis in humans: myth or reality? J Int Soc Sports Nutr 2017; 14(30): 30.
[http://dx.doi.org/10.1186/s12970-017-0184-9] [PMID: 28852372]

[151] Singh BK, Shaner DL. Biosynthesis of branched chain amino acids: from test tube to field the plant cell 1995; 7: 935-55.

[152] Chipman DM, Duggleby RG, Tittmann K. Mechanisms of acetohydroxyacid synthases. Curr Opin Chem Biol 2005; 9(5): 475-81.
[http://dx.doi.org/10.1016/j.cbpa.2005.07.002] [PMID: 16055369]

[153] Mitra A, Sarma SP. *Escherichia coli* ilvN interacts with the FAD binding domain of ilvB and activates the AHAS I enzyme. Biochemistry 2008; 47(6): 1518-31.
[http://dx.doi.org/10.1021/bi701893b] [PMID: 18193896]

[154] Gedi V, Yoon MY. Bacterial acetohydroxyacid synthase and its inhibitors--a summary of their structure, biological activity and current status. FEBS J 2012; 279(6): 946-63.
[http://dx.doi.org/10.1111/j.1742-4658.2012.08505.x] [PMID: 22284339]

[155] Lv Y, Kandale A, Wun SJ, *et al.* Crystal structure of *Mycobacterium tuberculosis* ketol-acid reductoisomerase at 1.0 Å resolution - a potential target for anti-tuberculosis drug discovery. FEBS J 2016; 283(7): 1184-96.
[http://dx.doi.org/10.1111/febs.13672] [PMID: 26876563]

[156] Chen C-Y, Chang Y-C, Lin B-L, *et al.* Use of cryo-EM to uncover structural bases of pH effect and cofactor bispecificity of ketol-acid reductoisomerase. J Am Chem Soc 2019; 141(15): 6136-40.
[http://dx.doi.org/10.1021/jacs.9b01354] [PMID: 30921515]

[157] Patel KM, Teran D. Crystal Structures of Staphylococcus aureus ketol-acid reductoisomerase in complex with two transition state analogues that have biocidal activity chemistry 2017; 23(72): 18289-95.

[158] Limberg G, Klaffke W, Thiem J. Conversion of aldonic acids to their corresponding 2-keto-3-dexy-analogs by the non-carbohydrate enzyme dihydroxy acid dehydratase (DHAD). Bioorg Med Chem 1995; 3(5): 487-94.
[http://dx.doi.org/10.1016/0968-0896(95)00072-O] [PMID: 7648198]

[159] Yan Y, Liu Q, Zang X, *et al.* Resistance-gene-directed discovery of a natural-product herbicide with a new mode of action. Nature 2018; 559(7714): 415-8.
[http://dx.doi.org/10.1038/s41586-018-0319-4] [PMID: 29995859]

[160] Tremblay LW, Blanchard JS. The 19 A structure of the branched chain amino acid transaminase (IlvE) from Mycobacterium tuberculosis Acta Crystallogr Sect F Biol Cryst Commun 2009; 65: 1071-77.

[161] Soper TS, Manning JM. Different modes of action of inhibitors of bacterial D-amino acid transaminase. A target enzyme for the design of new antibacterial agents. J Biol Chem 1981; 256(9): 4263-8.
[PMID: 7217082]

[162] Amorim Franco TM, Favrot L, Vergnolle O, Blanchard JS. Mechanism-Based Inhibition of the *Mycobacterium tuberculosis* Branched-Chain Aminotransferase by d- and l-Cycloserine. ACS Chem Biol 2017; 12(5): 1235-44.
[http://dx.doi.org/10.1021/acschembio.7b00142] [PMID: 28272868]

[163] Kohlhaw GB. Leucine biosynthesis in fungi: entering metabolism through the back door microbiology and molecular biology reviews 2003; 67: 1-5.

[164] Huisman FHA, Koon N, Bulloch EMM, *et al.* Removal of the c-terminal regulatory domain of α-isopropylmalate synthase disrupts functional substrate binding. Biochemistry 2012; 51(11): 2289-97.

[165] Zhang Z, Wu J, Lin W, *et al.* Subdomain II of α-isopropylmalate synthase is essential for activity J Biol Chem 2014; 289(40): 27966-78.

[166] de Carvalho LP, Blanchard JS. Kinetic and chemical mechanism of alpha-isopropylmalate synthase from *Mycobacterium tuberculosis*. Biochemistry 2006; 45(29): 8988-99.
[http://dx.doi.org/10.1021/bi0606602] [PMID: 16846242]

[167] Pandey P, Lynn AM, Bandyopadhyay P. Identification of inhibitors against α-Isopropylmalate Synthase of *Mycobacterium tuberculosis* using docking-MM/PBSA hybrid approach. Bioinformation 2017; 13(5): 144-8.
[http://dx.doi.org/10.6026/97320630013144] [PMID: 28690380]

[168] Manikandan K, Geerlof A, Zozulya AV, Svergun DI, Weiss MS. Structural studies on the enzyme complex isopropylmalate isomerase (LeuCD) from *Mycobacterium tuberculosis*. Proteins 2011; 79(1): 35-49.
[http://dx.doi.org/10.1002/prot.22856] [PMID: 20938981]

[169] Lee EH, Lee K, Hwang KY. Structural characterization and comparison of the large subunits of IPM isomerase and homoaconitase from *Methanococcus jannaschii*. Acta Crystallogr D Biol Crystallogr 2014; 70(Pt 4): 922-31.
[http://dx.doi.org/10.1107/S1399004713033762] [PMID: 24699638]

[170] Yasutake Y, Yao M, Kirita T, Tanaka I. Crystal structure of the *Pyrococcus horikoshii* isopropylmalate isomerase small subunit specificity of the enzyme. J Mol Bio 2004; 344(2): 325-3.

[171] Amorim Franco TM, Blanchard JS. Bacterial branched-chain amino acid biosynthesis: structures, mechanisms, and drugability. Biochemistry 2017; 56(44): 5849-65.
[http://dx.doi.org/10.1021/acs.biochem.7b00849] [PMID: 28977745]

[172] Hawkes TR, Cox JM, Fraser TE, Lewis T. A herbicidal inhibitor of isopropyl isomerase Z. Naturforsch 1993; 48c: 364-8.
[http://dx.doi.org/10.1515/znc-1993-3-440]

[173] Lee SG, Nwumeh R, Jez JM. Structure and mechanism of isopropylmalate dehydrogenase from *Arabidopsis thaliana*: insights on leucine and aliphatic glucosinolate biosynthesis. J Biol Chem 2016; 291(26): 13421-30.
[http://dx.doi.org/10.1074/jbc.M116.730358] [PMID: 27137927]

[174] Wittenbach VA, Teaney PW, Hanna WS, Rayner DR, Schloss JV. Herbicidal activity of an isopropylmalate dehydrogenase inhibitor. Plant Physiol 1994; 106(1): 321-8.
[http://dx.doi.org/10.1104/pp.106.1.321] [PMID: 12232331]

[175] Dumas R, Biou V, Halgand F, Douce R, Duggleby RG. Enzymology, structure, and dynamics of acetohydroxy acid isomeroreductase. Acc Chem Res 2001; 34(5): 399-408.
[http://dx.doi.org/10.1021/ar000082w] [PMID: 11352718]

[176] Kojima M, Kimura N, Miura R. Regulation of primary metabolic pathways in oyster mushroom mycelia induced by blue light stimulation: accumulation of shikimic acid. Sci Rep 2015; 5: 8630.
[http://dx.doi.org/10.1038/srep08630] [PMID: 25721093]

[177] Mir R, Jallu S, Singh TP. The shikimate pathway: review of amino acid sequence, function and three-dimensional structures of the enzymes. Crit Rev Microbiol 2015; 41(2): 172-89.
[http://dx.doi.org/10.3109/1040841X.2013.813901] [PMID: 23919299]

[178] Tohge T, Watanabe M, Hoefgen R, Fernie AR. Shikimate and phenylalanine biosynthesis in the green lineage. Front Plant Sci 2013; 4: 62.
[http://dx.doi.org/10.3389/fpls.2013.00062] [PMID: 23543266]

[179] Zhou L, Wu J, Janakiraman V, *et al.* Structure and characterization of the 3-deoxy-D-arabino-heptulosonate 7-phosphate synthase from Aeropyrum pernix. Bioorg Chem 2012; 40(1): 79-86.
[http://dx.doi.org/10.1016/j.bioorg.2011.09.002] [PMID: 22035970]

[180] Shumilin IA, Bauerle R, Wu J, Woodard RW, Kretsinger RH. Crystal structure of the reaction complex of 3-deoxy-D-arabino-heptulosonate-7-phosphate synthase from *Thermotoga maritima* refines the catalytic mechanism and indicates a new mechanism of allosteric regulation. J Mol Biol

2004; 341(2): 455-66.
[http://dx.doi.org/10.1016/j.jmb.2004.05.077] [PMID: 15276836]

[181]   Bender SL, Mehdi S, Knowles JR. Dehydroquinate synthase: the role of divalent metal cations and of nicotinamide adenine dinucleotide in catalysis. Biochemistry 1989; 28(19): 7555-60.
[http://dx.doi.org/10.1021/bi00445a009] [PMID: 2514789]

[182]   Mittelstädt G, Negron L, Schofield LR, Marsh K, Parker EJ. Biochemical and structural characterisation of dehydroquinate synthase from the New Zealand kiwifruit Actinidia chinensis. Arch Biochem Biophys 2013; 537(2): 185-91.
[http://dx.doi.org/10.1016/j.abb.2013.07.022] [PMID: 23916589]

[183]   Myrvold S, Reimer LM, Pompliano DL, Frost JW. Chemical inhibition of dehydroquinate synthase. J Am Chem Soc 1989; 111(5): 1861-6.

[184]   Chandran SS, Frost JW. Aromatic inhibitors of dehydroquinate synthase: synthesis, evaluation and implications for gallic acid biosynthesis. Bioorg Med Chem Lett 2001; 11(12): 1493-6.
[http://dx.doi.org/10.1016/S0960-894X(01)00065-8] [PMID: 11412967]

[185]   Light SH, Minasov G, Shuvalova L, et al. Insights into the mechanism of type i dehydroquinate dehydratase from structures of reaction intermediates. J Bio Chem 2011; 286(5): 3531-9.

[186]   Roszak AW, Robinson DA, Krell T, et al. The structure and mechanism of the type II dehydroquinate from Streptomyces coelicolor. Structure 2002; 10(4): 493-503.

[187]   Yao Y, Li Z-S. Structure-and-mechanism-based design and discovery of type II Mycobacterium tuberculosis dehydroquinate dehydratase inhibitors. Curr Top Med Chem 2014; 14(1): 51-63.
[http://dx.doi.org/10.2174/1568026613666131113150257] [PMID: 24236726]

[188]   Frederickson M, Parker EJ, Hawkins AR, Coggins JR, Abell C. Selective inhibition of type II dehydroquinases. J Org Chem 1999; 64(8): 2612-3.
[http://dx.doi.org/10.1021/jo990004q] [PMID: 11674325]

[189]   Sann CL, Abel C, Abell AD. A simple method for the preparation of 3-hydroxyiminodehydroquinate, a potent inhibitor of type II dehydroquinase. J Chem Soc Perkin Transaction 2002; 18: 2065-68.

[190]   Prazeres VF, Castedo L, Lamb H, Hawkins AR, González-Bello C. 2-substituted-3-dehydroquinic acids as potent competitive inhibitors of type II dehydroquinase. ChemMedChem 2009; 4(12): 1980-4.
[http://dx.doi.org/10.1002/cmdc.200900319] [PMID: 19856378]

[191]   Fonseca IO, Silva RG, Fernandes CL, de Souza ON, Basso LA, Santos DS. Kinetic and chemical mechanisms of shikimate dehydrogenase from Mycobacterium tuberculosis. Arch Biochem Biophys 2007; 457(2): 123-33.
[http://dx.doi.org/10.1016/j.abb.2006.11.015] [PMID: 17178095]

[192]   Ye S, Von Delft F, Brooun A, Knuth MW, Swanson RV, McRee DE. The crystal structure of shikimate dehydrogenase (AroE) reveals a unique NADPH binding mode. J Bacteriol 2003; 185(14): 4144-51.
[http://dx.doi.org/10.1128/JB.185.14.4144-4151.2003] [PMID: 12837789]

[193]   Díaz-Quiroz DC, Cardona-Félix CS, Viveros-Ceballos JL, et al. Synthesis, biological activity and molecular modelling studies of shikimic acid derivatives as inhibitors of the shikimate dehydrogenase enzyme of Escherichia coli. J Enzyme Inhib Med Chem 2018; 33(1): 397-404.
[http://dx.doi.org/10.1080/14756366.2017.1422125] [PMID: 29363372]

[194]   Avitia-Domínguez C, Sierra-Campos E, Salas-Pacheco JM, et al. Inhibition and biochemical characterization of methicillin-resistant Staphylococcus aureus shikimate dehydrogenase: an in silico and kinetic study. Molecules 2014; 19(4): 4491-509.
[http://dx.doi.org/10.3390/molecules19044491] [PMID: 24727420]

[195]   Hartmann MD, Bourenkov GP, Oberschall A, Strizhov N, Bartunik HD. Mechanism of phosphoryl transfer catalyzed by shikimate kinase from Mycobacterium tuberculosis. J Mol Biol 2006; 364(3): 411-23.

[http://dx.doi.org/10.1016/j.jmb.2006.09.001] [PMID: 17020768]

[196] Pereira JH, de Oliveira JS, Canduri F, *et al.* Structure of shikimate kinase from *Mycobacterium tuberculosis* reveals the binding of shikimic acid. Acta Crystallogr D Biol Crystallogr 2004; 60(Pt 12 Pt 2): 2310-9.
[http://dx.doi.org/10.1107/S090744490402517X] [PMID: 15583379]

[197] Segura-Cabrera A, Rodríguez-Pérez MA. Structure-based prediction of *Mycobacterium tuberculosis* shikimtae kinase inhibitors by high-throughput virtual screening. Bioorg Med Chem Lett 2008; 18(11): 3152-7.

[198] Kumar M, Verma S, Sharma S, Srinivasan A, Singh TP, Kaur P. Structure-based *in silico* design of a high-affinity dipeptide inhibitor for novel protein drug target Shikimate kinase of *Mycobacterium tuberculosis*. Chem Biol Drug Des 2010; 76(3): 277-84.
[PMID: 20626408]

[199] Gordon S, Simithy J, Goodwin DC, Calderón AI. Selective *Mycobacterium tuberculosis* shikimate kinase inhibitors as potential antibacterial. Perspect Medicin Chem 2015; 7: 9-20.

[200] Blanco B, Prado V, Lence E, *et al. Mycobacterium tuberculosis* shikimate kinase inhibitors: design and simulation studies of the catalytic turnover. J Am Chem Soc 2013; 135(33): 12366-76.
[http://dx.doi.org/10.1021/ja405853p] [PMID: 23889343]

[201] Priestman MA, Funke T, Singh IM, Crupper SS, Schönbrunn E. 5-Enolpyruvylshikimate-3-phosphate synthase from *Staphylococcus aureus* is insensitive to glyphosate. FEBS Lett 2005; 579(3): 728-32.
[http://dx.doi.org/10.1016/j.febslet.2004.12.057] [PMID: 15670836]

[202] Park H, Hilsenbeck JL, Kim HJ, *et al.* Structural studies of Streptococcus pneumoniae EPSP synthase in unliganded state, tetrahedral intermediate-bound state and S3P-GLP-bound state. Mol Microbiol 2004; 51(4): 963-71.
[http://dx.doi.org/10.1046/j.1365-2958.2003.03885.x] [PMID: 14763973]

[203] Healy_Fried ML, Funke T, Priestman MA, *et al.* Structural basis of glyphosate tolerance resulting from mutations of Pro101 in *Escherichia coli* 5-Enolpyruvylshikimate-3-phosphate synthase. J Bio Chem 2007; 282: 32949-55.

[204] Priestman MA, Healy ML, Becker A, *et al.* Interaction of phosphonate analogues of the tetrahedral reaction intermediate with 5-enolpyruvylshikimate-3-phosphate synyhase in atomic detail. Biochemistry 2005; 44: 3241-48.

[205] Kitzing K, Auweter S, Amrhein N, Macheroux P. Mechanism of chorismate synthase. Role of the two invariant histidine residues in the active site. J Biol Chem 2004; 279(10): 9451-61.
[http://dx.doi.org/10.1074/jbc.M312471200] [PMID: 14668332]

[206] Viola CM, Saridakis V, Christendat D. Crystal structure of chorismate synthase from Aquifex aeolicus reveals a novel beta alpha beta sandwich topology. Proteins 2004; 54(1): 166-9.
[http://dx.doi.org/10.1002/prot.10592] [PMID: 14705034]

[207] Tapas S, Kumar A, Dhindwal S, Preeti , Kumar P. Structural analysis of chorismate synthase from Plasmodium falciparum: a novel target for antimalaria drug discovery. Int J Biol Macromol 2011; 49(4): 767-77.
[http://dx.doi.org/10.1016/j.ijbiomac.2011.07.011] [PMID: 21801743]

[208] Davies GM, Barrett-Bee KJ, Jude DA, *et al.* (6S)-S-Fluoroshikimic Acid, an antibacterial agent acting on the aromatic biosynthetic pathway. Antimicrob Agents Chemother 1994; 38: 2.

[209] Thomas MG, Lawson Ch, Alalarson NM, *et al.* A Series of 2(Z)-2-Benzylidene--,7-didydroxybenzofuran-3[2H]-ones as inhibitors of chorismate synthase. Bioorg Med Chem Lett 2003; 13: 423-26.

[210] Burschowsky D, van Eerde A, Ökvist M, *et al.* Electrostatic transition state stabilization rather than reactant destabilization provides the chemical basis for efficient chorismate mutase catalysis. Proc Natl Acad Sci USA 2014; 111(49): 17516-21.

[http://dx.doi.org/10.1073/pnas.1408512111] [PMID: 25422475]

[211]   Ma J, Zheng X, Schnappauf G, Braus G, Karplus M, Lipscomb W-N. Yeast chorismate mutase in the R state: simulations of the active site. Proc Natl Acad Sci USA 1998; 95(25): 14640-5.
[http://dx.doi.org/10.1073/pnas.95.25.14640] [PMID: 9843942]

[212]   Hediger ME. Design, synthesis, and evaluation of aza inhibitors of chorismate mutase. Bioorg Med Chem 2004; 12(18): 4995-5010.
[http://dx.doi.org/10.1016/j.bmc.2004.06.037] [PMID: 15336279]

[213]   Khanapur M, Alvala M, Prabhakar M, *et al*. *Mycobacterium tuberculosis* chorismate mutase: A potential target for TB. Bioorg Med Chem 2017; 25(6): 1725-36.
[http://dx.doi.org/10.1016/j.bmc.2017.02.001] [PMID: 28202315]

[214]   Holland Ck, Berkovich DA, Kohn ML, Maeda H, Jez JM. Structural basis for substrate recognition and inhibition of prephenate aminotransferase from Arabidopsis. Plant J 2018; 94(2): 304-14.

[215]   Graindorge M, Giustini C, Kraut A, Moyet L, Curien G, Matringe M. Three different classes of aminotransferases evolved prephenate aminotransferase functionality in arogenate-competent microrganism. J Bio Chem 2014; 289: 3198-2008.

[216]   Legrand P, Dumas R, Seux M, *et al*. Biochemical characterization and crystal structure of Synechocystis arogenate dehydrogenase provide insights into catalytic reaction. Structure 2006; 14(4): 767-76.
[http://dx.doi.org/10.1016/j.str.2006.01.006] [PMID: 16615917]

[217]   Cho M-H, Corea ORA, Yang H, *et al*. Phenylalanine biosynthesis in *Arabidopsis thaliana*. Identification and characterization of arogenate dehydratases. J Biol Chem 2007; 282(42): 30827-35.
[http://dx.doi.org/10.1074/jbc.M702662200] [PMID: 17726025]

[218]   Mattaini KR, Sullivan MR, Vander Heiden MG. The importance of serine metabolism in cancer. J Cell Biol 2016; 214(3): 249-57.
[http://dx.doi.org/10.1083/jcb.201604085] [PMID: 27458133]

[219]   Thompson JR, Bell JK, Bratt J, Grant GA, Banaszak LJ. Vmax regulation through domain and subunit changes. The active form of phosphoglycerate dehydrogenase. Biochemistry 2005; 44(15): 5763-73.
[http://dx.doi.org/10.1021/bi047944b] [PMID: 15823035]

[220]   Unterlass JE, Wood RJ, Baslé A, *et al*. Structural insights into the enzymatic activity and potential substrate promiscuity of human 3-phosphoglycerate dehydrogenase (PHGDH). Oncotarget 2017; 8(61): 104478-91.
[http://dx.doi.org/10.18632/oncotarget.22327] [PMID: 29262655]

[221]   Rohde JM, Brimacombe KR, Liu L, *et al*. Discovery and optimization of piperazine-1-thiourea-based human phosphoglycerate dehydrogenase inhibitors. Bioorg Med Chem 2018; 26(8): 1727-39.
[http://dx.doi.org/10.1016/j.bmc.2018.02.016] [PMID: 29555419]

[222]   Hayashi H, Kagamiyama H. Transient-state kinetics of the reaction of aspartate aminotransferase with aspartate at low pH reveals dual routes in the enzyme-substrate association process. Biochemistry 1997; 36(44): 13558-69.
[http://dx.doi.org/10.1021/bi971638z] [PMID: 9354624]

[223]   Coulibaly F, Lassalle E, Baker HM, Baker EN. Structure of phosphoserine aminotransferase from *Mycobacterium tuberculosis*. Acta Crystallogr D Biol Crystallogr 2012; 68(Pt 5): 553-63.

[224]   Arora G, Tiwari P, Mandal RS, *et al*. High throughput screen identifies small molecule inhibitors specific for *Mycobacterium tuberculosis* phosphoserine phosphatase. J Biol Chem 2014; 289(36): 25149-65.
[http://dx.doi.org/10.1074/jbc.M114.597682] [PMID: 25037224]

[225]   Wang W, Cho HS, Kim R, *et al*. Structural characterization of the reaction pathway in phosphoserine phosphatase: crystallographic "snapshots" of intermediate states. J Mol Biol 2002; 319(2): 421-31.
[http://dx.doi.org/10.1016/S0022-2836(02)00324-8] [PMID: 12051918]

[226] Li X, Xun Z, Yang Y. Inhibition of phosphoserine phosphatase enhances the anticancer efficacy of 5-fluorouracil in colorectal cancer. Biochem Biophys Res Commun 2016; 477(4): 633-9.
[http://dx.doi.org/10.1016/j.bbrc.2016.06.112] [PMID: 27349874]

[227] Hawkinson JE, Acosta-Burruel M, Ta ND, Wood PL. Novel phosphoserine phosphatase inhibitors. Eur J Pharmacol 1997; 337(2-3): 315-24.
[http://dx.doi.org/10.1016/S0014-2999(97)01304-6] [PMID: 9430431]

[228] Yang Q, Yu K, Yan L, Li Y, Chen C, Li X. Structural view of the regulatory subunit of aspartate kinase from *Mycobacterium tuberculosis*. Protein Cell 2011; 2(9): 745-54.
[http://dx.doi.org/10.1007/s13238-011-1094-2] [PMID: 21976064]

[229] Yoshida A, Tomita T, Kuzuyama T, Nishiyama M. Mechanism of concerted inhibition of α2β2-type hetero-oligomeric aspartate kinase from *Corynebacterium glutamicum*. J Biol Chem 2010; 285(35): 27477-86.
[http://dx.doi.org/10.1074/jbc.M110.111153] [PMID: 20573952]

[230] Wang Z, Cole PA. Catalytic mechanism and regulation of protein kinases methods. Enzymol 2014; 548: 1-21.

[231] Bareich DC, Nazi I, Wright GD. Simultaneous *In vitro* assay of the first four enzymes in the fungal aspartate pathway identifies a new class of aspartate kinase inhibitor. Cell Chemical Biology 2003; 10: 967-73.

[232] Hadfield A, Shammas C, Kryger G, *et al.* Active site analysis of the potential antimicrobial target aspartate semialdehyde dehydrogenase. Biochemistry 2001; 40(48): 14475-83.
[http://dx.doi.org/10.1021/bi015713o] [PMID: 11724560]

[233] Faehnle CR, Le Coq J, Liu X, Viola RE. Examination of key intermediates in the catalytic cycle of aspartate-β-semialdehyde dehydrogenase from a gram-positive infectious bacteria. J Biol Chem 2006; 281(41): 31031-40.
[http://dx.doi.org/10.1074/jbc.M605926200] [PMID: 16895909]

[234] Cox RJ, Hadfield AT, Mayo-Martín MB. Difluoromethylene analogues of aspartyl phosphate: the first synthetic inhibitors of aspartate semi-aldehyde dehydrogenase. Chem Commun (Camb) 2001; 18(18): 1710-1.
[http://dx.doi.org/10.1039/b103988c] [PMID: 12240277]

[235] Evitt AS, Cox RJ. Synthesis and evaluation of conformationally restricted inhibitors of aspartate semialdehyde dehydrogenase. Mol Biosyst 2011; 7(5): 1564-75.
[http://dx.doi.org/10.1039/c0mb00227e] [PMID: 21369577]

[236] Alvarez E, Ramón F, Magán C, Díez E. L-cystine inhibits aspartate-β-semialdehyde dehydrogenase by covalently binding to the essential 135Cys of the enzyme. Biochim Biophys Acta 2004; 1696(1): 23-9.
[http://dx.doi.org/10.1016/j.bbapap.2003.09.002] [PMID: 14726201]

[237] Akai S, Ikushiro H, Sawai T, Yano T, Kamiya N, Miyahara I. The crystal structure of homoserine dehydrogenase complexed with l-homoserine and NADPH in a closed form. J Biochem 2019; 165(2): 185-95.
[http://dx.doi.org/10.1093/jb/mvy094] [PMID: 30423116]

[238] Jacques SL, Mirza IA, Ejim L, *et al.* Enzyme-assisted suicide: molecular basis for the antifungal activity of 5-hydroxy-4-oxonorvaline by potent inhibition of homoserine dehydrogenase. Cell Chemical Biology 2003; 10(10): 989-5.

[239] Jastrzebowska K, Gabriel I. Inhibitors of amino acids biosynthesis as antifungal agents. Amino Acids 2015; 47: 227-49.

[240] Krishna SS, Zhou T, Daugherty M, Osterman A, Zhang H. Structural basis for the catalysis and substrate specificity of homoserine kinase. Biochemistry 2001; 40(36): 10810-8.

[241] De Pascale G, Griffiths EJ, Shakya T, Nazi I, Wright GD. Identification and characterization of new

inhibitors of fungal homoserine kinase. ChemBioChem 2011; 12(8): 1179-82.
[http://dx.doi.org/10.1002/cbic.201100121] [PMID: 21538764]

[242] Garrido-Franco M, Ehlert S, Messerschmidt A, *et al.* Structure and function of threonine synthase from yeast. J Biol Chem 2002; 277(14): 12396-405.
[http://dx.doi.org/10.1074/jbc.M108734200] [PMID: 11756443]

[243] Murakawa T, Machida Y, Hayashi H. Product-assisted catalysis as the basis of the reaction specificity of threonine synthase. J Biol Chem 2011; 286(4): 2774-84.
[http://dx.doi.org/10.1074/jbc.M110.186205] [PMID: 21084312]

[244] Gahungu M, Arguelles-Arias A, Fickers P, *et al.* Synthesis and biological evaluation of potential threonine synthase inhibitors: Rhizocticin A and Plumbemycin A. Bioorg Med Chem 2013; 21(17): 4958-67.
[http://dx.doi.org/10.1016/j.bmc.2013.06.064] [PMID: 23891162]

[245] Kitabatake M, So MW, Tumbula DL, Söll D. Cysteine biosynthesis pathway in the archaeon Methanosarcina barkeri encoded by acquired bacterial genes? J Bacteriol 2000; 182(1): 143-5.
[http://dx.doi.org/10.1128/JB.182.1.143-145.2000] [PMID: 10613873]

[246] Ravilious GE, Jez JM. Structural biology of plant sulfur metabolism: from assimilation to biosynthesis. Nat Prod Rep 2012; 29(10): 1138-52.
[http://dx.doi.org/10.1039/c2np20009k] [PMID: 22610545]

[247] Pye VE, Tingey AP, Robson RL, Moody PCE. The structure and mechanism of serine acetyltransferase from *Escherichia coli*. J Biol Chem 2004; 279(39): 40729-36.
[http://dx.doi.org/10.1074/jbc.M403751200] [PMID: 15231846]

[248] Agarwal SM, Jain R, Bhattacharya A, Azam A. Inhibitors of *Escherichia coli* serine acetyltransferase block proliferation of Entamoeba histolytica trophozoites. Int J Parasitol 2008; 38(2): 137-41.
[http://dx.doi.org/10.1016/j.ijpara.2007.09.009] [PMID: 17991467]

[249] Rabeh WM, Cook PF. Structure and mechanism of *O*-acetylserine sulfhydrylase. J Biol Chem 2004; 279(26): 26803-6.
[http://dx.doi.org/10.1074/jbc.R400001200] [PMID: 15073190]

[250] Spyrakis F, Singh R, Cozzini P, *et al.* Isozyme-specific ligands for *O*-acetylserine sulfhydrylase, a novel antibiotic target. PLoS One 2013; 8(10): e77558.

[251] Salsi E, Bayden AS, Spyrakis F, *et al.* Design of *O*-acetylserine sulfhydrylase inhibitors by mimicking nature. J Med Chem 2010; 53(1): 345-56.
[http://dx.doi.org/10.1021/jm901325e] [PMID: 19928859]

[252] Franko N, Grammatoglou K, Campanini B, Costantino G, Jirgensons A, Mozzarelli A. Inhibition of *O*-acetylserine sulfhydrylase by fluoroalanine derivatives. J Enzyme Inhib Med Chem 2018; 33(1): 1343-51.
[http://dx.doi.org/10.1080/14756366.2018.1504040] [PMID: 30251899]

[253] Amori L, Katkevica S, Bruno A, *et al.* Design and synthesis of trans-2-substituted-cyclopropa-e-1-carboxylic acids as the first non-natural small molecule inhibitors of *O*-acetylserine sulfhydrylase Med. Chem Commun (Camb) 2012; 3: 1111-6.

[254] Romero I, Téllez J, Yamanaka LE, Steindel M, Romanha AJ, Grisard EC. Transsulfuration is an active pathway for cysteine biosynthesis in *Trypanosoma rangeli*. Parasit Vectors 2014; 7: 197.
[http://dx.doi.org/10.1186/1756-3305-7-197] [PMID: 24761813]

[255] Ripps H, Shen W. Review: taurine: a "very essential" amino acid. Mol Vis 2012; 18: 2673-86.
[PMID: 23170060]

[256] Tomasi ML, Li TWH, Li M, Mato JM, Lu SC. Inhibition of human methionine adenosyltransferase 1A transcription by coding region methylation. J Cell Physiol 2012; 227(4): 1583-91.
[http://dx.doi.org/10.1002/jcp.22875] [PMID: 21678410]

[257] Mato JM, Alvarez L, Ortiz P, Pajares MA. S-adenosylmethionine synthesis: molecular mechanisms and clinical implications. Pharmacol Ther 1997; 73(3): 265-80.
[http://dx.doi.org/10.1016/S0163-7258(96)00197-0] [PMID: 9175157]

[258] González B, Pajares MA, Hermoso JA, Guillerm D, Guillerm G, Sanz-Aparicio J. Crystal structures of methionine adenosyltransferase complexed with substrates and products reveal the methionine-ATP recognition and give insights into the catalytic mechanism. J Mol Biol 2003; 331(2): 407-16.
[http://dx.doi.org/10.1016/S0022-2836(03)00728-9] [PMID: 12888348]

[259] Taylor JC, Bock CW, Takusagawa F, Markham GD. Discovery of novel types of inhibitors of S-adenosylmethionine synthesis by virtual screening. J Med Chem 2009; 52(19): 5967-73.
[http://dx.doi.org/10.1021/jm9006142] [PMID: 19739644]

[260] Hu Y, Komoto J, Huang Y, *et al.* Crystal structure of S-adenosylhomocysteine hydrolase from rat liver. Biochemistry 1999; 38(26): 8323-33.
[http://dx.doi.org/10.1021/bi990332k] [PMID: 10387078]

[261] (a) Lee KM, Choi WJ, Lee Y, *et al.* X-ray crystal structure and binding mode analysis of human S-adenosylhomocysteine hydrolase complexed with novel mechanism-based inhibitors, haloneplanocin A analogues. J Med Chem 2011; 54(4): 930-8.
[http://dx.doi.org/10.1021/jm1010836] [PMID: 21226494] (b) Huang Y, Komoto J, Takata Y, *et al.* Inhibition of S-Adenosylhomocysteine hydrolase by acyclic sugar adenosine Analogue D-Eritadenine 2001; 277: 7477-82.

[262] Yang X, Hu Y, Yin DH, *et al.* Catalytic strategy of S-adenosyl-L-homocysteine hydrolase: transition-state stabilization and the avoidance of abortive reactions. Biochemistry 2003; 42(7): 1900-9.
[http://dx.doi.org/10.1021/bi0262350] [PMID: 12590576]

[263] Oulhaj A, Refsum H, Beaumont H, *et al.* Homocysteine as a predictor of cognitive decline in Alzhemer's disease. Int Geriatr Psychiatry 2010; 25(1): 82-90.

[264] Wilson-Miles E, Kraus JP. Cystathionine β-synthase: structure, function, regulation, and location of homocystinuria-causing mutations, being in charge of the formation of cystathione by conjugation of homocysteine with serine. J Biol Chem 2004; 279: 29871-4.
[http://dx.doi.org/10.1074/jbc.R400005200]

[265] Banerjee R, Evande R, Kabil O, Ojha S, Taoka S. Reaction mechanism and regulation of cystathionine β-synthase. Biochim Biophys Acta 2003; 1647(1-2): 30-5.
[http://dx.doi.org/10.1016/S1570-9639(03)00044-X] [PMID: 12686104]

[266] Meier M, Janosik M, Kery V, Kraus JP, Burkhard P. Structure of human cystathionine β-synthase: a unique pyridoxal 5'-phosphate-dependent heme protein. EMBO J 2001; 20(15): 3910-6.
[http://dx.doi.org/10.1093/emboj/20.15.3910] [PMID: 11483494]

[267] Ansari R, Mahta A, Mallack E, Luo JJ. Hyperhomocysteinemia and neurologic disorders: A review. J Clin Neurol 2014; 10(4): 281-8.
[http://dx.doi.org/10.3988/jcn.2014.10.4.281] [PMID: 25324876]

[268] Hellmich MR, Coletta C, Chao C, Szabo C. The therapeutic potential of cystathionine b-synthetase/hydrogen sulfide inhibition in cancer antioxidants & Redox Signaling 2015; 22: 424-8.

[269] Chan JJ, Chai Ch, Lim TW, *et al.* Cystathionine β-Synthase inhibition is a potential therapeutic approach to treatment of ischemic injury. ASN Neuro 2015; 7(2).

[270] Asimakopoulou A, Panopoulos P, Chasapis CT, *et al.* Selectivity of commonly used pharmacological inhibitors for cystathionine β synthase (CBS) and cystathionine γ lyase (CSE). Br J Pharmacol 2013; 169(4): 922-32.
[http://dx.doi.org/10.1111/bph.12171] [PMID: 23488457]

[271] Niu W, Chen F, Wang J, Qian J, Yan S. Antitumor effect of sikokianin C, a selective cystathionine β-synthase inhibitor, against human colon cancer *in vitro* and *in vivo*. MedChemComm 2017; 9(1): 113-20.

[http://dx.doi.org/10.1039/C7MD00484B] [PMID: 30108905]

[272]   Sun Q, Collins R, Huang S, *et al.* Structural basis for the inhibition mechanism of human cystathionine γ-lyase, an enzyme responsible for the production of H(2)S. J Biol Chem 2009; 284(5): 3076-85.
[http://dx.doi.org/10.1074/jbc.M805459200] [PMID: 19019829]

[273]   Corvino A, Severino B, Fiorino F, *et al.* Fragment-based de novo design of a cystathionine γ-lyase selective inhibitor blocking hydrogen sulfide production Fragment-based de novo design of a cystathionine γ-lyase selective inhibitor blocking hydrogen sulfide production. Nature Scientific Reports 6 2016; 34398: 1-11.

[274]   Stipanuk MH, Ueki I, Dominy JE Jr, Simmons CR, Hirschberger LL. Cysteine dioxygenase: A robust system for regulation of cellular cysteine levels. Amino Acids 2009; 37(1): 55-63.
[http://dx.doi.org/10.1007/s00726-008-0202-y]

[275]   Joseph CA, Maroney MJ. Cysteine dioxygenase: structure and mechanism. Chem Commun (Camb) 2007; 28(32): 3338-49.
[http://dx.doi.org/10.1039/b702158e] [PMID: 18019494]

[276]   Park E, Park SY, Dobkin C, Schuller-Levis G. Development of a novel cysteine sulfinic acid decarboxylase knockout mouse: dietary taurine reduces neonatal mortality. J Amino Acids 2014; 2014: 346809.
[http://dx.doi.org/10.1155/2014/346809] [PMID: 24639894]

[277]   Winge I, Teigen K, Fossbakk A, *et al.* Mammalian CSAD and GADL1 have distinct biochemical properties and patterns of brain expression. Neurochem Int 2015; 90: 173-84.
[http://dx.doi.org/10.1016/j.neuint.2015.08.013] [PMID: 26327310]

[278]   Radwanski ER, Last RL. Tryptophan biosynthesis and metabolism: biochemical and molecular genetics. Plant Cell 1995; 7(7): 921-34.
[PMID: 7640526]

[279]   Culbertson JE. hee Chung D, Ziebart KT, Espiritu E, Toner MD. Conversion of aminodeoxychorismate synthase into anthranilate synthase with janus mutations mechanism of pyruvate elimination catalyzed by chorismate enzymes. Biochemistry 2015; 54: 2372-84.
[PMID: 25710100]

[280]   Bera AK, Atanasova V, Dhanda A, Ladner JE, Parsons JF. Structure of aminodeoxychorismate synthase from *Stenotrophomonas maltophilia*. Biochemistry 2012; 51(51): 10208-17.
[http://dx.doi.org/10.1021/bi301243v] [PMID: 23230967]

[281]   Knöchel T, Ivens A, Hester G, *et al.* The crystal structure of anthranilate synthase from Sulfolobus solfataricus: functional implications. Proc Natl Acad Sci USA 1999; 96(17): 9479-84.
[http://dx.doi.org/10.1073/pnas.96.17.9479] [PMID: 10449718]

[282]   (a) Spraggon G, Kim C, Nguyen-Huu X, Yee MC, Yanofsky C, Mills SE. The structures of anthranilate synthase of Serratia marcescens crystallized in the presence of (i) its substrates, chorismate and glutamine, and a product, glutamate, and (ii) its end-product inhibitor, L-tryptophan. Proc Natl Acad Sci USA 2001; 98(11): 6021-6.
[http://dx.doi.org/10.1073/pnas.111150298] [PMID: 11371633] (b) Srivastava A, Sinha S. Uncoupling of an ammonia channel as a mechanism of allosteric inhibition in anthranilate synthase of Serratia marcescens: dynamic and graph theoretical analysis. Mol Biosyst 2016; 13(1): 142-55.
[http://dx.doi.org/10.1039/C6MB00646A] [PMID: 27833951]

[283]   Payne RJ, Toscano MD, Bulloch EMM, Abell AD, Abell C. Design and synthesis of aromatic inhibitors of anthranilate synthase. Org Biomol Chem 2005; 3(12): 2271-81.
[http://dx.doi.org/10.1039/b503802b] [PMID: 16010361]

[284]   Cookson TVM, Evans GL, Castell A, *et al.* Structures of Mycobacterium tuberculosis Anthranilate Phosphoribosyltransferase Variants Reveal the Conformational Changes That Facilitate Delivery of the Substrate to the Active Site Biochemistry 2015.
[http://dx.doi.org/10.1021/acs.biochem.5b00612]

[285]   Nurul Islam M, Hitchings R, Kumar S, *et al.* Mechanism of Fluorinated Anthranilate-Induced Growth Inhibition in *Mycobacterium tuberculosis.* ACS Infect Dis 2019; 5(1): 55-62.
[http://dx.doi.org/10.1021/acsinfecdis.8b00092] [PMID: 30406991]

[286]   Leopoldseder S, Claren J, Jürgens C, Sterner R. Interconverting the catalytic activities of (beta alpha)(8)-barrel enzymes from different metabolic pathways: sequence requirements and molecular analysis. J Mol Biol 2004; 337(4): 871-9.
[http://dx.doi.org/10.1016/j.jmb.2004.01.062] [PMID: 15033357]

[287]   Perveen S, Rashid N, Papageorgiou AC. Crystal structure of a phosphoribosyl anthranilate isomerase from the hyperthermophilic archaeon *Thermococcus kodakaraensis.* Acta Crystallogr F Struct Biol Commun 2016; 72(Pt 11): 804-12.
[http://dx.doi.org/10.1107/S2053230X16015223] [PMID: 27827353]

[288]   Bisswanger H, Kirschner K, Cohn W, Hager V, Hansson E. N-(5-Phosphoribosyl)anthranilate isomerase-indoleglycerol-phosphate synthase. 1. A substrate analogue binds to two different binding sites on the bifunctional enzyme from *Escherichia coli.* Biochemistry 1979; 18(26): 5946-53.
[http://dx.doi.org/10.1021/bi00593a029] [PMID: 391279]

[289]   Hennig M, Darimont BD, Jansonius JN, Kirschner K. The catalytic mechanism of indole-3-glycerol phosphate synthase: crystal structures of complexes of the enzyme from Sulfolobus solfataricus with substrate analogue, substrate, and product. J Mol Biol 2002; 319(3): 757-66.
[http://dx.doi.org/10.1016/S0022-2836(02)00378-9] [PMID: 12054868]

[290]   Zaccardi MJ, Yezdimer EM, Boehr DD. Functional identification of the general acid and base in the dehydration step of indole-3-glycerol phosphate synthase catalysis. J Biol Chem 2013; 288(37): 26350-6.
[http://dx.doi.org/10.1074/jbc.M113.487447] [PMID: 23900843]

[291]   Mazumder-Shivakumar D, Bruice TC. Molecular dynamics studies of ground state and intermediate of the hyperthermophilic indole-3-glycerol phosphate synthase. Proc Natl Acad Sci USA 2004; 101(40): 14379-84.
[http://dx.doi.org/10.1073/pnas.0406002101] [PMID: 15452341]

[292]   Shen H, Wang F, Zhang Y, *et al.* A novel inhibitor of indole-3-glycerol phosphate synthase with activity against multidrug-resistant *Mycobacterium tuberculosis.* FEBS J 2009; 276(1): 144-54.
[http://dx.doi.org/10.1111/j.1742-4658.2008.06763.x] [PMID: 19032598]

[293]   Zhou T, Wang F-F, Huang Q, Wang H-H, Shen H-B. Screening of novel inhibitors of indole--glycerol phosphate synthase from *Mycobacterium tuberculosis.* J Micr Infect 2015.

[294]   Ro H-S, Miles WE. Structure and Function of the Tryptophan Synthase a2b2 Complex. J Biol Chem 1999; 274: 36439-45.
[http://dx.doi.org/10.1074/jbc.274.51.36439] [PMID: 10593940]

[295]   Busch F, Rajendran C, Heyn K, Schlee S, Merkl R, Sterner R. Ancestral Tryptophan Synthase Reveals Functional Sophistication of Primordial Enzyme Complexes. Cell Chem Biol 2016; 23(6): 709-15.
[http://dx.doi.org/10.1016/j.chembiol.2016.05.009] [PMID: 27291401]

[296]   Xu Y, Abeles RH. Inhibition of tryptophan synthase by (1-fluorovinyl)glycine. Biochemistry 1993; 32(3): 806-11.
[http://dx.doi.org/10.1021/bi00054a010] [PMID: 8422385]

[297]   Abrahams KA, Cox JAG, Fütterer K, *et al.* Inhibiting mycobacterial tryptophan synthase by targeting the inter-subunit interface. Sci Rep 2017; 7(1): 9430.
[http://dx.doi.org/10.1038/s41598-017-09642-y] [PMID: 28842600]

[298]   Moggré G-J, Poulin MB, Tyler PC, Schramm VL, Parker EJ. Transition State Analysis of Adenosine Triphosphate Phosphoribosyltransferase. ACS Chem Biol 2017; 12(10): 2662-70.
[http://dx.doi.org/10.1021/acschembio.7b00484] [PMID: 28872824]

[299]   Champagne KS, Sissler M, Larrabee Y, Doublié S, Francklyn CS. Activation of the hetero-octameric

ATP phosphoribosyl transferase through subunit interface rearrangement by a tRNA synthetase paralog. J Biol Chem 2005; 280(40): 34096-104.
[http://dx.doi.org/10.1074/jbc.M505041200] [PMID: 16051603]

[300] Hove-Jensen B, Andersen KR, Kilstrup M, Martinussen J, Switzer RL, Willemoës M. Phosphoribosyl Diphosphate (PRPP). Microbiol Mol Biol Rev 2016; 81(1): 1-83.
[PMID: 28031352]

[301] Gohda K, Ohta D, Kozaki A, Fujimori K, Mori I, Kikuchi T. Identification of Novel Potent Inhibitors for ATP-Phosphoribosyl Transferase Using Three-Dimensional Structural Database Search Technique Quant. Struct-Act Relat 2001; 20: 143-7.
[http://dx.doi.org/10.1002/1521-3838(200107)20: 2< 143::AID-QSAR143> 3.0.CO;2-R]

[302] Javid-Majd F, Yang D, Ioerger TR, Sacchettini JC. The 1.25 A resolution structure of phosphoribosyl-ATP pyrophosphohydrolase from *Mycobacterium tuberculosis*. Acta Crystallogr D Biol Crystallogr 2008; 64(Pt 6): 627-35.
[http://dx.doi.org/10.1107/S0907444908007105] [PMID: 18560150]

[303] Henriksen ST, Liu J, Estiu G, Oltvai ZN, Wiest O. Identification of novel bacterial histidine biosynthesis inhibitors using docking, ensemble rescoring, and whole-cell assays. Bioorg Med Chem 2010; 18(14): 5148-56.
[http://dx.doi.org/10.1016/j.bmc.2010.05.060] [PMID: 20573514]

[304] Sivaraman J, Myers RS, Boju L, *et al.* Crystal Structure of Methanobacterium thermoautotrophicum Phosphoribosyl-AMP Cyclohydrolase HisI. Biochemistry 2005; 44(30): 10071-80.

[305] D'Ordine RL, Linger RS, Thai CJ, Davisson VJ. Catalytic zinc site and mechanism of the metalloenzyme PR-AMP cyclohydrolase. Biochemistry 2012; 51(29): 5791-803.
[http://dx.doi.org/10.1021/bi300391m] [PMID: 22741521]

[306] Gupta M, Prasad Y, Kumar Sharma S. Identification of Phosphoribosyl-AMP cyclohydrolase, as drug target and its inhibition in Brucella melitensis bv 1 16M using metabolic pathways analysis J Biomol Struct Dyn 2017; 35(2): 287-99.

[307] Due AV, Kuper J, Geerlof A, von Kries JP, Wilmanns M. Bisubstrate specificity in histidine/tryptophan biosynthesis isomerase from *Mycobacterium tuberculosis* by active site metamorphosis. Proc Natl Acad Sci USA 2011; 108(9): 3554-9.
[http://dx.doi.org/10.1073/pnas.1015996108] [PMID: 21321225]

[308] Söderholm A, Guo X, Newton MS, *et al.* Two-step ligand binding in a (βα)8 Barrel Enzyme: Substrate-bound structures shed new light on the catalytic cycle of HisA. J Biol Chem 2015; 290(41): 24657-68.
[http://dx.doi.org/10.1074/jbc.M115.678086] [PMID: 26294764]

[309] List F, Vega MC, Razeto A, Häger MC, Sterner R, Wilmanns M. Catalysis uncoupling in a glutamine aminotransferase bienzyme by unblocking the glutaminase active site. Chemistry & Biology 2012; 19: 1589-99.

[310] Chaudhuri BN, Lange SC, Myers RS, Davisson VJ, Smith JL. Toward understanding the mechanism of the complex cyclization reaction catalyzed by imidazole glycerolphosphate synthase: crystal structures of a ternary complex and the free enzyme. Biochemistry 2003; 42(23): 7003-12.
[http://dx.doi.org/10.1021/bi034320h] [PMID: 12795595]

[311] Chittur SV, Klem TJ, Shafer CM, Jo Davisson V. Mechanism for acivicin inactivation of triad glutamine amidotransferase. Biochemistry 2001; 40: 876-87.

[312] Glynn SE, Baker PJ, Sedelnikova SE, *et al.* Structure and mechanism of imidazoleglycerol-phosphate dehydratase. Structure 2005; 13(12): 1809-17.
[http://dx.doi.org/10.1016/j.str.2005.08.012] [PMID: 16338409]

[313] Cox JM, Hawkes TR, Bellini P, *et al.* Design and synthesis of inhibitors of imidazoleglycerol phosphate dehydratase as potential herbicides. Pestic Sci 1997; 50: 297-311.

[314] Sivaraman J, Li Y, Larocque R, Schrag JD, Cygler M, Matte A. Crystal structure of histidinol phosphate aminotransferase (HisC) from *Escherichia coli*, and its covalent complex with pyridoxal-5--phosphate and l-histidinol phosphate. J Mol Biol 2001; 311(4): 761-76.
[http://dx.doi.org/10.1006/jmbi.2001.4882] [PMID: 11518529]

[315] Marienhagen J, Sandalova T, Sahm H, Eggeling L, Schneider G. Insights into the structural basis of substrate recognition by histidinol-phosphate aminotransferase from *Corynebacterium glutamicum*. Acta Crystallogr D Biol Crystallogr 2008; 64(Pt 6): 675-85.
[http://dx.doi.org/10.1107/S0907444908009438] [PMID: 18560156]

[316] Ghodge SV, Fedorov AA, Fedorov EV, *et al.* Structural and mechanistic characterization of L-histidinol phosphate phosphatase from the polymerase and histidinol phosphatase family of proteins. Biochemistry 2013; 52(6): 1101-12.
[http://dx.doi.org/10.1021/bi301496p] [PMID: 23327428]

[317] Jha B, Kumar D, Sharma A, Dwivedy A, Singh R, Biswal BK. Identification and structural characterization of a histidinol phosphate phosphatase from *Mycobacterium tuberculosis*. J Biol Chem 2018; 293(26): 10102-18.
[http://dx.doi.org/10.1074/jbc.RA118.002299] [PMID: 29752410]

[318] Barbosa JARG, Sivaraman J, Li Y, *et al.* Mechanism of action and NAD+-binding mode revealed by the crystal structure of L-histidinol dehydrogenase. Proc Natl Acad Sci USA 2002; 99(4): 1859-64.
[http://dx.doi.org/10.1073/pnas.022476199] [PMID: 11842181]

[319] Ruszkowski M, Dauter Z. Structures of medicago truncatula L-Histidinol dehydrogenase show rearrangements required for nad+ binding and the cofactor positioned to accept a hydride. Nature Scientific Reports 7 2017; 10476: 1-13.

[320] Dancer JE, Ford MJ, Hamilton K, *et al.* Synthesis of potent inhibitors of histidinol dehydrogenase. Bioorg Med Chem Lett 1996; 6: 2131-6.
[http://dx.doi.org/10.1016/ 0960-894X(96)00384-8]

[321] Lunardi J, Martinelli LKB, Silva Raupp A, *et al.* *Mycobacterium tuberculosis* histidinol dehydrogenase: biochemical characterization and inhibition studies. RSC Advances 2016; 6: 28406-18.
[http://dx.doi.org/10.1039/C6RA03020C]

[322] Pahwa S, Kaur S, Jain R, Roy N. Structure based design of novel inhibitors for histidinol dehydrogenase from *Geotrichum candidum*. Bioorg Med Chem Lett 2010; 20(13): 3972-6.

[323] Marco-Marín C, Gil-Ortiz F, Pérez-Arellano I, Cervera J, Fita I, Rubio V. A novel two-domain architecture within the amino acid kinase enzyme family revealed by the crystal structure of *Escherichia coli* glutamate 5-kinase. J Mol Biol 2007; 367(5): 1431-46.
[http://dx.doi.org/10.1016/j.jmb.2007.01.073] [PMID: 17321544]

[324] Perez-Arellano I. Carmona-Alvarez, Gallego J, Cervera J. Molecular mechanism modulating glutamate kinase activity. Identification of the Proline Feedback Inhibitor Binding site J Mol Bio 2010; 404: 890-01.
[PMID: 20970428]

[325] Page R, Nelson MS, von Delft F, *et al.* Crystal structure of γ-glutamyl phosphate reductase (TM0293) from *Thermotoga maritima* at 2.0 A resolution. Proteins 2004; 54(1): 157-61.
[http://dx.doi.org/10.1002/prot.10562] [PMID: 14705032]

[326] Meng Z, Lou Z, Liu Z, *et al.* Crystal structure of human pyrroline-5-carboxylate reductase. J Mol Biol 2006; 359(5): 1364-77.
[http://dx.doi.org/10.1016/j.jmb.2006.04.053] [PMID: 16730026]

[327] Lemke CT, Howell PL. The structure of phosphorylated gsk-3b complexed with peptide, frattide, that inhibits b-catenin phosphorylation. Structure 2001; 9(12): 1143-52.

[328] Karlberg T, Collins R, van den Berg S, *et al.* Structure of human argininosuccinate synthetase. Acta

Crystallogr D Biol Crystallogr 2008; 64(Pt 3): 279-86.
[http://dx.doi.org/10.1107/ S0907444907067455] [PMID: 18323623]

[329] Jenkins GR, Tolleson WH, Newkirk DK, *et al.* Identification of fumonisin B1 as an inhibitor of argininosuccinate synthetase using fumonisin affinity chromatography and *in vitro* kinetic studies. J Biochem Mol Toxicol 2000; 14(6): 320-8.
[http://dx.doi.org/10.1002/1099-0461(2000)14:6<320::AID-JBT4>3.0.CO;2-9] [PMID: 11083085]

[330] Wu C-Y, Lee H-J, Wu S-H, Chen ST, Chiou SH, Chang GG. Chemical mechanism of the endogenous argininosuccinate lyase activity of duck lens delta2-crystallin. Biochem J 1998; 333(Pt 2): 327-34.
[http://dx.doi.org/10.1042/bj3330327] [PMID: 9657972]

[331] Chen X, Chen J, Zhang W, *et al.* Crystal structure and biochemical study on argininosuccinate lyase from *Mycobacterium tuberculosis*. Biochem Biophys Res Commun 2019; 510(1): 116-21.
[http://dx.doi.org/10.1016/j.bbrc.2019.01.061] [PMID: 30665717]

[332] Menyhárt J, Gróf J. Urea as a selective inhibitor of argininosuccinate lyase. Eur J Biochem 1977; 75(2): 405-9.
[http://dx.doi.org/10.1111/j.1432-1033.1977.tb11541.x] [PMID: 885138]

[333] Reboul CF, Porebski BT, Griffin MDW, *et al.* Structural and Dynamic Requirements for Optimal Activity of the Essential Bacterial Enzyme Dihydrodipicolinate Synthase. PLoS Comput Biol 2012; 8(6): 1-12.

[334] Dobson RCJ, Devenish SRA, Turner LA, *et al.* Role of arginine 138 in the catalysis and regulation of *Escherichia coli* dihydrodipicolinate synthase. Biochemistry 2005; 44(39): 13007-.

[335] Boughton BA, Griffin MDW, O'Donnell PA, *et al.* Irreversible inhibition of dihydrodipicolinate synthase by 4-oxo-heptenedioic acid analogues. Bioorg Med Chem 2008; 16(23): 9975-83.
[http://dx.doi.org/10.1016/j.bmc.2008.10.026] [PMID: 18977662]

[336] Blickling S, Renner Ch, Laber B, Pohlenz H-D, Holak TA. Reaction mechanism of *Escherichia coli* dihydrodipicolinate synthase investigated by X-ray crystallography and NMR spectroscopy. Biochemistry 1997; 36(1): 24-33.

[337] Cirilli M, Zheng R, Scapin G, Blanchard JS. The Three-Dimensional Structures of the *Mycobacterium tuberculosis* Dihydrodipicolinate Reductase−NADH−2,6-PDC and −NADPH−2,6-PDC Complexes. Structural and mutagenic analysis of relaxed nucleotide specificity Biochemistry 2003; 42: 10644-50.
[PMID: 12962488]

[338] W, Lee, S-H, Park, SG, Lee, HH, Park, HJ, Kim, HJ, Park, H, Park, JH, Lee Crystal structure of dihydrodipicolinate reductase (PaDHDPR) from Paenisporosarcina sp. TG-14: structural basis for NADPH preference as a cofactor. Sci Rep 2018; 8: 1-13.

[339] Paiva AM, Vanderwall DE, Blanchard JS, Kozarich JW, Williamson JM, Kelly TM. Inhibitors of dihydrodipicolinate reductase, a key enzyme of the diaminopimelate pathway of *Mycobacterium tuberculosis*. Biochimica et Bopphysica Acta 2001; 1545(1-2): 67-77.

[340] Beaman TW, Vogel KW, Drueckhammer DG, Blanchard JS, Roderick SL. Acyl group specificity at the active site of tetrahydridipicolinate N-succinyltransferase. Protein Sci 2002; 11(4): 974-9.
[http://dx.doi.org/10.1110/ps.4310102] [PMID: 11910040]

[341] Sagong H-Y, Kim K-J. Crystal Structure and Biochemical Characterization of Tetrahydrodipicolinate N-Succinyltransferase from *Corynebacterium glutamicum*. J Agric Food Chem 2015; 63(49): 10641-6.
[http://dx.doi.org/10.1021/acs.jafc.5b04785] [PMID: 26602189]

[342] Schnell R, Oehlmann W, Sandalova T, *et al.* Tetrahydrodipicolinate N-succinyltransferase and dihydrodipicolinate synthase from Pseudomonas aeruginosa: structure analysis and gene deletion. PLoS One 2012; 7(2): e31133.
[http://dx.doi.org/10.1371/journal.pone.0031133] [PMID: 22359568]

[343] Weyand S, Kefala G, Weiss MS. The three-dimensional structure of N-succinyldiaminopimelate aminotransferase from *Mycobacterium tuberculosis*. J Mol Biol 2007; 367(3): 825-38.

[http://dx.doi.org/10.1016/j.jmb.2007.01.023] [PMID: 17292400]

[344]   Cox RJ, Sherwin WA, Lam LKP, Vederas JC. Synthesis and Evaluation of Novel Substrates and Inhibitors of N-Succinyl-LL-diaminopimelate Aminotransferase (DAP-AT) from *Escherichia coli*. J Am Chem Soc 1996; 118(32): 7449-60.
[http://dx.doi.org/10.1021/ja960640v]

[345]   Nocek B, Reidl C, Starus A, *et al.* Structural Evidence of a Major Conformational Change Triggered by Substrate Binding in DapE Enzymes: Impact on the Catalytic Mechanism. Biochemistry 2018; 57(5): 574-84.
[http://dx.doi.org/10.1021/acs.biochem.7b01151] [PMID: 29272107]

[346]   Gillner D, Armoush N, Holz RC, Becker DP. Inhibitors of bacterial N-succinyl-L,L-diaminopimelic acid desuccinylase (DapE) and demonstration of *in vitro* antimicrobial activity. Bioorg Med Chem Lett 2009; 19(22): 6350-2.
[http://dx.doi.org/10.1016/j.bmcl.2009.09.077] [PMID: 19822427]

[347]   Pillai B, Cherney MM, Diaper CM, *et al.* Structural insights into stereochemical inversion by diaminopimelate epimerase: an antibacterial drug target. Proc Natl Acad Sci USA 2006; 103(23): 8668-73.
[http://dx.doi.org/10.1073/pnas.0602537103] [PMID: 16723397]

[348]   Sagong H-Y, Kima K-J. Structural basis for redox sensitivity in Corynebacterium glutamicum diaminopimelate epimerase: an enzyme involved in l-lysine biosynthesis Sci Rep 2017.
[http://dx.doi.org/10.1038/srep42318]

[349]   Diaper CM, Sutherland A, Pillai B, *et al.* The stereoselective synthesis of aziridine analogues of diaminopimelic acid (DAP) and their interaction with dap epimerase. Org Biomol Chem 2005; 3(24): 4402-11.
[http://dx.doi.org/10.1039/b513409a] [PMID: 16327902]

[350]   Son HF, Kim K-J. Structural basis for substrate specificity of meso-diaminopimelic acid decarboxylase from *Corynebacterium glutamicum*. Biochem Biophys Res Commun 2018; 495(2): 1815-21.
[http://dx.doi.org/10.1016/j.bbrc.2017.11.097] [PMID: 29233695]

[351]   Kelland JG, Arnold LD, Palcic MM, Pickard MA, Vederas JC. Analogs of Diaminopimelic Acid as Inhibitors of meso-Diaminopimelate Decarboxylase from Bacillus sphaericus and Wheat Germ.J Bio Chem 1986; 261(8): 13216-23.

[352]   Xu H, Andi B, Qian J, West AH, Cook PF. The alpha-aminoadipate pathway for lysine biosynthesis in fungi. Cell Biochem Biophys 2006; 46(1): 43-64.
[http://dx.doi.org/10.1385/CBB:46:1:43] [PMID: 16943623]

[353]   Toney MD. Aspartate aminotransferase: an old dog teaches new tricks. Arch Biochem Biophys 2014; 544: 119-27.
[http://dx.doi.org/10.1016/j.abb.2013.10.002] [PMID: 24121043]

[354]   Dajnowicz S, Johnston RC, Parks JM, *et al.* Direct visualization of critical hydrogen atoms in a pyridoxal 5-phosphate enzyme. Nature Communications 8 2017; 955: 1-9.

[355]   Cornell NW, Zuurendonk PF, Kerich MJ, Straight CB. Selective inhibition of alanine aminotransferase and aspartate aminotransferase in rat hepatocytes. Biochem J 1984; 220(3): 707-16.
[http://dx.doi.org/10.1042/bj2200707] [PMID: 6466297]

[356]   Rando RR. Irreversible inhibition of aspartate aminotransferase by 2-amino-3-butenoic acid. Biochemistry 1974; 13(19): 3859-63.
[http://dx.doi.org/10.1021/bi00716a006] [PMID: 4472160]

[357]   Lomelino CL, Andring JT, McKenna R, Kilberg MS. Asparagine synthetase: Function, structure, and role in disease. J Biol Chem 2017; 292(49): 19952-8.
[http://dx.doi.org/10.1074/jbc.R117.819060] [PMID: 29084849]

[358] Balasubramanian MN, Butterworth EA, Kilberg MS. Asparagine synthetase: regulation by cell stress and involvement in tumor biology. Am J Physiol Endocrinol Metab 2013; 304(8): E789-99.
[http://dx.doi.org/10.1152/ajpendo.00015.2013] [PMID: 23403946]

[359] Blaise M, Fréchin M, Oliéric V, *et al.* Crystal structure of the archaeal asparagine synthetase: interrelation with aspartyl-tRNA and asparaginyl-tRNA synthetases. J Mol Biol 2011; 412(3): 437-52.
[http://dx.doi.org/10.1016/j.jmb.2011.07.050] [PMID: 21820443]

[360] Koroniak L, Ciustea M, Gutierrez JA, Richards NGJ. Synthesis and characterization of an N-acylsulfonamide inhibitor of human asparagine synthetase. Org Lett 2003; 5(12): 2033-6.
[http://dx.doi.org/10.1021/ol034212n] [PMID: 12790521]

[361] Chen Y, Shertzer HG, Schneider SN, Nebert DW, Dalton TP. Glutamate cysteine ligase catalysis: dependence on ATP and modifier subunit for regulation of tissue glutathione levels. J Biol Chem 2005; 280(40): 33766-74.
[http://dx.doi.org/10.1074/jbc.M504604200] [PMID: 16081425]

[362] Biterova EI, Barycki JJ. Mechanistic details of glutathione biosynthesis revealed by crystal structures of Saccharomyces cerevisiae glutamate cysteine ligase. J Biol Chem 2009; 284(47): 32700-8.
[http://dx.doi.org/10.1074/jbc.M109.025114] [PMID: 19726687]

[363] Tadge A, Kang MH, Reynolds CP. The glutathione synthesis inhibitor buthionine sulfoxime synergistically enhanced melphalan activity against preclinical models. Blood Cancer J 2014; 4: 1-13.

[364] Gogos A, Shapiro L. Large conformational changes in the catalytic cycle of glutathione synthase. Structure 2002; 10(12): 1669-76.
[http://dx.doi.org/10.1016/S0969-2126(02)00906-1] [PMID: 12467574]

[365] Polekhina G, Board PG, Gali RR, Rossjohn J, Parker MW. Molecular basis of glutathione synthetase deficiency and a rare gene permutation event. EMBO J 1999; 18(12): 3204-13.
[http://dx.doi.org/10.1093/emboj/18.12.3204] [PMID: 10369661]

[366] Hara T, Kato H, Katsube Y, Oda J. A pseudo-michaelis quaternary complex in the reverse reaction of a ligase: structure of *Escherichia coli* B glutathione synthetase complexed with ADP, glutathione. and sulfate at 2.0 Å resolution. Biochemistry 1996; 35: 11967-74.

[367] (a) Sweet WL, Blanchard JS. Human erythrocyte glutathione reductase: chemical mechanism and structure of the transition state for hydride transfer. Biochemistry 1991; 30(35): 8702-9.
[http://dx.doi.org/10.1021/bi00099a031] [PMID: 1888731] (b) Deonarain MP, Berry A, Scrutton NS, Perham RN. Alternative Proton Donors/Acceptors in the Catalytic Mechanism of the Glutathione Reductase of *Escherichia coli*: The Role of Histidine-439 and Tyrosine-99. Biochemistry 1989; 28: 9602-07.

[368] Mittl PRE, Schulz GE. Structure of glutathione reductase from *Escherichia coli* at 1.86 A resolution: comparison with the enzyme from human erythrocytes. Protein Sci 1994; 3(5): 799-809.
[http://dx.doi.org/10.1002/pro.5560030509] [PMID: 8061609]

[369] Urig S, Fritz-Wolf K, Réau R, *et al.* Undressing of phosphine gold(I) complexes as irreversible inhibitors of human disulfide reductases. Angew Chem Int Ed Engl 2006; 45(12): 1881-6.
[http://dx.doi.org/10.1002/anie.200502756] [PMID: 16493712]

[370] Seefeldt T, Zhao Y, Chen W, *et al.* Characterization of a novel dithiocarbamate glutathione reductase inhibitor and its use as a toll to modulate intracellular glutathione J Bio Chem 2009; 284: 2729-37.

SUBJECT INDEX

A

Acceptor 34, 53, 97, 154
  amino 155
  electron 34
Acetohydroxyacid 101, 103, 120
  reductoisomerase 120
  synthase inhibitors 103
  synthase reaction 101
Acetolactate 101, 103, 104, 120
  synthase 120
Acetyl-CoA 34, 37, 57, 58, 61, 65, 66, 67, 68, 71, 123, 175, 176, 177
  enolate 66
  production 61
Acid 6, 9, 10, 20, 22, 44, 66, 76, 79, 82, 84, 86, 87, 108, 156, 171, 192, 197, 200, 210, 238, 243, 251, 255, 263, 272, 275, 276
  2-amino-4-oxo-5-hydroxypentanoic 171
  2-oxo amino 156
  4-methoxy-trans-but-3-enoic 276
  4-oxoheptadienedionic 255
  4-oxoheptenedioic 255
  6-hydrazoheptane-1,7-dioic 263
  agaricic 9, 10
  α-ketopimelic 255
  amino-4-methoxy-trans-but-3-enoic 275
  aminooxyacetic 192
  aspterric 108
  basic amino 251
  carboxylic amino 272
  citric 66
  converting cysteine sulfinic 197
  glutamic 200
  guanidino amino 243
  histidine amino 210
  indazole-3-carboxilic 6
  koningic 20, 22
  lipoic 44
  lysophosphatidic 79, 82
  natural amino 171
  oleic 76, 86, 87
  phophatidic 86

phosphatidic 79, 84
phosphoenol pyruvic 123
pyrrolidine heterocyclic amino 238
Acivicin 226, 227
  analogue 226
Aconitase 34, 37, 38, 39, 115, 116
  activity 39
  family 115
  human cytosolic 38
  inhibitors 38
Active site region 12, 18, 21, 226
  superimposed 12
Active site residues 48, 65, 73, 259
  representative 259
Active site threonine 253
Activity 4, 18, 25, 30, 31, 49, 80, 90, 108, 154, 165, 175, 196, 278, 225, 239
  antifungal 165, 175
  high serine biosynthetic 154
  inhibitory 20, 90, 108, 278
  mitochondrial 49
Acylglycerophosphate acyltransferase reaction 85
Alanine 97, 98, 182
  fluoro 182
  target aminoacid 98
  trifluoro 182
Alanine biosynthesis 97
Alcohol 177, 186, 211, 235
  oxidation 211
Aldimide 95, 96, 98, 147, 174, 275
  external 98, 174, 275
  internal 98, 147
Aldol type reaction 65, 126
Alkoxide 11, 53, 166, 177
Alkylating agents 6, 287
Alkyl transposition 104
Allosteric 10, 24, 31, 32, 52, 214
  binding site 213
  effector 24
  enzyme 10
  sites 52, 214
Alloxan concentration 39
Alzheimer disease 188

www.ingramcontent.com/pod-product-compliance
Lightning Source LLC
Chambersburg PA
CBHW050808220326
41598CB00006B/152